HIGH-INTENSITY
ULTRASONIC FIELDS

ULTRASONIC TECHNOLOGY
A Series of Monographs

General Editor
Lewis Balamuth
Ultrasonic Systems, Inc., Farmingdale, N. Y.

HIGH-INTENSITY ULTRASONIC FIELDS

Edited by
L.D. Rozenberg
Acoustics Institute
Academy of Sciences of the USSR
Moscow, USSR

Translated from Russian by
James S. Wood

⨁ **PLENUM PRESS** • **NEW YORK–LONDON** • **1971**

The original Russian text, published by Nauka Press for the Acoustics Institute, Academy of Sciences of the USSR, in 1968, has been corrected for the present edition. The English translation is published under an agreement with Mezhdunarodnaya Kniga, the Soviet book export agency.

Л. Д. РОЗЕНБЕРГ

МОЩНЫЕ УЛЬТРАЗВУКОВЫЕ ПОЛЯ
MOSHCHNYE UL'TRAZVUKOVYE POLYA

Library of Congress Catalog Card Number 78-128509
SBN 306-30497-X

© 1971 Plenum Press, New York
A Division of Plenum Publishing Corporation
227 West 17th Street, New York, N.Y. 10011

United Kingdom edition published by Plenum Press, London
A Division of Plenum Publishing Company, Ltd.
Davis House (4th Floor), 8 Scrubs Lane, Harlesden, NW10 6SE, England

Dedicated to
Nikolai Nikolaevich Andreev

Preface

This is the second volume in the series, "Physics and Engineering of High-Intensity Ultrasound,"* and is devoted to the specific characteristics of high-intensity sound fields.

The characteristics of such fields constitute the subject of nonlinear acoustics, and the book submitted herewith does not aspire to be an exhaustive treatise in that area. As the reader is aware, nonlinear processes are typified by interrelatedness and the inadmissibility of superposition, so that one should properly investigate the whole complex of phenomena in the large. However, the state of the art of research in this direction is such that we cannot yet pursue that approach and must be content to treat isolated aspects of nonlinear sound fields. This approach, of course, while only approximate, nevertheless has distinct advantages. Our orientation in this area is still, so to speak, "linear," and by the investigation of individual effects we are better able to grasp their physical significance, i.e., they are more comprensible to us. Consequently, the investigation of individual aspects of nonlinear sound fields, which can lead us to qualitative and sometimes quantitative insight, is not only of practical importance, but also fits better into our accustomed "linear" way of thinking.

As far as the individual aspects are concerned, they have been set apart not only for the extent to which they can be utilized to construct the physical foundations of ultrasonic technology, but also from considerations of their timeliness and the contributions of Soviet scientists, particularly members of the Acoustics Insti-

---

*Reference here is to the series of Russian works, of which "Sources of High Intensity Ultrasound" was the first translated as part of the present series — Publisher.

v

tute of the Academy of Sciences of the USSR. Thus, the book does
not contain material on the nonlinear interaction of nonlinear absorption of high-intensity acoustic oscillations in solids, etc. Although the main body of the text generally encompasses the results
of the author of the given part of the book, the materials of other
Soviet and foreign researchers working in the particular field are
also incorporated in the interest of clarity and completeness.

Part I presents a survey of the results of investigations conducted by the author in the nonlinear absorption of sound in liquids
and gases. Of special interest are the effects on nonlinear absorption in convergent and divergent cylindrical and spherical
fronts.

Part II is based on results obtained by the author largely
during his tenure at the Acoustics Institute. It is given over to the
rather complex and beguiling problem of the radiation pressure,
an area that is often imprecisely and at times incorrectly handled.

Part III comprises a survey of our knowledge concerning
acoustic streaming. Even though the advances made in this area
at the Acoustics Institute are relatively limited, the subject is
still important enough to warrant its inclusion as an independent
part of the book.

The last three parts (IV, V, and VI) are concerned with a problem that is exceedingly complex, multifaceted, and often vague,
despite the great many papers that have been published on it, namely ultrasonic cavitation. As the reader is aware, cavitation is the
most active factor in a large majority of practical applications;
moreover, it is extremely important as a unique physical process
in itself. There is a certain arbitrariness in the breakdown of the
subject into three parts. All three parts are based on investigations conducted at the Acoustics Institute and marked by many con-
ceptually new quantitative and qualitative materials. For example,
Part IV represents a probing analysis of the equations describing
the motion of an individual bubble and an assessment of the instability effect. The results are compared with the data of high-
speed motion pictures of cavitation bubbles. Part V describes the
recently discovered effect of the multiplication of cavitation bubbles, a procedure developed for determining the acoustic energy
losses in the formation of cavitation bubbles, and an investigation
of the very interesting effects of the static pressure on the be-

havior of cavitation bubbles, primarily on the enhancement of cavitation erosion.

Part VI is devoted to the vital, though so far, comparatively neglected problem of the behavior of the cavitation zone in general. The work in this area is currently in the state of development, but the results to date already permit a clear conceptualization of the phenomena evolving in a cavitation zone and occassionally certain inferences useful for practical applications.

The pioneer in the formulation and development of problems in nonlinear acoustics in the large is Academician N. N. Andreev, on whose initiative some of the earliest investigations in this area were undertaken.

The editor is grateful to V. S. Grigor'ev and G. A. Ostroumov, who assumed the task of reviewing the manuscript of the book and offered many valuable comments. Also deserving acknowledgment is the considerable service rendered by V. A. Akulichev in preparing the manuscript for publication.

L. D. Rozenberg

Contents

PART II — ACOUSTIC RADIATION PRESSURE

Z. A. Gol'dberg

PART V — EXPERIMENTAL INVESTIGATIONS
OF ULTRASONIC CAVITATION

M. G. Sirotyuk

PART VI — THE CAVITATION ZONE

L. D. Rozenberg

PART I

ABSORPTION OF FINITE-AMPLITUDE WAVES

K. A. Naugol'nykh

Introduction

Normally in acoustics one is concerned with sound waves of small amplitude in the sense that the perturbations elicited by these waves in the equilibrium state of the medium are small. The propagation of such waves is described in terms of approximate equations derived by linearization of the hydrodynamic equations and equation of state. This, the so-called linear-acoustical approximation, proves inadequate for the case of sound waves of large intensity, which are being encountered on a growing scale in present-day engineering.

Large sound intensities are used in high-power ultrasonic equipment and are generated in the operation of jet and rocket engines. Lately large-amplitude hypersound has been produced in solids by the scattering of powerful light beams from lasers by lattice vibrations.

The propagation of intense sound waves is accompanied by a whole series of effects characterized by their dependence on wave amplitude. The description of these effects requires that the nonlinear terms of the hydrodynamic equations be included. The presence of so-called nonlinear effects significantly alters the propagation pattern of the intense sound wave and, in particular, the nature of its absorption. This part of the book is devoted to an analysis of the indicated problem of the absorption of strong waves.

Other nonlinear effects are sound pressure, also called the acoustic radiation pressure, acoustic streaming, the scattering of sound by sound, etc. Certain relevant problems are considered in subsequent parts of the book.

Chapter 1

Plane Waves of Finite Amplitude

§ 1. Absorption of a Low-Intensity Sound Wave

During the propagation of a sound wave, its energy is converted into latent energy of thermal motion by irreversible processes attributed to the viscosity and thermal conductivity of the medium, whereupon gradual absorption of the wave takes place.

The process of sound absorption in liquids and gases is described by the hydrodynamic equations taking into account viscosity and thermal conductivity. If one seeks the solution of the linearized hydrodynamic equations for the one-dimensional case in the form of a plane harmonic wave of the type $\exp(kx - \omega t)$, the wave number k turns out to be complex; its real part determines the wavelength, and its imaginary part describes the absorption coefficient: $k = k_1 + i\alpha$,

$$e^{kx-\omega t} = e^{-\alpha x} e^{i\,(k_1 x - \omega t)}. \tag{1}$$

Here α is expressed by the formula (see, e.g., [1])

$$\alpha = \frac{\omega^2}{2\rho c^3} \left[\frac{4}{3}\eta + \eta' + \varkappa \left(\frac{1}{c_v} - \frac{1}{c_p} \right) \right], \tag{2}$$

from which it is apparent that the absorption coefficient varies as the square of the wave frequency ω and a linear combination of the shear viscosity coefficient η, dilatational viscosity coefficient η', and thermal conductivity coefficient $\varkappa$. The rest of the notation is

5

conventional: ρ is the density of the medium, c is the velocity of sound, and c_p, c_v are the specific heats at constant pressure and volume, respectively.

Comparison with experiment shows that in monatomic gases and liquids whose internal degrees of freedom are not excited at the given temperature, the experimental values of the absorption coefficient exhibit good agreement with the theoretical results according to Eq. (2). In a number of monatomic gases such as carbon dioxide or hydrogen and in certain liquids (water or benzene), the frequency behavior of the absorption departs from the square-law dependence predicted by Eq. (2). These absorption anomalies are attributed to relaxation effects. During the compression or expansion of the medium elicited by the sound wave (as in any other rapid change of state), the thermodynamic equilibrium of the medium can be upset, whereupon irreversible equilibrium recovery processes develop in the medium, accompanied by the dissipation of energy and leading to the anomalous attenuation of sound [1, 2].

The absorption α_r due to relaxation processes is expressed by the formula [3]:

$$\alpha_r = \frac{c_\infty^2 - c_0^2}{2c_0^3} \frac{\omega^2 \tau}{1 + \omega^2 \tau^2},\tag{3}$$

in which case the real part of the wave number is equal to

$$k_1 = \frac{\omega}{c_0} \left(1 - \frac{c_\infty^2 - c_0^2}{2c_0^2} \frac{\omega^2 \tau^2}{1 + \omega^2 \tau^2} \right).\tag{4}$$

Here τ is the relaxation time; c_∞ and c_0 are the velocities of sound at high frequencies ($\omega \tau \gg 1$) and at low frequencies ($\omega \tau \ll 1$), respectively.

It is evident from relations (3) and (4) that relaxation processes lead to dispersion of the velocity of sound, i.e., they lead to a frequency dependence of the phase velocity $c = \omega/k_1$ and to a specific frequency dependence of the absorption per wavelength (Fig. 1).

Note that according to (1), the shape of the wave does not change during its absorption, but remains harmonic, suffering only a decrease in amplitude.

By contrast, the shape of an acoustic signal of arbitrary form changes during the absorption process, becoming smoother than

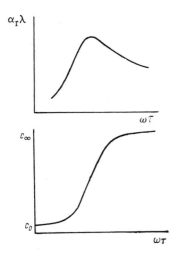

Fig. 1. Frequency dependences of the absorption coefficient per wavelength $\alpha_r \lambda$ and propagation velocity c of a sound wave in the relaxation domain.

before. Different spectral components of the signal are damped differently because the absorption coefficient increases as the square of the frequency. Thus, a wave initially in the shape of a rectangular pulse acquires a Gaussian profile at large distances. Its "width" is on the order of magnitude $(ax)^{\frac{1}{2}}$, i.e., it grows as the square root of the path traversed by the wave, while the "amplitude" of the wave (its peak value) falls off inversely as $x^{\frac{1}{2}}$ (this is discussed in further detail in [1]). Here a denotes the coefficient of ω^2 in expression (2):

$$a = \frac{b}{2\rho c^3}; \quad b = \frac{4}{3}\eta + \eta' + \varkappa\left(\frac{1}{c_v} - \frac{1}{c_p}\right). \tag{5}$$

§2. Qualitative Description of the Propagation of Finite-Amplitude Waves

The relations presented in the preceding section describe the absorption of sound of fairly low intensity, when it is admissible to linearize the hydrodynamic equation and equation of state.

With an increase in the sound intensity, this approximation becomes inapplicable, and effects are observed which are described by the nonlinear terms of the hydrodynamic equations and equation of state and which significantly alter the pattern of sound wave propagation and, in particular, its absorption process [1, 2, 4-8].

We analyze these effects in the example of periodic waves having an initially sinusoidal form. This case is clearly the most important for ultrasonic practice. We first consider a plane traveling wave radiated in a medium with viscosity and thermal conductivity by an oscillating plane in a semi-infinite half-space.

Let the wave have a sinusoidal form near the radiator. If the wave intensity is large enough, then, during propagation, its form changes due to the difference in the velocities of different portions of the wave profile (we interpret the profile of a wave as the distribution of certain variables such as the pressure, velocity, or density in the direction of propagation).

Points of the profile that correspond to large compression move faster than points corresponding to smaller densities. As a result, the steepness of the wave fronts increases, resulting eventually in the occurrence of a discontinuity in each period of the wave and the formation of a sawtooth wave.

On the other hand, the influence of viscosity and thermal conductivity tends to smooth the wave profile, diminishing the velocity and temperature gradients. Consequently, during the propagation of a harmonic wave (near the radiator), the steepness of the fronts will increase until the effects of the nonlinear and dissipative processes cancel one another or, as customarily stated, until stabilization of the wave shape takes place.

This stabilization is relative, in other words, it does not mean that with further propagation the wave shape does not change, because with damping of the wave amplitude the nonlinear effects are attenuated, and the wave profile is gradually smoothed out, returning once again to a sinusoidal shape. Therefore, a wave in which the effects of nonlinear and dissipative processes cancel is regarded as a wave of relatively stable form, by which is implied a slower variation of the profile of such a wave than the profile of an arbitrary wave of the same amplitude and frequency.

The profile of a wave of relatively stable form is determined by the relation between the nonlinear and dissipative effects. If the wave intensity is large enough, nonlinear effects dominate, and the shape of the wave is sharply altered. In this case an initially sinusoidal wave becomes a weak periodic shock wave, and its profile acquires a sawtooth configuration.

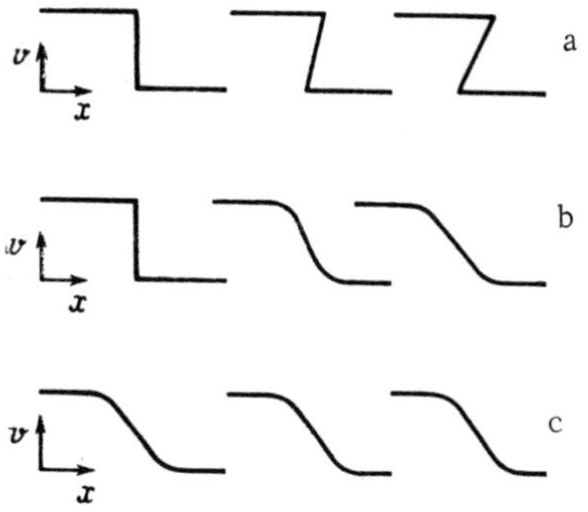

Fig. 2. Variation of the profile of a
single compression shock. a) Due to
nonlinear effects; b) due to dissipative
effects; c) under the combined influence of both effects.

In the case of small intensities, the opposite effect is observed;
the influence of dissipative processes prevails, the wave is damped
out before the nonlinear effects have time to build up, and its propagation is adequately described by Eq. (1).

The various ways in which nonlinear and dissipative effects
act to vary the wave shape are illustrated in Fig. 2, which is borrowed from [5]. The figure shows the variation of the profile of
a single compression shock with regard for nonlinear and dissipative effects, as well as for the simultaneous allowance for these
effects when stabilization of the wave occurs.

The distortion of the wave shape may also be treated as a
variation of its spectral composition; in the propagation of an initially monochromatic wave the high-frequency harmonics grow,
attain a maximum in the interval where the variations of the wave
shape are strongest, and then decay. The resulting continuous
transfer of energy from the fundamental harmonic to the strongly
absorbing high-frequency components causes intense attenuation
of the wave.

It may also be stated that an increase of the wave absorption
is caused by the increasing steepness of the wave fronts due to the
stronger dissipation of energy as the velocity and temperature
gradients grow larger. As a result, the wave absorption is dependent on the distance from the radiator, being small near the
source, attaining a maximum in the region where the wave has a
sawtooth shape, and then decreasing once again.

Here, in general, it is important to distinguish between dissipation of the total energy of the wave and the absorption of its fun-

damental harmonic or, in other words, the amplitude coefficient of absorption of the fundamental harmonic as opposed to the intensity coefficient of absorption of the wave [10, 11].

The following qualitative picture of the propagation of a finite-amplitude wave is substantiated by mathematical analysis and experimental data.

§ 3. Fundamental Equations

The propagation of finite-amplitude waves is described by the system of hydrodynamic equations

$$\rho\left(\frac{\partial \mathbf{v}}{\partial t} + \mathbf{v}\nabla\mathbf{v}\right) = -\nabla p + \eta\Delta\mathbf{v}' + \left(\frac{\eta}{3} + \eta'\right)\nabla \operatorname{div} \mathbf{v}, \tag{6}$$

$$\frac{\partial \rho}{\partial t} + \operatorname{div}\rho\mathbf{v} = 0, \tag{7}$$

the heat-transfer equation

$$\rho T\left(\frac{\partial s}{\partial t} + (\mathbf{v}\nabla s)\right) = \frac{\eta}{2}\left(\frac{\partial v_i}{\partial x_k} + \frac{\partial v_k}{\partial x_i} - \frac{2}{3}\,\delta_{ik}\,\frac{\partial v_l}{\partial x_l}\right)^2 + \eta'\,(\operatorname{div} v)^2 + \varkappa\operatorname{div}\nabla T, \tag{8}$$

and the equation of state

$$p = p\,(\rho,\,s). \tag{9}$$

Here **v** is the hydrodynamic velocity of the medium, p is the pressure, T is the temperature, and s is the entropy per unit mass. Equations (6)-(9) are nonlinear, and their integration presents considerable difficulties.

However, if we confine our perspective to the analysis of sound of reasonably limited intensity, such that the Mach numbers characterizing the motion are small, $M = v_0/c_0 \ll 1$ (where v_0 is the particle velocity amplitude of the wave), it becomes admissible to introduce a number of approximations, which facilitate the solution of the problem significantly. This does not mean that such simplifications exclude the consideration of cases when the nonlinear effects are very large. As a matter of fact, the nonlinear corrections to the solutions of linear acoustics contain, in addition to terms of the order $M = v_0/c_0$, terms describing cumulative effects that are proportional to the quantity Mkr, as in the case of a plane wave traveling along the r axis.

Therefore, nonlinear effects are significantly felt not only in the obvious case of wave amplitudes so large that $M \gtrsim 1$, but also in the far more prevalent and acoustically more important case when $M \ll 1$ and yet cumulative effects are large (as in the case, for example, of a traveling plane wave when $Mkr \approx 1$). In fact, such situations, when the number M is again a small parameter of the problem, but cumulative nonlinear effects need to be taken into account, constitute the chief topic of the present part of the book. In these cases it is possible to simplify considerably the set of initial hydrodynamic equations and the equation of state on the basis of the smallness of M and the absorption per wavelength α/k, permitting the development of an extremely effective approximative nonlinear theory of the propagation of finite-amplitude sound.

In particular, the equation of state may be written in the form

$$p - p_0 = c_0^2 (\rho - \rho_0) + \frac{1}{2} \left(\frac{\partial c^2}{\partial \rho} \right)_s (\rho - \rho_0)^2 + \frac{\partial p}{\partial s} (s - s_0) + \ldots \qquad (10)$$

Limiting our analysis to the one-dimensional case, we transform Eq. (6) by means of (8), combining the viscosity term in that equation with the last term of the state equation (10), i.e., the nonadiabatic term [12, 13], whereupon Eq. (6) takes on the form

$$\rho \left[\frac{\partial v}{\partial t} + (v \nabla) \, v \right] = - \nabla p_s + b \Delta v, \qquad (11)$$

where

$$b = \frac{4}{3} \eta + \eta' + \varkappa \left(\frac{1}{c_v} - \frac{1}{c_p} \right);$$

$$p_s - p_0 = c_0^2 (\rho - \rho_0) + \frac{1}{2} \left(\frac{\partial c^2}{\partial \rho} \right)_s (\rho - \rho_0)^2. \qquad (12)$$

The latter relation is often written in the alternative form [9]:

$$p_s - p_0 = A \frac{\rho - \rho_0}{\rho_0} + \frac{B}{2} \left(\frac{\rho - \rho_0}{\rho_0} \right)^2, \qquad (13)$$

where

$$A = \rho_0 c_0^2; \quad B = \rho_0^2 \left(\frac{\partial c^2}{\partial \rho} \right)_s,$$

TABLE 1

Substance	T, °C	γ from [9]	γ from [14]	Substance	T, °C	γ from [9]	γ from [14]
H_2O	0	5.2		C_6H_6	25	7.5	
	20	6.1	7.1		40	9.5	
	40	6.4		Glycerin	20	9.8	
	60	6.7		Transformer oil	20		7.5
	80	7.0		Acetone	20		11.0
	100	7.1					
C_2H_5OH	0	11.5					
	20	11.3	11.8	Gasoline A-70	20		12.2
	40	11.8		Chloroform	20		12.5
	60	11.9					
CS_2	10	7.4		Monatomic gas		1.67	
	40	7.1		Diatomic gas		1.40	
CCl_4	10	9.1					
	20	9.7	11.4				
	40	10.3					

and the ratio

$$\frac{B}{A} = \frac{\rho_0}{c_0}\left(\frac{\partial c^2}{\partial \rho}\right)_{\rho,\, s} = 2\rho_0 c_0\left(\frac{\partial c}{\partial \rho}\right)_T + \frac{2\chi c_0 T}{c_p}\left(\frac{\partial c}{\partial T}\right)_p, \qquad (14)$$

where χ, the compressibility of the medium, is adopted as a characteristic of the degree of nonlinearity of the equation of state of the medium (12) [9]. The ratio B/A is customarily replaced by the quantity

$$\gamma = \frac{B}{A} + 1 = \left(\frac{\partial c^2}{\partial \rho}\right)_s \frac{\rho_0}{c_0^2} + 1, \qquad (15)$$

which is equal to the specific heat ratio $\gamma = c_p/c_v$ in the case of gases and to the exponent n in the empirical equation of state for condensed media [2]:

$$p = A\,(\rho/\rho_0)^n - B. \qquad (16)$$

The values of γ for certain substances are presented in Table 1, which has been compiled from the results of [9, 14]. In [9] the values of γ have been computed from Eq. (14) on the basis of the experimental density and temperature dependences of the velocity of sound, and in [14] they have been measured directly in terms of the magnitude of the second harmonic in a finite-amplitude wave.

Also, the systems (7), (11), (12) can be reduced in the given approximation to a single equation on the assumption that the absorption is small, $\alpha/k \approx M$ [15-17]. We explicate the process of this transformation using plane waves as an example.

Introducing the function $w = \int dp_s/\rho$ and the velocity potential φ we rewrite the equation of continuity (7) in the form

$$\frac{\partial w}{\partial t} + \frac{\partial \varphi}{\partial r} \frac{\partial w}{\partial r} + c^2 \frac{\partial^2 \varphi}{\partial r^2} = 0. \tag{17}$$

From the Euler equation (11) we have

$$w = -\frac{\partial \varphi}{\partial t} - \frac{1}{2}\left(\frac{\partial \varphi}{\partial t}\right)^2 + \frac{b}{\rho_0}\frac{\partial^2 \varphi}{\partial r^2}, \tag{18}$$

and from Eq. (12) we obtain

$$c^2 = c_0 + (\gamma - 1)\, w. \tag{19}$$

Now, substituting (18) and (19) into Eq. (17), we find, correct to squared terms on the small parameter M,

$$c_0^2 \varphi'' - \ddot{\varphi} + \frac{b}{\rho_0}\dot{\varphi}'' + c_0 (\gamma + 1)\varphi'\varphi'' = 0. \tag{20}$$

Transforming in this equation from the variables r, t to the variables $y = t - (r/c_0)$, r and returning from the potential to the velocity $v = \partial\varphi/\partial r$, we approximately obtain

$$\frac{\partial v}{\partial r} - \frac{\varepsilon}{c_0^2}\, v\,\frac{\partial v}{\partial y} = a\,\frac{\partial^2 v}{\partial y^2}, \tag{21}$$

where $\varepsilon = (\gamma + 1)/2$.

This equation is striking in two respects: first, despite its approximate character, in the cases of interest, it describes with sufficient accuracy both nonlinear and dissipative processes; second, the equation is of the same type as the Burgers equation, which admits transformation to a linear equation of the diffusion type (18).

Introducing the dimensionless variables $u = v/v_0$, $\sigma = r/L$, where, as will become apparent presently, it is convenient to interpret L as the quantity

$$L = \frac{1}{k\varepsilon M} = \frac{c_0^2}{\varepsilon \omega v_0},$$ (22)

we can rewrite the fundamental equation of the nonlinear theory of sound propagation in a form which will be more useful

$$\frac{\partial u}{\partial \sigma} - u \frac{\partial u}{\partial (\omega y)} = \Gamma^{-1} \frac{\partial^2 u}{\partial (\omega y)^2},$$ (23)

where

$$\Gamma = 2\varepsilon \, \mathrm{Re} = \frac{\varepsilon}{\pi} \frac{v_0 \lambda}{b/\rho_0} = 2\varepsilon \frac{p_0'}{b\omega};$$

p_0' is the pressure amplitude of the wave.

The dimensionless parameter Γ, which is proportional to the ratio of the nonlinear term in Eq. (23) to the term accounting for the influence of viscosity and thermal conductivity, is, along with M, the most important characteristic of the propagation of high-intensity sound characterizing the relative contributions of non-linear and dissipative effects. This parameter, which was first introduced in [10] as a criterion of the influence of nonlinear effects, is proportional to the usual hydrodynamic Reynolds number (Re) and differs from it only by the factor 2ε, which accounts for the nonlinearity of the equation of state; it is often referred to in the literature either as the "Reynolds number" [5] or the "acoustic Reynolds number" [16].

Note that Γ is equivalent to the parameter $(\alpha_0 L)^{-1}$ used in [19], is related to the coefficient k in [20] by the expression $k = (\Gamma/4) \times [1 - \exp(-2\alpha x)]$, and is similar to the parameter $(-2A)$ of [15] (see [17]).

Equation (23) affords a good starting point for the analysis of the propagation of a finite-amplitude wave, as it enables us to deduce a number of simple approximate solutions. These solutions may be divided into two groups according to two distinct approaches to the problem of finite-amplitude wave propagation.

The first group of solutions, to be considered in §4, describes the profile of the finite-amplitude wave and its variation during propagation. The second group of solutions may be regarded as the result of the Fourier series expansion of the solutions in the first group. These solutions, which are discussed in §5, describe the propagation of a finite-amplitude wave from the spectral point of view, characterizing the amplitude variation of the separate harmonics during wave propagation.

Within each of these main groups are several solutions that differ in their domains of applicability. One of them is applicable only for large values of the parameter Γ, and the other only for small values; some solutions are valid only near the source, others only in the region of wave stabilization.

§ 4. Formation of First-Order Discontinuties in a Sound Wave

We wish to consider the propagation of a plane sound wave of sufficient intensity, such that the values of the parameter Γ are large: $\Gamma \gg 1.$* We assume that the wave initially has a sinusoidal form:

$$v = v_0 \sin \omega t \quad \text{at} \quad r = 0. \tag{24}$$

For large values of Γ, the right-hand side of Eq. (23) may be neglected; the rest of the equation is easily integrated if y is considered as a function of v and σ.

The resulting solution found by Soluyan and Khokhlov [16] may be written in the form

$$\omega y = -\sigma \frac{v}{v_0} + f\left(\frac{v}{v_0}\right), \tag{25}$$

where $f(v/v_0)$ is an arbitrary function.

Ascertaining this function from the boundary condition (24), we obtain a solution describing an initially sinusoidal traveling plane wave† :

$$\omega y = \arcsin \frac{v}{v_0} - \sigma \frac{v}{v_0}. \tag{26}$$

---

* We emphasize once again that the condition ot small numbers M stipulated above is always assumed true: $M \ll 1$. Note that the conditions $M \ll 1$ and $\Gamma \gg 1$ are not conflicting and are in fact true in real situations. For instance, if the pressure amplitude of a wave propagating in water is $p_0' = 25$ atm ($c_0 = 1.5 \cdot 10^5$ cm /sec, $\varepsilon = 4$, $\alpha/v^2 = 22 \cdot 10^{-17}$ sec$^2 \cdot$ cm^{-1}) and the frequency $v = 10^6$ cps, then $\Gamma \approx 10^2$, and $M \approx 10^{-3}$.

† Equation (26) may be thought of as the zero-order term in the expansion of the solution of Eq. (1) in reciprocal powers of Γ. The ensuing first-order term in this expansion has been found by Blackstock [17].

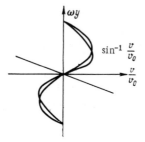

Fig. 3. Plot of the profile of
a finite-amplitude wave.

This expression is amenable to graphical analysis. Plotting the ratio v/v_0 on the horizontal axis and ωy on the vertical, it is a simple matter to represent Eq. (26) as a sum of two functions: $\sin^{-1}(v/v_0)$ and the linear function $-\sigma(v/v_0)$, the slope of which increases with the distance σ traversed by the wave (Fig. 3). Rotating Fig. 3 counterclockwise through 90°, we obtain the dependence we need, $v = v(\omega y)$, at various distances from the source. The resulting shapes of the wave profile are shown in Fig. 4. It is clear that near the source ($\sigma \ll 1$) the distortions of the sinu-

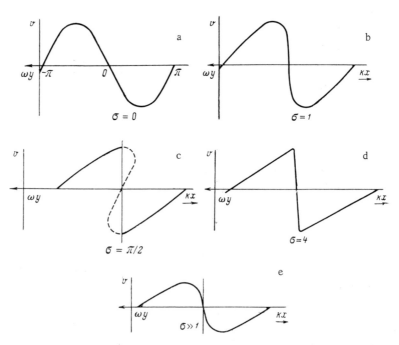

Fig. 4. Variation of the profile of a finite-amplitude wave during propagation.

soidal form are small (Fig. 4a). During propagation the waves gradually increase, so that at a distance $\sigma = 1$, the wave profile in the vicinity of the point v = 0 becomes vertical (Fig. 4b), then multivalued (Fig. 4c), a situation that is physically inadmissible and indicates the formation of a discontinuity (shock wave).

The distance to the formation of the discontinuity in dimensionless units is readily determined from Eq. (22). For $\sigma = 1$ (instant of formation of the discontinuity), r = L, i.e., the quantity $L = 1/k\varepsilon M$, chosen as the unit of length, is then the discontinuity buildup distance.

In water, for example, at an intensity of 10 W/cm^2, we find $M \approx 2.4 \cdot 10^{-4}$ and $L/\lambda \approx 1.6 \cdot 10^2$ (where λ is the wavelength).

After formation of the discontinuity, the pattern of motion is complicated in general, the wave losing its identity as a unidirectional traveling wave due to reflection at the surface of the discontinuities. However, for moderate wave amplitudes, which is the case that interests us, when $M \ll 1$, these effects are negligible.

Thus, the relations connecting the variation of the velocity, pressure, and density in a shock wave and in the solution (26) turn out to be the same, correct to second-order terms inclusively on the number M [1]. Therefore, in the second approximation, on both sides of the discontinuity the sound wave remains simple, traveling in one direction, and the boundary conditions at the discontinuity are satisfied without the introduction of waves reflected from the discontinuity. From the condition for matching of the continuous solution with the relations at the discontinuity, we obtain a simple rule governing the position of the discontinuity from the known continuous solution [1]. In our case this rule leads to the requirement of equal areas bounded by the dashed curve in Fig. 4c (to the left and right of the discontinuity), so that the discontinuity must be situated at the point $\omega y = 0$. With this result, we readily deduce from (26) the following relation determining the magnitude of the discontinuity v_p:

$$\sin^{-1} \frac{v_p}{v_0} - \sigma \frac{v_p}{v_0} = 0,$$

which may be rewritten

$$\frac{v_p}{v_0} = \sin \sigma \frac{v_p}{v_0}. \tag{27}$$

Looking once again at Fig. 3, we interpret relation (27) as the
condition for the point on the wave profile corresponding to the
velocity v_p to reach the discontinuity front situated on the axis
$\omega y = 0$. Analyzing relation (27) graphically with the aid of con-
structions analogous to Fig. 3, but for different values of σ, we
readily verify that after the onset of the discontinuity at $\sigma = 1$
(see Fig. 4b), its magnitude grows until at $\sigma = \pi/2$ the maximum
value of the particle velocity v_0 reaches the discontinuity,* whereupon the magnitude of the discontinuity diminishes, and the wave
assumes a sawtooth shape (Fig. 4d). The law of diminution of the
discontinuity becomes particularly simple in the domain of large
σ, when the right-hand side of (27) can be expanded in the vicinity
of the point $\sigma(v/v_0) \approx \pi$.

Thus, setting $\sigma(v_p/v_0) \approx \pi - \delta$, where $\delta \ll \pi$, we have
$\sin \sigma(v_p/v_0) = \sin(\pi - \delta) = \sin \delta \approx \delta = \pi - \sigma(v_p/v_0)$; inserting
this result into (27), we obtain [16, 21]

$$\frac{v_p}{v_0} = \frac{\pi}{1 + \sigma}. \tag{28}$$

Comparison with expression (27) reveals that the error in-
curred by the use of the approximate relation (28) does not exceed
2% even for $\sigma > 3.6$ [21]. The foregoing portrayal of the transition of a sine wave into a sawtooth has been observed experimentally in the propagation of ultrasonic waves in gases (Mendousse
[15]; Werth [22]) and in liquids (Burov and Krasil'nikov [23]; Romanenko [24]). A sample oscillogram of the wave shape in water in
the vicinity of $\sigma \approx \pi/2$ for $\Gamma \approx 40$, obtained by Romanenko [24], is
shown in Fig. 5.

The onset of a discontinuity produces strong attenuation of the
wave, the amplitude of the latter decreasing in accordance with
(28). Physically this damping is attributable to irreversible compression processes in the shock wave and the attendant increase
in the entropy and dissipation of energy.

These processes are implicitly included in the derivation of
Eq. (28) by virtue of the relations determining the magnitude of the
jump in the hydrodynamic variables at a first-order discontinuity.
The foregoing is confirmed by the fact that the calculation of the

---

* This is easily verified by direct substitution of the solution $v_p = v_0$ for $\sigma = \pi/2$ into
Eq. (27).

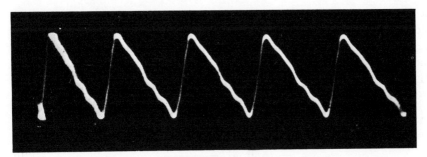

Fig. 5. Oscillogram of the wave shape in water in the vicinity of $\sigma \simeq \pi/2$ for $\Gamma = 40$. The frequency $f \approx 0.8$ Mc; $p_0 \approx 80$ atm.

absorption of a sawtooth wave on the basis of a direct computation of the energy dissipation of the wave per period at the first-order discontinuity yields a result analogous to (28) [25] and differing therefrom only in the absence of the factor π in the numerator of the right-hand side. This disparity is due to the fact that Eq. (28) describes the absorption of a wave having an initially sinusoidal ($\sigma = 0$) form in terms of the amplitude v_0 of that wave and the distance traversed by it. The following equation, obtained in [25] (see also [8]);

$$\frac{v_p}{v_0} = \frac{1}{1+\sigma} \tag{28a}$$

characterizes the decrease of the discontinuity of a periodic shock wave that has an initial sawtooth form (at $\sigma = 0$) with a peak value of the hydrodynamic velocity equal to v_0.

As apparent from (28), a finite-amplitude sawtooth, unlike a small-amplitude wave, is damped nonexponentially, the degree of damping increasing with increasing wave amplitude (inasmuch as σ is proportional to the wave amplitude).

Directly linked with this fact is a specific kind of "saturation" effect; with an increase in the initial wave amplitude v_0, the peak value of the particle velocity of the sawtooth, coinciding with v_p, increases more and more slowly at a fixed point of space, asymptotically approaching the limit v_{pm}, which is no longer dependent on the initial amplitude v_0:

$$v_{pm} = \frac{\pi c_0}{\varepsilon k x}. \tag{29}$$

It is instructive to compare the solution (26) with the exact solution of the hydrodynamic equations for an ideal medium, which describe the propagation of a so-called simple wave. This solution has been obtained by Poisson [26], Riemann [27], and others (see [1, 28], for example, for a brief description of these papers).

For the case of a harmonic boundary condition, the solution may be written in the form

$$\frac{v}{v_0} = \sin \omega \left[t - \frac{x}{c\,(v) + v} \right], \qquad (30)$$

where

$$c\,(v) = c_0 + \frac{(\gamma - 1)}{2}\, v, \qquad (31)$$

provided the equation of state is adopted in the form (12). The above relation implies that the wave propagation velocity differs for different parts of the profile, and this is the reason for the variation in wave shape during propagation.

In the given case of small numbers M, the second term in the argument of (30) can be expanded on a small correction to the velocity of sound, resulting in the expression

$$\frac{v}{v_0} = \sin \omega \left(t - \frac{r}{c_0} + \sigma\, \frac{v}{v_0} \right), \qquad (32)$$

which exactly coincides with the solution (26). A comparison of (30) and (32) clarifies the nature of the approximations used in the derivation of (26), which were discussed earlier; the variations of the velocity of sound are assumed to be small (M ≪ 1), but this does not prohibit the investigation of large phase corrections during propagation of the wave.

We note that the spatial velocity of a point on the wave profile corresponding to a definite value of the velocity v differs from the propagation velocity of small-amplitude sound by the amount

$$c\,(v) + v - c_0 = v + \frac{(\gamma - 1)}{2}\, v, \qquad (33)$$

which comprises two terms. One [the first term on the right-hand side of (33)] describes the transport of the wave by the moving

medium, while the other is attributable to the nonlinearity of the
equation of state.

For gases the first term normally yields the greater contribution: thus, in the case of air it is five times the second term.
In liquids the nonlinearity of the state equation is the more significant: in water, for example, the second term is three times
the first one.

We next consider a sawtooth wave. During its subsequent
propagation, the relative contribution of nonlinear effects is reduced by attenuation, causing the resulting discontinuities to gradually smooth out.

This process of smoothing of the sawtooth is described by the
simple solution of (23), which takes account of the influence of
dissipative processes and makes it possible to determine the form
of one period of the wave [16, 17]:

$$\frac{v}{v_0} = \frac{1}{1+\sigma}\left[-\omega y + \pi th\,\frac{\omega y}{\Delta}\right] - \pi \leqslant \omega y \leqslant \pi, \qquad (34)$$

where

$$\Delta = \frac{2}{\pi}\frac{1+\sigma}{\Gamma} = \frac{2}{\tilde{\Gamma}} = \frac{1}{2\varepsilon\,\widetilde{Re}} \qquad (35)$$

is the dimensionless thickness of the shock wave; $\tilde{\Gamma}$ differs from
Γ in that the constant velocity v_0 is replaced in the former by the
instantaneous value of the peak velocity $v_p = v_0\pi/(1 + \sigma)$: $\tilde{\Gamma} =
2\rho c_0\varepsilon v_p/b\omega$.

It can be verified by direct substitution into (1) that (30) is an
exact solution of that equation. Equation (34) may be regarded as
the result of matching the solution for the sloping rectilinear portion of the sawtooth profile to the conventional equation describing
the structure of a first-order discontinuity [second term in the
brackets of (34)] (see, e.g., [1]). The factor in front of the brackets
of (34) accounts for the attenuation of the wave with distance.

The wave profile constructed from Eq. (34) is illustrated in
Fig. 6 for two values of Γ. It is apparent that the thickness of the
discontinuity depends on Γ and decreases as the latter increases,
reverting to zero as $\Gamma \to \infty$. In this case from (34) we obtain the

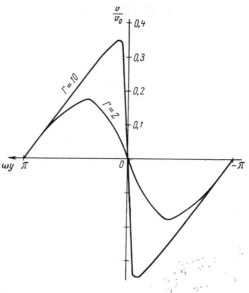

Fig. 6. Profile of one period of a finite-amplitude
wave for $\Gamma = 2$ and $\Gamma = 10$.

following solution describing a sawtooth:

$$\frac{v}{v_0} = \frac{1}{1 + \sigma} [-\omega y + \pi], \tag{36}$$

the peak value of which, v_p, attained at $\omega y = 0$, decays in exact correspondence with (28). Moreover, the solution (34) indicates the limits of applicability of the results of the preceding section based on neglect of the thickness of the discontinuity.

Clearly, this approximation is applicable under the condition

$$\frac{\Delta}{2\pi} = \frac{1 + \sigma}{2\pi^2 \Gamma} = \frac{1}{2\pi \tilde{\Gamma}} \ll 1. \tag{37}$$

Note that, according to (33), the parameter $\tilde{\Gamma}$ may still be regarded in order of magnitude as the ratio of the wavelength of an initially sinusoidal wave to the thickness of the resulting discontinuity. Accordingly, the value $\Gamma = \tilde{\Gamma}$ ($\sigma = \pi/2$) gives the maximum of that ratio.

Solution (34), in turn, becomes inapplicable when the thickness of the discontinuity is comparable with the wavelength, i.e., wherever $\Delta/2\pi \approx 1$, which is true at distances $r \gtrsim 1/\alpha$.

§5. Variation of the Spectral Composition of a Finite-Amplitude Wave during Propagation

The distortion of an initially sinusoidal wave during propagation may be treated as a variation of its spectral composition. Thus, the expansion of the solution (26) on a small phase correction indicates a growth of the high-frequency harmonics:

$$v = v_0 \sin \omega y + \frac{r}{L} \frac{v_0}{2} \sin 2\omega y + \cdots \tag{38}$$

Here L is the distance to the formation of discontinuity and is defined by Eq. (22).*

We observe that solutions similar to (38) have been obtained by several authors through direct integration of the hydrodynamic equations, using the method of successive approximations (Lamb [29]; Fubini [30]). The series (38) converges rapidly only at small distances relative to L from the radiator.

More efficient in this respect is the solution obtained by the expansion of expression (32), which is equivalent to the solution (26), in a Fourier series [16].

Thus, the expression

$$v = v_0 \sin \left[\omega t - kr + \sigma \frac{v}{v_0} \right], \quad \sigma = \frac{r}{L} \tag{39}$$

may be regarded as a transcendental equation in v. As noted by Blackstock [28], the exact solution of an analogous equation has been obtained by Bessel in an investigation of the Kepler problem. In our case it can be written as follows [28]:

$$\frac{v}{v_0} = \sum_{n=1}^{\infty} \frac{2}{n\sigma} I_n (n\sigma) \sin n (\omega t - kr), \tag{40}$$

where $I_n(n\sigma)$ is an n-th-order Bessel function. This expression is applicable for $0 \lessgtr \sigma \lessgtr 1$. In view of the fact that this solution has

---

* An expansion analogous to (38) to fourth-order terms inclusively has been obtained by Blackstock [28] for the problem corresponding to specification of a harmonic displacement [rather than the velocity, as in (26)] at the origin. It differs significantly from (38) only in the immediate vicinity of the source, i.e., in the domain where $\sigma \ll M$, due to the finite displacement of the radiating surface.

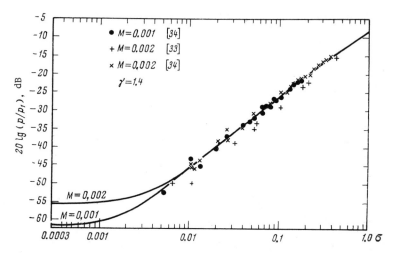

Fig. 7. Amplitude of the second harmonic versus distance. The solid curves represent the results of a calculation in [28].

been obtained independently by Fubini [30] (see also [20, 28, 31, 32]), Blackstock has suggested that it be called henceforth the Bessel–Fubini solution.

The growth of the harmonics during the propagation of a large-intensity wave in gases has been observed in several experiments [33, 34], the results of which are illustrated in Fig. 7, taken from [28]. The solid curves represent the theoretical dependence of the amplitude of the second harmonic on the distance traveled by the wave, according to Eq. (40) in the interval $1 \gtrsim \sigma \gg M$ and according to an expansion analogous to (38), but corresponding to specification of a harmonic displacement at the origin, in the interval $1 \gg \sigma \gtrsim 0$. The experimental data are in good agreement with the theoretical results.

The Bessel–Fubini solution describes the variation of the spectrum of a high-intensity wave in the interval characterized by progressive variation of the shape of an initially sinusoidal wave to the point of formation of a discontinuity over the extent of each wavelength and by monotone growth of all the harmonics.

A more general solution, taking account of the formation of a discontinuity and therefore applicable in the region $\sigma > 1$, has

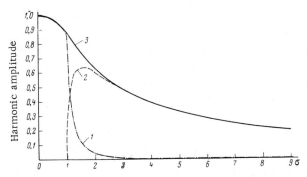

Fig. 8. Amplitude of the first harmonic versus distance.

been deduced by Blackstock [21]:

$$\frac{v}{v_0} = \sum_{n=1}^{\infty} B_n \sin n\omega y, \qquad (41)$$

where B_n is the sum of two terms, the first of which dominates in the interval $0 \lesssim \sigma \lesssim 1$ and coincides with the amplitude of the n-th harmonic in the Bessel–Fubini solution, while the second affords the principal contribution in the interval $\sigma > 1$ and assumes a simple form for $\sigma \gg 1$, so that B_n is determined by the relation

$$B_n = \begin{cases} \dfrac{2}{n\sigma} I_n(n\sigma), & 0 \lesssim \sigma \lesssim 1, \\[2mm] \dfrac{2}{n(1+\sigma)}, & \sigma \gg 1. \end{cases}$$

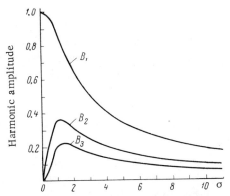

Fig. 9. Amplitude of the first, second, and third harmonics versus distance for $\Gamma \gg 1$.

The relative contribution of these two terms and the resultant amplitude B_1 of the first harmonic are presented in Fig. 8; here curve 1 corresponds to the Bessel–Fubini solution, curve 2 to a sawtooth, and curve 3 to their sum. Figure 9 illustrates the dependences of the dimensionless amplitudes B_n of the first three harmonics on the distances traversed by the wave (Figs. 8 and 9 are borrowed from [21]). It is clear that after formation of the discontinuity, in the vicinity of $\sigma \approx 1$ (see Fig. 4b), the harmonics cease their growth and instead gradually decay.

The solution (41) is applicable in the interval where the thickness of the discontinuity is negligibly small in comparison with the wavelength, $\Delta/2\pi \ll 1$, and becomes less accurate as the shock fronts wash out. In this case the variation of the spectral composition of the wave is described by the solution obtained by the Fourier series expansion of expression (34), this solution coinciding in the interval of applicability of (34) with the familiar solution of Fay [35]:

$$\frac{v}{v_0} = \frac{2}{\Gamma} \sum_{n=1}^{\infty} \frac{\sin n\omega y}{\sinh \dfrac{n(1+\sigma)}{\Gamma}} \cdot \tag{42}$$

This solution describes the gradual decay of the harmonics of a finite-amplitude wave in the domain where gradual smoothing of the wave profile occurs (see Figs. 4d and 4e). It is applicable for arbitrary values of Γ, but only in the half-space $\sigma > \pi/2$, and exactly coincides with the solution (41) in the domain of large values of Γ and σ.

At a sufficient distance from the radiator, where considerable smoothing of the previously formed discontinuities has already occurred and the wave shape again approaches a sine curve, Eq. (42) approximately implies

$$\frac{v}{v_0} = 2\Gamma^{-1} \left[e^{-\alpha x} \sin \omega y + e^{-2\alpha x} \sin 2\omega y + \cdots \right]. \tag{43}$$

It is worth noting that the coefficient of absorption of the second harmonic is twice, rather than four times, the absorption coefficient for the first harmonic and therefore increases linearly rather than as the square of the frequency, clearly as a result of the continuous transfer of energy from low- to high-frequency harmonics.

This means, for example, that the second harmonic, which has a frequency 2ω and is described by the second term of (43), decays more slowly than a small-amplitude wave of the same frequency and at large distances can have a greater intensity than a wave generated by a source of frequency 2ω [17].

Rewriting the first term of (43) in the form

$$v \approx \frac{b\omega}{\rho_0 c_0 \sigma} e^{-\alpha x} \sin \omega y, \qquad (44)$$

we readily verify that in the given domain $\sigma \gg 1$, the wave amplitude no longer depends on the particle velocity amplitude v_0 at the source. This happens because with increasing wave intensity the absorption also increases, so that the quantity of energy transported to large distances does not increase.

The foregoing is also illustrated by the graphs of Fig. 10, borrowed from [17]. They show the excess absorption experienced by the first harmonic, as characterized by the quantity $-20 \log [B_1/\exp (-\alpha x)]$, relative to the small-amplitude absorption.

It is evident that an increase in the source amplitude of the wave causes an increase in the excess absorption.

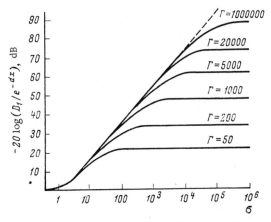

Fig. 10. Excess absorption of the first harmonic relative to the small-amplitude absorption as a function of the distance traversed by the wave for various values of Γ.

The cited graphs have been plotted on the basis of the exact solution of Eq. (23), which coincides with the solution (42) in the case of large Γ, but differs therefrom in that, unlike the Fay solution, it is equally applicable for small values of σ, i.e., near the source, so that it is possible to analyze the entire process in the buildup and smoothing of the discontinuities.

This solution was first obtained by Mendousse [15] (see also [16, 17]) and may be written in the form

$$\frac{v}{v_0} = \frac{2}{\Gamma} \frac{\partial \log \zeta}{\partial y}, \tag{45}$$

where

$$\zeta = \sum_{n=0}^{\infty} (-1)^n I_n\left(\frac{\Gamma}{2}\right) \exp\left(-n^2 \frac{\sigma}{\Gamma}\right) \cos n\omega y.$$

The complex quality of this solution renders it difficult to utilize directly. However, as remarked in [17], it is possible to

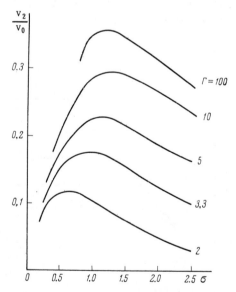

Fig. 11. Amplitude of the second harmonic versus distance for various values of Γ_1, according to a calculation in [19].

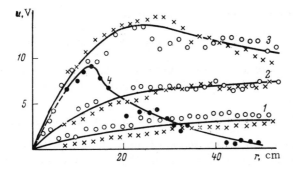

Fig. 12. Experimental dependence of the second harmonic on the distance during propagation of a 1.5-MHz wave [36].

perform a harmonic analysis of this solution and to compute the dependence of the coefficients B_n in the expansion

$$\frac{v}{v_0} = \sum_{n=1}^{\infty} B_n \sin n\omega y \qquad (46)$$

on the distance σ for various values of the parameter Γ.

The results of such a calculation for B_2, which exhibit good agreement with the data of [19] obtained by another method, are shown in Fig. 11. It is clear that during wave propagation, the amplitude of the harmonic gradually increases, reaching a maximum and then decreasing.

With an increase in Γ, the nature of the dependence changes very slightly, but the curve as a whole is displaced toward larger values of v, asymptotically approaching the limiting dependence $B_n(\sigma)$ defined by the solution (41). These results are confirmed by the experimental data of [36-42]; this is clear, for example, from Fig. 12, borrowed from [36], which shows the dependence of the second harmonic on the distance during propagation of a wave of frequency 1.5 Mc in water (curves 1-3) and transformer oil (curve 4). Curve 1 corresponds to an initial intensity $I_0 = 0.5$ W/cm^2, curve 2 to $I_0 = 2$ W/cm^2, and curve 3 to $I_0 = 7.6$ W/cm^2.

For small values of Γ the nonlinear distortions of the wave are inconsequential, and in this case it is permissible to use an approximate solution of the form [10]

$$\frac{v}{v_0} = e^{-ax} \sin \omega y + \frac{v_0}{2} \Gamma \left(e^{-2ax} - e^{-4ax} \right) \sin 2\omega y, \qquad (47)$$

which only takes account of the second harmonic of the wave and describes its gradual increase and subsequent decay.

§ 6. Absorption of a Finite-Amplitude Wave

The variation of the shape of the finite-amplitude wave during propagation affects its absorption. The absorption of a finite-amplitude wave having an initially sinusoidal form depends on the distance. Near the source, where the shape of the wave is nearly sinusoidal, the absorption is small and is described by the standard expressions of linear acoustics, then it increases, and in the region of largest wave distortion ($\sigma \approx \pi/2$), it reaches a maximum, after which it decreases. The absorption intensity at a given point of space depends on the wave amplitude, increasing with it. This may be interpreted in particular as the result of an increase in the energy dissipation at the first-order discontinuity as the pressure jump increases there. A quantitative description of the absorption of a finite-amplitude sine wave can be obtained in principle on the basis of the solutions presented in the preceding sections.

As an example, we consider the absorption of a wave in the most important practical domain $\sigma \gtrsim \pi/2$, where nonlinear effects are especially pronounced. From the physical point of view the wave absorption is most reasonably characterized by the degree of energy dissipation that it suffers. In acoustical practice, however, the amplitude absorption coefficients are of interest. We therefore begin with a consideration of the attenuation of the fundamental harmonic of a finite-amplitude wave.

Its amplitude v_1 in the domain $\sigma \gtrsim \pi/2$ decreases with distance according to the law [see (42)]

$$\frac{v_1}{v_0} = \frac{2}{\Gamma} \frac{1}{\sinh \frac{1 + \sigma}{\Gamma}}. \qquad (48)$$

Differentiating this relation, we readily find the amplitude absorption coefficient α_1' for the first harmonic [8]:

$$\alpha_1' = \frac{1}{v_1} \frac{dv_1}{dr} = \alpha \sqrt{1 + \widetilde{\Gamma}_1}, \qquad (49)$$

where $\tilde{\Gamma}_1 = 2\rho c_0 v_1 \varepsilon / b\omega$ is the instantaneous value of the parameter and, unlike the constant Γ_1, varies with decreasing v_1.

It is apparent from (49) that the nature of the dependence of the absorption coefficient on Γ_1 differs in the limiting cases of large and small values of the parameter; hence it is possible to arbitrarily distinguish, as did Gol'dberg [10], two regimes of finite-amplitude wave absorption, namely, the sawtooth regime and the sine wave regime.

We first consider the former, which occurs for $\Gamma_1 \gg 1$. In this case, as evident from (49),

$$\frac{\alpha_1'}{\alpha} = \frac{\tilde{\Gamma}_1}{2} = \varepsilon \widetilde{Re}_1, \tag{50}$$

where $\widetilde{Re}_1 = p_1' / b\omega$ is the instantaneous value of the Reynolds number. In other words, the absorption coefficient for the fundamental harmonic can greatly exceed (provided $\Gamma_1 \gg 1$) the small-amplitude absorption coefficient, its value increasing linearly as the wave amplitude.

Bearing in mind that for a sawtooth wave

$$\frac{v_p}{v_1} = \frac{\pi}{2}, \tag{51}$$

where v_p is the peak value of the particle velocity and v_1 is the amplitude of the first harmonic, we readily verify that

$$\frac{\alpha_1'}{\alpha} = \frac{4\rho_0 c_0 \varepsilon v_p}{\pi b\omega} = \frac{\alpha}{\pi} \tilde{\Gamma}, \tag{52}$$

where $\tilde{\Gamma} = 2\rho c_0 \varepsilon v_p / b\omega$.

Thus, the fundamental harmonic absorption coefficient is expressed in terms of the peak value of the sawtooth velocity. Making use of Eq. (28), we obtain an expression for the coefficient characterizing the attenuation of the peak value of the sawtooth:

$$\alpha' \equiv -\frac{1}{v_p} \frac{dv_p}{dr}. \tag{53}$$

It turns out that the value is determined by a formula that exactly coincides with (52).

Moreover, the higher-frequency harmonics are damped by exactly the same law as the peak value of the sawtooth and its first harmonic, provided their order number is not too large, so that $\Gamma_1^2 \gg n^2$.

This can be asserted by means of the solution (42), from which, in particular, we obtain the following expression for the harmonic absorption coefficient:

$$\frac{\alpha_n'}{\alpha} = \sqrt{n^2 + \Gamma_1^2}. \tag{54}$$

For $\Gamma_1^2 \gg n^2$

$$\frac{\alpha_n'}{\alpha} \approx \widetilde{\Gamma}_1 = \frac{2}{\pi}\,\widetilde{\Gamma}. \tag{55}$$

The identical character of the attenuation of not too high harmonics indicates that during the propagation of a sawtooth, its shape changes very little; hence the wave intensity absorption coefficient, which characterizes the rate of energy dissipation, is approximately equal in this case to twice the amplitude absorption coefficient.

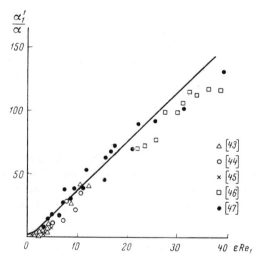

Fig. 13. Theoretical (solid curve) and experimental dependences of the first harmonic absorption coefficient on the Reynolds number.

But if the wave intensity is small, $\Gamma_1 \ll 1$, then from (49) we obtain an expression for the absorption coefficient in the sine wave regime

$$\frac{\alpha_1'}{\alpha} \approx 1 + \frac{1}{2}\tilde{\Gamma}_1^2, \tag{56}$$

from which it is apparent that the growth of the first harmonic absorption is small (since $\Gamma_1 \ll 1$). In this case the absorption coefficient for the first harmonic differs from the absorption coefficient for the higher harmonics [see (54)] and, accordingly, is no longer simply equal to half the intensity absorption coefficient. Here we discover that in the absorption of a finite-amplitude wave whose shape changes during the course of propagation, the diminution of the wave energy and the attenuation of its harmonics generally follow different laws.

Summarizing the foregoing and returning to Eq. (49), we note that the absorption coefficient for the fundamental harmonic in the stabilization domain $\sigma \gtrsim \pi/2$ at first grows as the square, then linearly as Γ_1 and, hence, as the wave amplitude.

This is distinctly seen in the graph of Fig. 13, which has been plotted on the basis of Eq. (49). Also shown in the figure are the results of an experimental measurement of the absorption coefficient in water according to the data of various authors [43-47].

We see that the experimental and theoretical results are in fairly good agreement. A certain scatter of the points is clearly attributable to certain experimental problems (generation of a plane wave, the need for local, rather than space-averaged measurements of the absorption, etc.).

The problems of experimental procedure in studies of the propagation of a finite-amplitude wave have been investigated in detail in Zarembo and Krasil'nikov's book [6], and we shall not discuss them at this point.

Chapter 2

Spherical and Cylindrical Waves
of Finite Amplitude

§1. Formation of First-Order
Discontinuities in Spherical and
Cylindrical Waves

The propagation of spherical and cylindrical waves of finite amplitude from the qualitative point of view is largely akin to the propagation of plane waves. In that case, nonlinear effects cause variations in the shape of the propagating wave, and this can lead to the onset of shock waves, accompanied by the intense absorption of sound. In the quantitative aspect, however, there are important distinctions, which are manifested, in particular, in the different rate of growth of nonlinear distortions during the propagation of spherical and cylindrical waves, due to the amplitude variation of such waves during propagation as the result of their divergence (or convergence). For the description of the propagation of spherical and cylindrical finite-amplitude waves, the set of hydrodynamic equations and the state equation are conveniently reduced, as in the analysis of medium-intensity plane waves, to a single approximate equation of the form [48-51]

$$\frac{\partial v}{\partial r} + n\frac{v}{r} - \frac{\varepsilon}{c_0^2}v\frac{\partial v}{\partial y} = a\frac{\partial^2 v}{\partial y^2}, \tag{57}$$

where n = 1, $\frac{1}{2}$, 0 for spherical, cylindrical, and plane waves, respectively. For n = 1 or $\frac{1}{2}$, Eq. (57) is only applicable in the domain $kr \gg 1$, $y = t \mp (r - r_0)/c_0$ for divergent or convergent waves, respectively. The rest of the notation is the same as in Eq. (21).

Equation (57) is conveniently written in the alternative form

$$\frac{\partial W}{\partial z} - W \frac{\partial W}{\partial \omega y} = \Gamma^{-1} B\,(z)\, \frac{\partial^2 W}{\partial\,(\omega y)^2}\,, \tag{58}$$

where

$$W = \begin{cases} \dfrac{v}{v_0}\dfrac{r}{r_0} \\[2mm] \dfrac{v}{v_0}\sqrt{\dfrac{r}{r_0}}\,, \\[2mm] \dfrac{v}{v_0} \end{cases} z = \begin{cases} \sigma_0\left|\ln\dfrac{r}{r_0}\right| \\[2mm] 2\sigma_0\left|\sqrt{\dfrac{r}{r_0}}-1\right|, \\[2mm] \sigma \end{cases} B\,(z) = \begin{cases} e^{z/\sigma_0}, & n = 1 \\[2mm] 1+\dfrac{z}{2\sigma_0}, & n = \dfrac{1}{2} \\[2mm] 1, & n = 0 \end{cases} \tag{59}$$

$$\sigma = \varepsilon M k r, \quad \Gamma = (\gamma + 1)\,\rho c_0 v_0/b\omega.$$

Note that in terms of W(z, ωy), Eq. (58) for spherical and cylindrical waves is similar to Eq. (23) for plane waves, except that the coefficient B is dependent on the spatial coordinates. This means that the spherical (or cylindrical) wave may be regarded as a plane wave in a medium whose viscosity varies in the direction of wave propagation [48-49].

Consider the propagation of a divergent or convergent traveling spherical (or cylindrical) wave of finite amplitude, harmonic at the point $r = r_0$:

$$\frac{v}{v_0} = \sin \omega y. \tag{60}$$

If the intensity of the wave is sufficient ($\Gamma \gg 1$), then in the first stage of wave propagation the right-hand side of Eq. (58) may be neglected. The remaining left-hand side

$$\frac{\partial W}{\partial z} - W \frac{\partial W}{\partial \omega y} = 0 \tag{61}$$

exactly coincides with the left-hand side of Eq. (23), permitting us at once to write down the solution of the stated problem:

$$\omega y = - zW + \arcsin W, \tag{62}$$

where z and W are determined by Eqs. (59). This solution, like the solution (26) in the plane wave case, describes the progressive distortion of the propagating wave (see Figs. 3 and 4). Here the change in shape of the profile of the spherical (cylindrical) wave, as characterized by the coefficient z in the first term of the right-hand side of (61), is proportional to the logarithm (square root) of the distance traversed by the wave, as opposed to the plane-wave case, where the dependence is linear.

The distortion of the wave profile leads to the formation of a discontinuity at the point z determined from the condition z = 1, so that the position of the discontinuity is determined by the conditions:

for a spherical wave,

$$\sigma_0 \left| \ln \frac{r}{r_0} \right| = 1, \ n = 1; \tag{63}$$

for a cylindrical wave,

$$2\sigma_0 \left| \sqrt{\frac{r}{r_0}} - 1 \right| = 1, \ n = \frac{1}{2}; \tag{64}$$

for a plane wave,

$$\sigma = 1, \ n = 0. \tag{65}$$

From (63) we can determine the coordinate of the discontinuity in a divergent spherical wave

$$r_1^{(-)} = r_0 e^{1/\sigma_0} \tag{66}$$

and in a convergent spherical wave

$$r_1^{(+)} = r_0 e^{-1/\sigma_0}. \tag{67}$$

In order to highlight the distinction between these expressions, we write out the equations deduced from them for the path traveled by the wave to the buildup of the discontinuity

$$l^{(-)} = (r_1^{(-)} - r_0) = r_0 (e^{1/\sigma_0} - 1); \tag{68}$$

$$l^{(+)} = (r_0 - r_1^{(+)}) = r_0 \left(1 - \frac{1}{e^{1/\sigma_0}} \right). \tag{69}$$

It is clear that, with a decrease in $\sigma_0 = \varepsilon M k r_0$ (which can, for example, indicate a reduction in the amplitude of the radiated wave or a reduction in the radius of the radiating sphere), the length of the path to the formation of discontinuity increases sharply in a divergent wave, whereas in a convergent wave it never exceeds r_0.

For example, let a sphere of radius $r_0 = 3$ cm radiate a wave into water at a frequency $\nu = 10^6$ cps and pressure amplitude $p_0' = 25$ atm; here $\Gamma \approx 10^2$, $M \approx 10^{-3}$, and $\sigma_0 \approx 5 \cdot 10^4$. Then in a divergent initially sinusoidal wave the discontinuity is formed at a distance $l \approx 5r_0$. If the amplitude of the radiated wave is reduced by one half, this distance becomes approximately equal to $37r_0$. For a convergent spherical traveling wave, which can be approximately realized by means of a focusing system, we obtain, respectively,

$$l^{(+)} \approx \frac{5}{6} r_0 \quad \text{and} \quad l^{(+)} \approx \frac{37}{38} r_0.$$

Moreover, from (64) we obtain the coordinate of the discontinuity in a divergent cylindrical wave

$$r_1^{(-)} = r_0 \left[1 + \frac{1}{2\sigma_0}\right]^2 \tag{70}$$

and in a convergent cylindrical wave

$$r_1^{(+)} = r_0 \left[1 - \frac{1}{2\sigma_0}\right]^2. \tag{71}$$

It is interesting that, as apparent from (71), the formation of a discontinuity in a convergent cylindrical wave is only possible for values of $\sigma_0 \geq \frac{1}{2}$; for smaller values one obtains negative values of r_1, which are meaningless and indicate in this case that the formation of a discontinuity is impossible in general.

After the formation of a discontinuity at the point $z = 1$, the process of distortion of the wave profile continues, resulting in the formation of a sawtooth wave at the point $z_2 = \pi/2$ (see Fig. 4), its amplitude decaying according to a law deduced from (62) (see Fig. 3), with the following simple form in the domain $z > \pi$

$$W_p = \frac{\pi}{1 + z}. \tag{72}$$

It is a simple matter on the basis of (72) to obtain a solution describing sawtooth spherical, cylindrical, and plane waves:

$$W = \frac{1}{1+z}(-\omega y + \pi).$$ (73)

It is verifiable by direct substitution that this solution satisfies Eq. (61).

For clarification we rewrite (73) in explicit form

for a spherical wave,

$$\frac{vr}{v_0 r_0} = \frac{1}{1 + \sigma_0 \left| \ln \dfrac{r}{r_0} \right|}(-\omega y + \pi), \quad n = 1,$$ (74)

for a cylindrical wave,

$$\frac{v}{v_0}\sqrt{\frac{r}{r_0}} = \frac{1}{1 + 2\sigma_0 \left| \sqrt{\dfrac{r}{r_0}} - 1 \right|}(-\omega y + \pi), \quad n = \frac{1}{2},$$ (75)

and for a plane wave

$$\frac{v}{v_0} = \frac{1}{1+\sigma}(-\omega y + \pi), \quad n = 0.$$ (76)

This clearly exhibits the nonexponential behavior of the attenuation of sawtooth waves with distance.

Expressions (74) and (75) are valid either for divergent or for convergent waves.

The applicability of the solutions (74)-(76) is limited to the domain of existence of the sawtooth, where the thickness of the discontinuity is negligible relative to the wavelength. This condition prevails if the wave intensity is large enough, i.e., $\tilde{\Gamma} > 1$. As the wave propagates, its intensity diminishes along with gradual smoothing and diffusion of the shock fronts. This process is described in the plane-wave case by Eq. (34), which gives the shape of one period of the wave and is an exact solution of Eq. (23).

It turns out that the process of diffusion or spreading of the discontinuities in a spherical or cylindrical wave can also be described by a solution similar to (34), but, unlike the latter, it only approximately satisfies Eq. (58) [49-51]. This solution may be written

$$W = \frac{1}{1+z}\left[-\omega y + \pi \tanh \frac{\omega y}{\Delta}\right], \tag{77}$$

where the values of z and W are determined by expressions (3), and the dimensionless thickness of the shock front is equal to

$$\Delta = \frac{2}{\pi} \frac{1+z}{\Gamma}\left(\frac{r}{r_0}\right)^n; \tag{78}$$

$n = 1$, $\frac{1}{2}$, 0 for spherical, cylindrical, and plane waves, respectively.

This solution is applicable under the conditions [49-50]:

$$\begin{array}{l} \sigma_0^2 \, \Gamma \gg \sigma \text{ in the spherical wave case } (n = 1); \\ 2\sigma_0 \Gamma \gg 1 \text{ in the cylindrical wave case } (n = \frac{1}{2}). \end{array} \tag{79}$$

As apparent from (78), the thickness Δ of the shock front does not remain stationary for two reasons.

First, the shock front spreads due to the reduction in its amplitude as a result of absorption, so that Δ increases as $|\ln(r/r_0)|$ in a spherical wave and as $\sim 2\sigma_0 |1 - (r/r_0)^{\frac{1}{2}}|$ in a cylindrical wave.

This mechanism of smoothing of the discontinuities is completely analogous to the corresponding mechanism acting in the plane-wave case (where Δ grows linearly as the path traversed by the wave). It is identically manifested in the case of divergent as well as convergent waves [49-51].

Second, and most important, Δ varies due to the presence of the factor r/r_0 [or $(r/r_0)^{\frac{1}{2}}$ in the cylindrical case, due to the amplitude variation caused by geometrical convergence, which acts differently in the case of divergent, $r/r_0 > 1$, and convergent, $r/r_0 < 1$, waves]. This produces qualitative differences between the propagation of divergent and convergent waves. Thus, in the propagation of a divergent wave, after the formation of a sawtooth profile in the vicinity of the point $z \approx z_2 = \pi/2$, there occurs a monotone spreading of the formed discontinuities due to the reduction in wave amplitude both as a result of absorption and as a result of divergence, so that, at a certain distance $z = z_3$, the dimensionless thickness Δ of the shock front becomes comparable with π.

This means that the process of smoothing of the fronts leads to inversion of the sawtooth back to a sine wave and the invalidation of the solution (78). The position of z_3 is determined from the condition $\Delta \approx \pi$. According to (78), this condition leads in the spherical-wave case to an equation for the dimensionless coordinate σ_3 of the formation of the sine wave

$$\sigma_3 \ln \frac{\sigma_3}{\sigma_0} \approx \frac{\pi^2}{2} \Gamma. \tag{80}$$

The analogous equation in the case of a divergent cylindrical wave is written

$$\sigma_3 \approx \pi^2 \frac{\Gamma}{2}. \tag{81}$$

This establishes the far limit of applicability of the solution (78) describing the propagation of a sawtooth wave.

As evident from (66) and (68), with a decrease in σ_0 the coordinate of the formation of discontinuity rapidly increases, approaching σ_3, this effect shrinking and eventually (at $\sigma_1 \approx \sigma_3$) annihilating completely the domain of existence of the sawtooth.

In the case of a convergent wave the factors dictating the amplitude variation of the wave due to absorption and geometric considerations have the opposite effect. This causes the dimensionless thickness of the shock front after the formation of discontinuity to increase somewhat at first, reach a maximum, and then begin to diminish (Fig. 14). The maximum attainable thickness of the front is determined by the value of the parameter σ_0, increasing as it increases. For large enough values of the parameter, such that $\sigma_0 \gg \Gamma$, the maximum thickness of the shock front can exceed π, implying that in the interval between the points $\sigma^{(1)}$ and $\sigma^{(2)}$, at which $\Delta \approx \pi$, the wave shape is nearly sinusoidal.

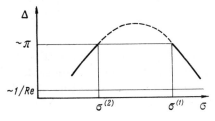

Fig. 14. Dimensionless thickness of the shock front versus distance for a convergent wave.

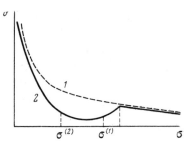

Fig. 15. Amplitude of a convergent spherical wave versus distance. 1) Linear theory; 2) influence of non-linear effects.

In other words, the absorption of the sawtooth formed at the point z_1 turns out to be so strong as to cause a local reduction in the amplitude of the convergent wave during its propagation and a concomitant increase in the thickness of the shock front. This is illustrated in Fig. 15, in which curve 1 represents the amplitude variation of a convergent spherical wave without regard for absorption, and curve 2 represents the same with regard to absorption of the sawtooth wave. Thus, by means of (74) we readily verify that, for example, in a convergent spherical wave

$$\frac{dv_p}{dr} \sim -1 + \frac{\sigma_0}{1 + \sigma_0 \ln \frac{r_0}{r}}, \tag{82}$$

from which it is evident that, for sufficient values of σ_0, the derivative dv_p/dr can become positive in a certain interval of r values, and this indicates a reduction in the wave amplitude v_p in that interval.

§ 2. Variation of the Spectral Composition of Spherical and Cylindrical Waves

The variation of the spectral composition of spherical and cylindrical waves due to nonlinear distortions during propagation are easily described on the basis of the solutions presented in the preceding section.

We first consider a wave of sufficiently large intensity ($\Gamma \gg 1$), harmonic at the point $r = r_0$. The variation of the profile of such a wave leads to the formation of a discontinuity at the point z_1, and its subsequent attenuation is described by the solution (62), which may be rewritten

$$W = \sin{(\omega y + zW)}. \tag{83}$$

Treating this expression as a transcendental equation in W(ωy, z), we can find its solution in Fourier series form as in the plane-wave case:

$$W = \sum_{n=1}^{\infty} B_n \sin{n\omega y}, \tag{84}$$

where B_n is the sum of two terms, the first of which dominates in the interval $0 \lesssim z \lesssim 1$ and corresponds to the amplitude of the n-th harmonic in the Bessel – Fubini solution in the plane wave case, while the second yields the principal contribution in the interval z > 1 and assumes a simple form for z $\gg$ 1 (see Fig. 7) [50]. This enables us to write

$$B_n = \begin{cases} \dfrac{2}{nz} I_n(nz), & 0 \lesssim z \lesssim 1, \\[2mm] \dfrac{2}{n(1+z)}, & z \gg 1, \end{cases} \tag{85}$$

where $I_p(nz)$ is an n-th order Bessel function.

With (85) we investigate the variation of the harmonic composition of the wave over its initial interval of propagation in the domain z $\ll$ 1. Expanding $I_p(nz)$ in the interval of small values of the argument, we obtain

$$W \approx \sin{\omega y} + \frac{z}{2} \sin{2\omega y} + \ldots \tag{86}$$

Inserting here the values of z and W from (59), we write the following relations in explicit form (see also [1, 13, 52]):

for a spherical wave,

$$\frac{vr}{v_0 r_0} = \sin{\omega y} + \frac{\sigma_0}{2} \left| \ln{\frac{r}{r_0}} \right| \sin{2\omega y} + \cdots; \tag{87}$$

for a cylindrical wave (divergent and convergent),

$$\frac{v}{v_0} \sqrt{\frac{r}{r_0}} = \sin{\omega y} + \sigma_0 \left| \sqrt{\frac{r}{r_0}} - 1 \right| \tag{88}$$

for a plane wave,

$$\frac{v}{v_0} = \sin{\omega y} + \frac{1}{2} \frac{r}{L} \sin{2\omega y}. \tag{89}$$

It is clear that the growth of the harmonics occurs at different rates in spherical, cylindrical, and plane waves, a fact that is attributable to the disparate character of the amplitude variations of these waves with distance due to geometrical divergence (or convergence).

The amplitudes of the harmonics increase until the instant of formation of discontinuity at $z = 1$, then diminish (see Fig. 9). The maximum amplitude of the second harmonic (at $z = 1$), as apparent from (86), amounts to half the amplitude of the wave at the fundamental frequency. The solution (85) is applicable as long as the wave intensity is large enough, $\tilde{\Gamma} \gg 1$, so that the width of the discontinuities is negligibly small. In the interval where this condition does not hold, the diffusion of the shock fronts and the corresponding attenuation of the high-frequency components of the spectrum become appreciable.

This process is described in the plane-wave case by the familiar Fay solution, which may be regarded as the result of expanding the solution (77) in a Fourier series with $n = 0$ (plane wave).

In the case of spherical and cylindrical waves, it is also possible to carry out an analogous transformation of the solution (77), yielding the more general expression

$$W = \frac{2}{\Gamma}\left(\frac{r}{r_0}\right)^n \sum_{n=1}^{\infty} \frac{\sin n\omega y}{\sinh\left[n\frac{1+z}{\Gamma}\left(\frac{r}{r_0}\right)^n\right]} , \tag{90}$$

where, as before, $n = 1$, $\frac{1}{2}$, 0 for spherical, cylindrical, and plane waves, respectively.

It is important, however, to note that, unlike the corresponding solution for plane waves ($n = 0$), the solution (90) for spherical and cylindrical waves ($n = 1$, $\frac{1}{2}$) is approximate by virtue of the error of expression (77) in these cases and the limitation of its validity to the domain in which conditions (79) hold, so that it is impossible to analyze the entire process of inversion of the sawtooth to a sine wave.

Finally, if the wave intensity is small from the outset and the nonlinear effects are small, during propagation only the second harmonic is significant.

In this case the propagation of a finite-amplitude wave may be described by the solutions obtained by the Krylov–Bogolyubov method. The approximate solutions obtained for spherical [13] and cylindrical [49] waves in the interval $kr \gg 1$ show that during the propagation of a wave initially monochromatic at the point $r = r_0$, its spectral decomposition includes the second harmonic, which grows, reaches a maximum, then gradually decays:

$$\frac{vr}{v_0 r_0} = e^{-\alpha \,(r-r_0)} \sin \omega y - \sigma_0 \left| \ln \frac{r}{r_0} \right| e^{-2\alpha \,(r-r_0)} \sin 2\omega y. \tag{91}$$

This expression is applicable if the relation

$$\frac{1}{2}\sigma_0 \left| \sqrt{\frac{r}{r_0}} - 1 \right| e^{-2\alpha \,(r-r_0)} \ll 1 \tag{92}$$

is fulfilled, which, as apparent from (91), expresses the condition of smallness of the second harmonic relative to the first.

The corresponding equations for a cylindrical wave have the form

$$\frac{v}{v_0}\sqrt{\frac{r}{r_0}} = e^{-\alpha \,(r-r_0)} \sin \omega y + \sigma_0 \left| \sqrt{\frac{r}{r_0}} - 1 \right| e^{-2\alpha \,(r-r_0)} \sin 2\omega y; \tag{93}$$

$$\frac{1}{2}\sigma_0 \left| \sqrt{\frac{r}{r_0}} - 1 \right| e^{-2\alpha \,(r-r_0)} \ll 1. \tag{94}$$

§ 3.   Absorption of Spherical and Cylindrical Waves of Finite Amplitude

The energy dissipation of spherical and cylindrical finite-amplitude waves occurs as in the plane-wave case, with the same kind of wave profile, increasing with the steepness of the wave fronts.

With an increase in the sound intensity, the steepness of the wave fronts increases, and this in turn gives rise to a growth of the wave absorption. Analyzing the dependence of the absorption coefficient on the wave amplitude, we can, as mentioned earlier, arbitrarily recognize two regimes of the absorption of a finite-amplitude wave in correspondence with the two limiting expressions for the absorption coefficient in the plane-wave case, for $\Gamma \ll 1$ and for $\Gamma \gg 1$. The first regime corresponds to a sine wave, when the variations of the wave shape are small; the wave is damped out before nonlinear effects can evolve in it. The second regime corresponds to a sawtooth, when the nonlinear effects lead to the formation of discontinuities in the vicinity of $z \approx z_1 = 1$. The quantitative description of the absorption of spherical and cylindrical finite-amplitude waves in the first regime and in the transition region is difficult in view of the inapplicability in this case (for $\Gamma \lesssim 1$) of the solution (90), which is basic to the calculation of the plane wave absorption coefficient, in which case the corresponding solution is valid for any Γ. It is possible, however, to introduce the following approximate estimate of the absorption coefficient in the interval $\Gamma \ll 1$. It can be shown on the basis of the equations of Chap. 1 that, in the domain of small Γ, the absorption coefficient for the first harmonic of a plane wave is proportional to the square of the amplitude ratio of the second to the first harmonic:

$$\frac{\alpha_1'}{\alpha} \approx 1 + \frac{1}{2}\left(\frac{v}{v_1}\right)^2.$$

$$(95)$$

This relation between the degree of wave distortion characterized by the relative magnitude of the second harmonic and the absorption coefficient must be fulfilled in the case of spherical and cylindrical waves as well, making it possible by means of (91) and (93) to assess the absorption of these waves:

$$\frac{\alpha_1'}{\alpha} \approx 1 + \frac{1}{2}\left[\sigma_0\left|\ln\frac{r}{r_0}\right|\right]^2 e^{-2\alpha\,(r-r_0)};$$

$$(96)$$

$$\frac{\alpha_1'}{\alpha} \approx 1 + \frac{1}{2}\left[2\sigma_0\left|\sqrt{\frac{r}{r_0}}-1\right|\right]^2 e^{-2\alpha\,(r-r_0)}.$$

$$(97)$$

The validity of these expressions can also be confirmed directly, carrying out a calculation analogous to that in [10] for the plane-wave case.

In the considerably more important case from the practical point of view of large intensities, when a sawtooth is formed, the value of the absorption coefficient characterizing the reduction in wave amplitude due to energy dissipation is readily determined by differentiation of (72) (see also [13, 25, 53-55])

$$\alpha' \equiv -\frac{1}{W_p}\frac{dW_p}{dr} = \frac{\varepsilon k v_p}{\pi c_0};\qquad(98)$$

$$\frac{\alpha'}{\alpha} = \frac{1}{\pi}\tilde{\Gamma} = \frac{1}{2}\tilde{\Gamma}_1,\qquad(99)$$

where $\tilde{\Gamma} = 2\rho c v_p \varepsilon / b\omega$; $\tilde{\Gamma}_1 = 2\rho c v_1 \varepsilon / b\omega$; $v_1 = v_p\frac{2}{\pi}$.

As expected, the amplitude absorption coefficient for a saw-tooth turns out to be the same for plane, spherical, and cylindrical waves. In the domain where the variations of the wave shape are inconsequential, the intensity absorption coefficient is simply equal to twice the value of α'.

We need to point out certain characteristics of the absorption of a finite-amplitude wave. In cases when the first term in the denominators of (74)-(76) is negligible relative to the second term (this is admissible, for example, at sufficient distances from the source), the attenuation of the peak value in a sawtooth is described by the expressions:

for a plane wave,

$$v_p \approx \frac{\pi c_0}{\varepsilon k r},\quad n = 0;\qquad(100)$$

for a spherical wave,

$$v_p \approx \frac{\pi c_0}{\varepsilon k r \ln\frac{r}{r_0}},\quad n = 1;\qquad(101)$$

for a cylindrical wave,

$$v_p \approx \frac{\pi c_0}{\varepsilon k 2\sqrt{r}\,(\sqrt{r}-\sqrt{r_0})},\, n = \frac{1}{2}.\qquad(102)$$

It is interesting to compare these attenuation laws with those for a single triangular pulse with a shock front having the form [1]:

for a plane wave,

$$v_p \sim \frac{1}{\sqrt{r}},\quad n = 0;\qquad(103)$$

for a spherical wave,

$$v_p \sim \frac{1}{r \sqrt{\ln \frac{r}{r_0}}}, \, n=1;$$ (104)

for a cylindrical wave,

$$v_p \sim \frac{1}{r^{3/4}}, \quad n = \frac{1}{2}.$$ (105)

It is clear that the attenuation of the pulse is slower than that of the sawtooth wave. The reason for this disparity can be explained as follows. As already indicated, the absorption of a sawtooth wave may be regarded formally as a clipping of the vertices of its profile as a result of the rear points of the wave profile catching up with its front (see Fig. 2). It is obvious that the effectiveness of this clipping process depends on how rapidly the points of the profile overtake the wave front, i.e., on the ratio of the velocities of points on the profile and the velocity of the front. But the velocity of the front of a first-order discontinuity is equal to [1]

$$c_0 + \varepsilon \frac{v_1 + v_1}{2},$$ (106)

where v_1 and v_2 are the velocities of the medium before and after discontinuity.

Consequently, the velocity of the front of a single pulse with a velocity jump equal to v_p is

$$c_0 + \varepsilon \frac{v_p}{2}.$$ (107)

The velocity of the front of a sawtooth wave of equivalent amplitude is equal to

$$c_0 + \varepsilon \frac{-v_p + v_p}{2} = c_0.$$ (108)

Therefore, the pulse front moves ahead more rapidly than the sawtooth waves. As apparent from (100)-(102) (see also the description of difficult for points on the profile to overtake the front, and the process of clipping of the vertex occurs more slowly in this case, resulting in weaker attenuation of the single pulse relative to the sawtooth wave.

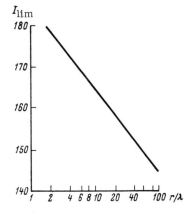

Fig. 16. Intensity limit for a plane wave versus distance.

Another peculiarity is worth noting in the propagation of sawtooth waves. As apparent from (100)-(102) (see also the description of Fig. 10), the amplitude of a sawtooth wave in the domain $\sigma \gg 1$ no longer depends on the oscillation amplitude of the radiator since the absorption increases with the wave intensity. In other words, no matter how large the intensity of the radiated wave, its intensity at a given point of space (in the domain $\sigma \gg 1$) cannot become larger than some limiting value admitted by the medium.

These limiting intensities are easily derived from expressions (100)-(102) bearing in mind that for a sawtooth wave

$$I = \rho_0 \frac{v_p^2}{3} c_0;$$ (109)

$$I_{\text{lim-pl}} = \frac{\rho_0}{3} \frac{\pi^2 c_0^3}{\varepsilon^2 k^2 r^2};$$ (110)

$$I_{\text{lim-sph}} = \frac{\rho_0}{3} \frac{\pi^2 c_0^3}{\varepsilon^2 k^2 r^2 \left(\ln \dfrac{r}{r_0} \right)^2};$$ (111)

$$I_{\text{lim-cyl}} = \frac{\rho_0}{3} \frac{\pi^2 c_0^3}{\varepsilon^2 k^2 r 4 (\sqrt{r} - \sqrt{r_0})^2}.$$ (112)

For illustration a graph taken from [53] is presented in Fig. 16, showing the dependence of the limiting intensity on the distance from the radiator for the case of a plane wave [Eq. (110)].

Chapter 3

Dependence of the Gain
of a Focusing System
on the Sound Intensity

§ 1. Introduction

Various types of focusing systems are widely used for the generation of high-intensity ultrasound. The gain of such systems is determined from diffraction considerations, from which it is expressed in terms of the wavelength and geometrical parameters of the system [56]. Normally, in calculations of the gain, the wave absorption is neglected, since the inclusion of absorption in the linear-acoustical approximation merely adds insignificant corrections when the sound frequency is not too great. Thus, in the operation of a focusing system in water at frequencies up to 5 or 6 Mc, the inclusion of absorption within the framework of linear acoustics only changes the gain by a few percent.

It is important to note, however, that, in focusing systems, even at relatively small sound intensities, appreciable nonlinear effects can occur near the radiating surface of the system. They are manifested, in particular, in a change in the shape of the wave during its propagation. A wave of sinusoidal form near the surface of the focusing system becomes a sawtooth on the path to the focus, with a considerable increase in the attendant absorption.

With an increase in intensity, nonlinear effects become more strongly pronounced, the wave turns into a sawtooth earlier, and the absorption coefficient of the sawtooth increases.

These two factors cause an increase in the wave absorption in the vicinity of the focus and a decrease in the gain of the focusing system as the wave intensity is increased,* effects that have in fact been verified experimentally [58].

Relying on the preceding notions concerning the transition of an initially sinusoidal wave into a sawtooth and making use of the laws derived for the attenuation of spherical and cylindrical saw-tooth waves, we can easily make an approximate calculation of the reduction in gain of a focusing system. The corresponding calculations for the use of spherical and cylindrical concentrators will be presented in the ensuing sections [59].

§ 2. Spherical Concentrator

Consider a radiator whose surface has the configuration of a spherical surface segment of radius F with an exit aperture of diameter 2d, as shown schematically in Fig. 17.

Neglecting diffraction effects and low-amplitude absorption, whose inclusion, as already noted, usually contributes only minor corrections, we assume that there propagates from the surface of the concentrator to the surface of a sphere of radius r_f a convergent spherical wave approximately described by the expression

$$vr = Fv_0 \sin \omega y, \qquad\qquad (113)$$

where r is the distance from the focus of the concentrator and v_0 is the velocity amplitude at the radiating surface for r = F.

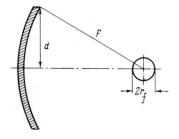

Fig. 17. Diagram pertinent to the analysis of a spherical concentrator.

---

*The conceptual possibility of such an effect has been indicated in [57].

We define r_f so that the particle velocity amplitude at the point r_f

$$v_f = \frac{Fv_0}{r_f} \tag{114}$$

will be equal to the velocity amplitude at the focus as predicted by diffraction theory.

If the radiator operates so that the oscillation amplitude is the same over its whole surface, the velocity gain of the focusing system has the form [56]

$$K_v = \frac{kF}{2}\sin^2\alpha_m, \tag{115}$$

where α_m is the aperture angle of the system.

By definition the velocity gain is the ratio of the velocity v_f at the focus (directed along the axis of the focusing system) to the velocity v_0 at the radiator surface, so that $K_v = v_f/v_0$, whereupon, according to (114), we obtain

$$r_f = \frac{F}{K_v}. \tag{116}$$

During propagation, the wave changes its form due to nonlinear effects, and if the wave intensity is sufficient, a sawtooth is formed. The site of formation of the sawtooth wave is determined by Eq. (67), which may be written in the form

$$r_1 = Fe^{-1/\sigma_0}, \tag{117}$$

where $\sigma_0 = \varepsilon kMF$.

The propagation of the sawtooth over the interval from r_1 to the focus $r = r_f$ is accompanied by powerful absorption, described by Eq. (74). From this formula we readily determine the peak value v_{pf} of the particle velocity at the focus*:

$$v_{pf} = \pi K_v v_0 (1 + \sigma_0 \ln K_v)^{-1}. \tag{118}$$

---

* The minor discrepancy of this and the ensuing equations from the corresponding results of [4] is attributable to the slightly different nature of the approximations invoked for calculation of the absorption of the sawtooth on the interval from r_1 to r_f. It is readily established, however, that this difference lessens as the role of nonlinear effects increases and vanishes altogether when those effects are large enough to render the first term in the denominator of (118) negligible relative to the second term.

As evident from (118), with an increase in the particle velocity amplitude v_0 at the radiator surface, the particle velocity at the focus increases more slowly, asymptotically approaching a constant value, which is determined by the relation

$$v_{pf\text{-lim}} = \frac{\pi c_0 K_v}{\varepsilon k F \ln K_v}, \tag{119}$$

i.e., saturation occurs, as always happens in the propagation of sawtooth waves. Consequently, there is a limiting sound intensity, independent of the intensity at the radiator, that can be obtained at the focus of the given system.†

Recognizing that the intensity I of the sawtooth wave is related to the peak value v_p of the particle velocity by the relation

$$I = \frac{\rho_0 v_p^3}{3} c_0, \tag{120}$$

it is a simple matter to determine the limiting intensity at the focus:

$$I_{f\text{-lim}} = \frac{\rho_0 c_0}{3} \left[\frac{\pi c_0 K_v}{\varepsilon k F \ln K_v} \right]^2. \tag{121}$$

Clearly, the limiting intensity at the focus of the system increases with the wavelength $\lambda = 2\pi/k$ and decreases as the focal length F is increased.

The attenuation of the sawtooth on the interval from r_1 to r_f causes a reduction in the gain of the system.

Without regard for attenuation, the velocity gain K_v, which is equal by definition to the ratio of the velocity amplitude at the focus to the velocity amplitude at the radiator surface, is determined by relation (115), and the gain K_v' with regard for nonlinear attenuation is expressed by the equation deduced from (118):

$$\frac{K_v'}{K_v} = [1 + \sigma_0 \ln K_v]^{-1}. \tag{122}$$

Clearly, it diminishes with increasing wave amplitude $[\sigma_0 = \varepsilon(v_0/c_0)kr_0]$. We now calculate the variation of the intensity gain. For small intensities, when nonlinear effects are unappreciable,

---

* This fact was brought to the author's attention by L. D. Rozenberg.

in the case of small concentrator aperture angles, the intensity gain is defined by the expression

$$K_I = (K_v)^2. \tag{123}$$

But if the nonlinear effects are appreciable and a wave with intensity

$$I_0 = \frac{\rho_0 v_0^2}{2} c_0 \tag{124}$$

near the radiator surface becomes a sawtooth on the path to the focus, with a peak value v_{pf} of the particle velocity at the focus as determined by Eq. (118), then its intensity at the focus, in accordance with (120), is equal to

$$I_f = \frac{\rho_0 v_{pf}^2}{3} c_0. \tag{125}$$

Then the intensity gain with regard for nonlinear attenuation can be determined as

$$K_I' = \frac{2}{3} \left(\frac{v_{pf}}{v_0}\right)^2. \tag{126}$$

Equations (118), (119), (21), and (122) are applicable under the condition

$$z\,(F) = \sigma \ln K_v \geqslant 0. \tag{127}$$

Thus, it is apparent from (122) that, for $z(F) < 0$, it turns out that $K_v' > K_v$; but this is impossible, as absorption can only reduce the gain. A more detailed analysis reveals that the condition $z(F) \leq 0$ corresponds to the situation when nonlinear effects are insignificant, i.e., the wave cannot attain a sawtooth form on the interval from F to r_f. In this case $K_v' = K_v$, and $v_f = K_v v_0$ (see Fig. 18, which schematically represents the dependence of K_v' on v_0). There-

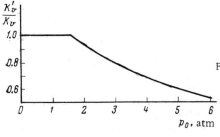

Fig. 18. Gain of a focusing system versus the wave amplitude.

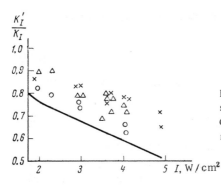

Fig. 19. Decrease of the gain of a focusing system with increasing sound intensity. Comparison between theory and experimental data of [58].

fore, the magnitude of the parameter $z(F)$ can be used to estimate the influence of nonlinear effects on the gain of the focusing system. If $z(F) \leq 0$, the distortions of the wave are insignificant, it stays sinusoidal, and the gain is determined by the usual expression (115). If $z(F) > 0$, nonlinear effects cause an appreciable variation in the wave shape, its absorption grows, and the gain decreases, in correspondence with (122). These results are corroborated by the experimental data of [58, 59].

In [58] the gain has been measured of a spherical lens with focal length $F = 10.5$ cm. The measurements were carried out by the calorimetric method in the continuous mode at a frequency of 2 MHz. The results of this study in the domain where $\sigma_0 \ln K_v > 0$ are shown in Fig. 19. The solid curve has been plotted on the basis of Eqs. (118) and (127).

An investigation of the sound fields in the vicinity of the focus of several spherical concentrators has been investigated in [59], in which the presence of nonlinear absorption effects for $\sigma_0 \ln K_v > 0$ and their absence for $\sigma_0 \ln K_v \leq 0$ were observed in correspondence with the theoretical estimates.

§3. Cylindrical Concentrator

We now consider a concentrator whose radiating surface represents a segment of a cylindrical surface of radius F with exit aperture width $2d$ (see Fig. 17). The analysis is entirely analogous to that in the preceding section. Neglecting diffraction effects and small-amplitude absorption, we assume that there propagates from the surface of the concentrator to the surface of a cylinder of radius r_f a convergent cylindrical wave approximately described by the expression

$$v\sqrt{r} = v_0 \sqrt{F} \sin\omega y. \tag{128}$$

We define r_f so that the particle velocity amplitude

$$v_f = \sqrt{\frac{F}{r_f}}\, v_0 \tag{129}$$

at the distance r_f will be equal to the velocity amplitude at the focus, as calculated by the usual expression for the system gain, $v_f = K_v v_0$, where K_v is determined from the relation [56]

$$K_v = \left(\frac{2kF}{\pi}\right)^{1/2} \sin\alpha_m, \tag{130}$$

and the meaning of α_m is clear from Fig. 17.

We then obtain

$$r_f = \frac{F}{K_v^2}. \tag{131}$$

Nonlinear effects in the convergent cylindrical wave can lead to the formation of a discontinuity, so that the wave turns into a sawtooth at a point r_1, determined by condition (71)

$$r_1 = F\left(1 - \frac{1}{2\sigma_0}\right)^2. \tag{132}$$

Bearing in mind the attenuation of a convergent cylindrical sawtooth wave [see (75)], we find the peak value v_{pf} of the particle velocity at the focus:

$$v_{pf} = \frac{\pi K_v v_0}{1 + 2\sigma_0 (K_v - 1)}. \tag{133}$$

From this we readily find an expression for the limiting value to which the velocity at the focus of the system tends asymptotically with increasing potential on the radiator:

$$v_{pf\text{-lim}} = \frac{\pi c_0 K_v}{2\,\varepsilon\, kF(K_v - 1)}. \tag{134}$$

The gain, which is equal by definition to the ratio of the velocity at the focus to the velocity at the radiator surface, assumes the form with regard to the attenuation of the sawtooth wave

$$\frac{K_v'}{K_v} = \frac{\pi}{1 + 2\sigma_0 (K_v - 1)}; \tag{135}$$

Eqs. (133)-(135) are applicable under the condition

$$z (F) = 2\sigma_0 [K_v - 1] \geqslant 0. \tag{136}$$

The quantity z(F) may be treated as a parameter characterizing the contribution of nonlinear effects in the concentrator.

If $z(F) \leq 0$, the nonlinear effects are negligible, and the gain is determined by expression (130).

If $z(F) > 0$, the attenuation of the sawtooth generated in this case leads to a reduction of the gain in correspondence with expression (135).

Chapter 4

Absorption of Finite-Amplitude Waves
in Relaxing Media and in Solids

§ 1. Sound Absorption in Relaxing Media

The buildup times of thermodynamic equilibrium in a medium are normally rather small relative to the time scales typical of acoustical processes; hence, the medium remains in equilibrium during the propagation of sound in it.

There are situations, however, in which slow processes are possible in the medium, such as the excitation of retarded degrees of freedom or the occurrence of chemical reactions between various components of the medium, where such reactions are unable to keep pace with the rapid changes of state induced by acoustic oscillations.

The compression and rarefaction of the medium during the transmission of a sound wave upset the thermodynamic equilibrium, and irreversible processes of the approach to equilibrium take place, accompanied by an increase in the entropy and dissipation of energy, which result in the relaxation absorption of sound (see, e.g., [1, 2]). The intensities of these processes depend, of course, on the wave frequency, or, more precisely, on its relation to the characteristic equilibrium rise time, i.e., to the relaxation time τ.

At low frequencies $(\omega \tau \ll 1)$, the change of state occurs so slowly that the medium stays in equilibrium. At high frequencies $(\omega \tau \gg 1)$, the internal degrees of freedom act as if "frozen," i.e., they cannot change their state, and the relaxation processes again prove inconsequential.

At frequencies $\omega\tau \approx 1$, the equilibrium recovery processes are most vigorous, and the relaxation coefficient of small-amplitude sound absorption is maximized (see Fig. 1).

The nature of the propagation of a finite-amplitude wave in a relaxing medium is also frequency-dependent. In fact, the conditions of propagation of a finite-amplitude wave at high and low frequencies are profoundly dissimilar. If $\omega\tau \gg 1$, the fundamental wave and all its high-frequency harmonics formed due to nonlinear effects fall into the frequency domain where the relaxation absorption coefficient is monotone decreasing.

If $\omega\tau \ll 1$, the fundamental wave falls outside the relaxation absorption domain, where the high-frequency components characterized by frequencies comparable with the relaxation frequency are strongly absorbed.

We wish to consider the propagation of a large-intensity sound wave in a relaxing medium in the limiting cases of low $(\omega\tau \ll 1)$ and high $(\omega\tau \gg 1)$ frequencies, assuming that the sound dispersion induced by relaxation processes is small:

$$m = \frac{c_\infty^2 - c_0^2}{c_0^2} \ll 1, \tag{137}$$

and having as our prime concern the conditions for formation of periodic shock waves, whose onset results in the strong attenuation of sound [60, 61].

In nonequilibrium systems, the various thermodynamic variables (such as the pressure p) are no longer determined by the specification of any two thermodynamic variables. Their determination requires the designation of certain auxiliary parameters, in the simplest case, one parameter ξ characterizing the given nonequilibrium state (physically ξ may be determined, for example, by the concentration of reacting components or number of excited atoms, etc.), so that

$$p = p\,(\rho,\ s,\ \xi). \tag{138}$$

For small deviations from equilibrium, ξ may be determined from the so-called reaction equation (see [1, 2])

$$\tau \frac{\partial \xi}{\partial t} = \xi_0 - \xi \tag{139}$$

(ξ_0 is the equilibrium value of ξ), and thus (138) can be used to find the equation of state of the relaxing medium. Combining this equation with the hydrodynamic equations, we obtain a set of two approximate equations describing the propagation of a finite-amplitude wave in a relaxing medium [5]

$$\frac{\partial v}{\partial x} - \alpha v \frac{\partial v}{\partial y} = - \frac{B\tau}{2\rho_0 c_0^2} \frac{\partial^2 \xi}{\partial y^2}; \quad \tau \frac{\partial \xi}{\partial y} + \xi = - \frac{m\rho_0 c_0}{B} v, \tag{140}$$

where $B = (\partial p / \partial \xi) \rho_0$ is a small quantity of the order $M = v_0/c_0$.

For the consideration of the propagation of a low-frequency wave ($\omega \tau \ll 1$), we first present the solution of the auxiliary problem of the structure of a unit stationary compression shock of amplitude $2v_0$ [62, 63]

$$\frac{y + y_0}{v_0} = \ln \frac{(v_0 + v)^{\Gamma_p - 1} v_0^2}{(v_0 - v)^{\Gamma_p + 1}}, \tag{141}$$

where

$$\Gamma_p = \frac{(\gamma + 1)}{m} \frac{v_0}{c_0} \tag{142}$$

is a parameter characterizing the relation of the nonlinear and dissipative effects during wave propagation in a relaxing medium. The analysis of this solution shows that the profile of the shock can be gently sloping and symmetric (Fig. 20a) if $\Gamma_p \ll 1$; steep and asymmetric (Fig. 20b) if $\Gamma_p < 1$; or even discontinuous (Fig. 20c) if $\Gamma_p > 1$. The gradual variation of the wave profile over the discontinuity (see Fig. 20c) is related to the fact that the compressed matter behind the shock wave rapidly enters the nonequilibrium state, and the slow equilibrium restoration process is accompanied by a variation of the hydrodynamic variables (in this case, v) behind the wave front.

Combining the solution (141) with the solution describing the linear part of the wave profile, we obtain a solution describing the profile of one period of a sawtooth wave in the relaxing medium

$$\frac{v}{v_0} = \frac{1}{1 + \sigma} (- \omega y + \pi v (\omega y)). \tag{143}$$

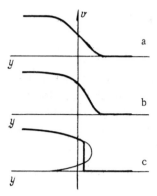

Fig. 20. Profile of a compression shock in a relaxing medium.

It differs from the solution (34) for a nonrelaxing medium only in
the second term in the parentheses, where in place of the hyper-
bolic tangent describing the structure of a first-order discontinuity
in a medium having ordinary viscosity, we have the expression
$v(\omega y)$ from the solution (142). The one-period waveforms obtained
on the basis of this solution are shown in Fig. 21a for $\Gamma_p < 1$ and
in Fig. 21b for $\Gamma_p > 1$. Note that the influence of relaxation processes in the low-frequency case ($\omega \tau \ll 1$) merely induces asym-
metry in the wave profile, the nature of the absorption relative
to the absorption of a sawtooth remaining unchanged in the nonrelaxing medium.

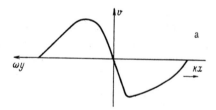

Fig. 21. Profile of one period of a finite-
amplitude wave in a relaxing medium for
$\omega\tau \ll 1$.

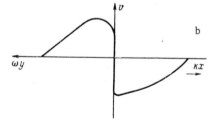

For high frequencies ($\omega\tau \gg 1$), the set of equations (140) approximately reduces to a single equation [60, 61]:

$$\frac{\partial v}{\partial x} - \alpha\left(v + \frac{m}{2\varkappa c_0}\right)\frac{\partial v}{\partial y} + \frac{m}{2c_0\tau} = 0,$$

the solution of which, describing the propagation of a sine wave, may be written

$$\omega\left(t - \frac{x}{c_\infty}\right) = \sin^{-1} W - \sigma W, \qquad (144)$$

where

$$W = \frac{v}{v_0} e^{sx}; \quad s = \frac{m}{2c_0\tau}; \quad \sigma = \frac{\alpha\omega v_0}{s}(1 - e^{-sx}). \qquad (145)$$

Structurally this expression coincides with Eq. (26), which means that we can at once utilize the results obtained in the analysis of that solution. The distortions induced by nonlinear effects in the wave profile produce a discontinuity at the point $\sigma = 1$, so that the wave becomes a sawtooth at the point $\sigma = \pi/2$. It is interesting to note, however, that, as opposed to nonrelaxing media, where σ is monotone-increasing in proportion to the path traversed by the wave, in the given case $\sigma(x)$ tends asymptotically to a limiting value

$$\sigma_\infty = \frac{\gamma+1}{2} M\omega\tau\left[\frac{2c_0^2}{c_\infty^2 - c_0^2}\right]. \qquad (146)$$

This introduces the critical initial amplitude

$$\frac{v_{0\,cr}}{c_0} = \frac{2}{\gamma+1}\frac{c_\infty^2 - c_0^2}{c_0^2}\frac{1}{\omega\tau}, \qquad (147)$$

such that for $v_0 < v_{0\,cr}$ (corresponding to the condition $\Gamma_p < 1$) $\sigma_\infty < 1$ and, hence, generally, a discontinuity is not formed.

The specific nature of the dependence of σ and W on x also creates a special kind of absorption regime for the finite-amplitude wave in a relaxing medium for $\omega\tau \gg 1$, this regime being determined for $\sigma_\infty \gg 1$ by the approximate relation

$$\frac{v}{v_0} \approx \frac{\pi e^{-sx}}{1 + \frac{\alpha\omega v_0}{s}(1 - e^{-sx})}. \qquad (148)$$

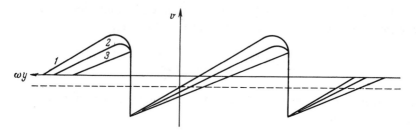

Fig. 22. Profile of a finite-amplitude wave in a relaxing medium for $\Gamma_p > 1$.
1) $\omega\tau \ll 1$; 2) $\omega\tau \approx 1$; 3) $\omega\tau \gg 1$.

It is important to note, however, that in the domain of small sx, this attenuation law essentially coincides with the ordinary relation (28) characterizing the attenuation of a plane sawtooth wave, a fact that is easily verified by the expansion of the exponentials in (148) on the small parameter sx:

$$\frac{v}{v_0} \approx \frac{\pi}{1 + \alpha\omega v_0 x}. \tag{149}$$

Using the foregoing results for the limiting cases of low and high frequencies, we can state some qualitative considerations for the intermediate frequency case, when $\omega\tau \approx 1$, as has been done in detail in [60] for arbitrary values of Γ_p. Without reiterating these arguments in their entirety, we merely give the most important results with respect to the immediate problems of the absorption of finite-amplitude waves.

If $\Gamma_p \ll 1$, the nonlinear effects are strongly pronounced in the absorption of sound in a relaxing medium, irrespective of the frequency. But if $\Gamma_p > 1$, the distortions of the wave cause the formation of a sawtooth wave, whose profile for one period is represented by curve 1 in Fig. 22 for $\omega\tau \ll 1$ [see (143)] and by curve 3 for $\omega\tau \gg 1$ [see (144)]. The areas bounded above by this curve and below by the horizontal axis are equal, indicating the presence of a constant velocity component.

The onset of this constant component (its magnitude is indicated by the dashed line) is equivalent, for the transition from $\omega\tau \ll 1$ to $\omega\tau \gg 1$, to a change of the wave propagation velocity from c_0 to c_∞. The attenuation of sawtooth waves in the cases $\omega\tau \ll 1$ and $\omega\tau \gg 1$ is described by expressions (143) and (148), respectively.

It is obvious that, in the intermediate case $\omega\tau \approx 1$, the wave profile has the form represented by curve 2 in Fig. 22, and its absorption is described by formulas of the type (143) or (148).

§ 2. The Absorption of Finite-Amplitude Sound in Solids

The absorption of sound in an isotropic solid due to viscosity and thermal conductivity is described by equations similar to the corresponding expressions for gases and liquids. The absorption coefficient turns out to be proportional to the frequency ω squared and is, for a longitudinal wave,

$$\alpha_l = \frac{\omega^2}{2\rho c_l^3}\left[\left(\frac{4}{3}\eta + \eta'\right) + \frac{\varkappa T^2\beta^2\rho^2\left(c_l^2 - \frac{4}{3}c_t^2\right)}{c_p^2}\right]; \qquad (150)$$

for a transverse wave [1],

$$\alpha_t = \frac{\eta\omega^2}{2\rho c_t^3}. \qquad (151)$$

Here η and η' are the shear and dilatational viscosity coefficients, β is the coefficient of thermal expansion, c_l, c_t are the longitudinal and transverse wave velocities, $\varkappa$ is the thermal conductivity coefficient, c_p is the specific heat at constant pressure, and ρ is the density of the medium.

These equations, like expressions (2) and (3) for sound absorption in gases and liquids, are only applicable for calculations of the absorption of waves of sufficiently small amplitude amenable to description by linear theory. If, however, the wave intensity is large, the nonlinear effects become appreciable.

In the case of a longitudinal wave, these effects are analogous to the distortion of the wave profile during propagation in a gas or liquid due to the differences in the propagation velocities of the separate parts of the profile [64], a result that is confirmed by the several experimental studies of Gedroits, Zarembo, and Krasil'-nikov [65, 66], Rollins [67], and Shklovskaya-Kordi [68]. The propagation and absorption of a wave may be described within the framework of the considered approximations by a method entirely analogous to that in Chap. 1 [69]. The role of the nonlinear parameter γ in this case is taken by the exponent in the empirical equation of state for a condensed medium in the form [2]

$$P = A\left[\left(\frac{\rho}{\rho_0}\right)^{\gamma} - 1\right], \tag{152}$$

where A is a constant.

For metals we find a value $\gamma \approx 46 \pm 6$ [2, 7]; we recall for comparison that for water $\gamma = 7$. Note that this parameter can be expressed in terms of the coefficients of the cubic terms in the expansion of the free energy of a solid on the invariants of the strain tensor [67]; these coefficients are sometimes called the "third-order elastic moduli" [70].

It should be pointed out, however, that the transition of an ultrasonic wave to a sawtooth in a solid and the appreciable growth in its absorption by comparison with the small-amplitude case have yet to be observed experimentally; this is a result of the eminent difficulties involved in the generation of a finite-amplitude ultrasonic wave in relatively "rigid" solids, so that the experimentally attainable Reynolds numbers (or Γ) have been small. Thus, in [68] the displacement amplitude during the propagation of sound in an aluminum sample at $4.5 \cdot 10^6$ cps attained $\sim 10^{-7}$ cm, which for an absorption coefficient $\alpha_l \approx 8 \cdot 10^{-3}$ cm^{-1} corresponds to values of $\Gamma \approx 10^{-1}$.

Recently, in connection with investigations of the interaction of high-intensity optical radiation with matter, researchers have discovered the generation of hypersonic waves by a laser beam focused onto a solid [71]. The intensity of the waves is so great that, according to estimates in [72], the corresponding values of Γ even at room temperatures are close to unity and increase by three or four orders of magnitude with a reduction to liquid helium temperatures, a result that seems to indicate the presence of appreciable nonlinear effects in this case.

In a transverse wave, the nature of the nonlinear effects is different, its shape remaining unaltered in propagation, as verified by the following argument, which is valid within the framework of the second approximation [64]. The distortion of a longitudinal wave is engendered by the dissimilarity of the velocities of different points on the wave profile. This dissimilarity is due, first, to the dependence of the velocity of sound on the density (and, hence, on the degree of compression of the medium) and, second, to the entrainment of the wave by the moving medium.

In the propagation of a transverse wave, there are no compressions and rarefactions of the medium, and the particle velocity is orthogonal to the velocity of propagation. Consequently, the velocity of all points of the transverse wave profile is the same, and it is not distorted during propagation. For this reason, the results obtained above concerning the absorption of finite-amplitude waves prove inapplicable in the case of transverse waves.

In real solids there are other possible sound absorption mechanisms that result in a strong dependence of the absorption on the amplitude of the sound wave and which are related, for example, to variations in the structure of the medium (see [73, 74]), but the discussion of these most intriguing problems is beyond the scope of our article.

In conclusion I wish to convey my heartfelt appreciation to Prof. D. T. Blackstock for being so kind as to furnish me with copies of his papers, as well as to N. G. Kozhelupova for assisting with the preparation of the manuscript for publication.

References

1. L. D. Landau and E. M. Lifshits, Mechanics of Continuous Media, Moscow (1954).
2. Ya. B. Zel'dovich and Yu. P. Raizer, Physics of Shock Waves and High-Temperature Hydrodynamic Effects, Moscow (1965).
3. I. G. Mikhailov, V. A. Solov'ev, and Yu. Syrnikov, Fundamentals of Molecular Acoustics, Izd. "Nauka" (1964).
4. L. K. Zarembo and V. A. Krasil'nikov, Certain aspects of the propagation of finite-amplitude ultrasonic waves in liquids, Usp. Fiz. Nauk, 18(4):688 (1959).
5. M. J. Lighthill, Viscosity effects in sound waves of finite amplitude, in: Survey in Mechanics, Cambridge (1956).
6. L. K. Zarembo and V. A. Krasil'nikov, Introduction to Nonlinear Acoustics, Izd. "Nauka" (1966).
7. G. A. Ostroumov, Lectures on Nonlinear Acoustics, Izd. Leningrad. Univ. (1966).
8. K. A. Naugol'nykh, Absorption of sound waves of finite amplitude, Akust. Zh., 4(2):115 (1958).
9. R. T. Beyer, Physical Acoustics (W. P. Mason, ed.), Vol. 2B, Academic Press, New York—London (1965).
10. Z. A. Gol'dberg, Propagation of finite-amplitude plane waves, Akust. Zh., 3(4):322 (1957).

11.   D. T. Blackstock, Theoretical analysis of propagation of sound waves containing shocks, J. Acoust. Soc. Am., 36(5):1032 (1964).

12.   Z. A. Gol'dberg, Propagation of plane sound waves of finite amplitude in a viscous heat-conducting medium, Akust. Zh., 5(1):118 (1959).

13.   K. A. Naugol'nykh, Propagation of spherical sound waves of finite amplitude in a viscous heat-conducting medium, Akust. Zh., 5(1):80 (1959).

14.   V. V. Shklovskaya-Kordi, Acoustical method of determining the internal pressure in liquids, Akust. Zh., 9(11):107 (1963).

15.   J. S. Mendousse, Nonlinear dissipative distortion of progressive sound waves at moderate amplitudes, J. Acoust. Soc. Am., 25(1):51 (1953).

16.   S. I. Soluyan and R. V. Khokhlov, Propagation of finite-amplitude sound waves in a dissipative medium, Vest. MGU, Ser. III, Fizika i Astronomiya, 3:52 (1961).

17.   D. T. Blackstock, Thermoviscous attenuation of plane, periodic finite amplitude sound waves, J. Acoust. Soc. Am., 36(3):534 (1964).

18.   J. D. Cole, On a quasi-linear parabolic equation occurring in aerodynamics, Quart. Appl. Math., 9(3):225 (1951).

19.   B. D. Cook, New procedure for computing finite amplitude distortion, J. Acoust. Soc. Am., 34(7):941 (1962).

20.   W. Keck and R. T. Beyer, Frequency spectrum of finite amplitude ultrasonic waves in liquids, Phys. Fluids, 3:346 (1960).

21.   D. T. Blackstock, Connection between the Fay and Fubini solutions for plane sound waves of finite amplitude, J. Acoust. Soc. Am., 39(6):1019 (1966).

22.   G. C. Werth, Attenuation of repeated shock waves in tubes, J. Acoust. Soc. Am., 25:821 (1953).

23.   V. A. Burov and V. A. Krasil'nikov, Direct observation of the distortion of intense ultrasonic waves in a liquid, Dokl. Akad. Nauk SSSR, 118:920 (1958).

24.   E. V. Romanenko, Experimental Investigation of the Propagation of Finite-Amplitude Waves in a Liquid (Candidate's Dissertation), Akust. Inst., Moscow (1962).

25.   I. Rudnick, On the attenuation of a repeated sawtooth shock wave, J. Acoust. Soc. Am., 25(5):1012 (1953).

26.   S. D. Poisson, Mémoire sur la théorie du son [Note on the theory of sound], J. École Polit. (Paris), 7:364 (1808).

27.   B. Riemann, Über die Fortpflanzung der Luftwellen endlicher Schwingungsweite [Propagation of Airborne Waves of Infinite Amplitude], Göttingen. Abhandl., 8 (1860).

28.   D. T. Blackstock, Propagation of plane sound waves of finite amplitude in nondissipative fluids, J. Acoust. Soc. Am., 34(9) (1962).

29.   H. Lamb, Dynamic Theory of Sound, Arnold, London (1931).

30.   G. Fubini, Pressione di radiatione acustica i onde di grande ampierra [Acoustic radiation pressure in large-amplitude waves], Alta Frequenza, 4:530 (1935).

31.   L. E. Hargrove, Fourier series for the finite amplitude sound waveform in a dissipationless medium, J. Acoust. Soc. Am., 32:511 (1960).

32.   D. T. Blackstock, Convergence of the Keck—Beyer perturbation solution for plane waves of finite amplitude in a viscous fluid, J. Acoust. Soc. Am., 39(2):411 (1966).

33.   A. L. Thuras and R. T. Jenkins, Extraneous frequencies generated in air carrying intense sound waves, J. Acoust. Soc. Am., 6:173 (1935).

34.   O. N. Geertseen, A Study of Finite Amplitude Distortion of a Sound Wave in Air, Univ. California Techn. Rept., III (1951).

35.   R. D. Fay, Plane sound waves of finite amplitude, J. Acoust. Soc. Am., 3:222 (1931).

36.   L. K. Zarembo, V. A. Krasil'nikov, and V. V. Shklovskaya-Kordi, Distortion of a finite-amplitude ultrasonic waveform in a liquid, Dokl. Akad. Nauk SSSR, 109(3):484 (1956).

37.   I. G. Mikhailov and V. A. Shutilov, Distortion of the finite-amplitude ultrasonic waveform in various liquids, Akust. Zh., 6(3):340 (1960).

38.   R. P. Rya, A. G. Lutsh, and R. T. Beyer, Measurement of the distortion of finite ultrasonic waves in liquids by a pulse method, J. Acoust. Soc. Am., 34(1):31 (1952).

39.   K. L. Zankel and E. A. Hiedemann, Simple demonstration of the presence of second harmonics in progressive ultrasonic waves of finite amplitude, J. Acoust. Soc. Am., 30(6):582 (1958).

40.   L. L. Myasnikov, On the "flipping" of a large-amplitude sound wave, Zh. Tekh. Fiz., 8:1896 (1938).

41.   R. T. Beyer and V. Narasimhau, Note on finite amplitude waves in liquids, J. Acoust. Soc. Am., 29(4):532 (1957).

42.   F. E. Fox and W. A. Wallace, Absorption of finite amplitude sound waves, J. Acoust. Soc. Am., 26(6):994 (1954).

43.   L. K. Zarembo, Absorption of Finite Amplitude Ultrasonic Waves in a Liquid (Candidate's Dissertation), MGU, Moscow (1958).

44.   L. K. Zarembo, V. A. Krasil'nikov, and V. V. Shklovskaya-Kordi, Absorption of finite-amplitude ultrasonic waves in liquids, Dokl. Akad. Nauk SSSR, 109(4):731 (1956).

45.   O. M. Towlet and R. B. Lindsay, Absorption and velocity of ultrasonic waves of finite amplitude in liquids, J. Acoust. Soc. Am., 27(3):530 (1955).

46.   V. A. Burov and V. A. Krasil'nikov, Absorption of ultrasonic waves of large intensity in water, Dokl. Akad. Nauk SSSR, 124(3):571 (1959).

47.   K. A. Naugol'nykh and E. V. Romanenko, Propagation of finite-amplitude waves in a liquid, Akust. Zh., 4(2):200 (1958).

48.   K. A. Naugol'nykh, S. I. Soluyan, and R. V. Khokhlov, Spherical waves of finite amplitude in a viscous thermally conducting medium, Akust. Zh., 9(1):54 (1963).

49.   K. A. Naugol'nykh, S. I. Soluyan, and R. V. Khokhlov, Cylindrical waves of finite amplitude in a dissipative medium, Vest. MGU, Ser. III, 4:65 (1962).

50.   D. T. Blackstock, On plane, spherical, and cylindrical sound waves of finite amplitude in lossless fluids, Tech. Rep. AF 49 (638), General Dynamics, Rochester, New York (1965).

51.   R. Khokhlov, K. Naugol'nykh, and S. Soluyan, Waves of moderate amplitudes in absorbing media, Acustica, 14(5):248 (1964).

52.   H. S. Heaps, Waveform of finite amplitude derived from equations of hydrodynamics, J. Acoust. Soc. Am., 34:355 (1962).

53.   I. Rudnick, On the attenuation of high-amplitude waves of stable sawtooth form, J. Acoust. Soc. Am., 30(4):339 (1958).

54.   P. J. Westervelt, Self-scattering of high-intensity sound, Proc. Third Internat. Congress Acoustics, Stuttgart (1959), ed. by L. Cremer, Amsterdam (1961).

55.   K. A. Naugol'nykh, Propagation of Finite-Amplitude Sound Waves (Candidate's Dissertation), Akust. Inst. (1959).

56.   L. D. Rozenberg, Sound Focusing Systems, Izd. AN SSSR, Moscow—Leningrad (1949).

57.   A. K. Burov, Generation of high-intensity ultrasonic oscillations for the treatment of malignant tumors in animals and humans, Dokl. Akad. Nauk SSSR, 106(2):239 (1956).

58.   D. V. Khaminov, Dependence of the gain of a sound-focusing system on the intensity of ultrasound in water, Akust. Zh., 3(3):294 (1957).

59.   K. A. Naugol'nykh and E. V. Romanenko, Dependence of the gain of a focusing system on the sound intensity, Akust. Zh., 5(2):191 (1959).

60.   S. I. Soluyan and R. V. Khokhlov, Finite-amplitude acoustic waves in a relaxing medium, Akust. Zh., 8(2):220 (1962).

61.   R. Khokhlov and S. Soluyan, Propagation of acoustic waves of moderate amplitude through dissipative and relaxing media, Acustica, 14(5):242 (1964).

62.   A. L. Polyakov, S. I. Soluyan, and R. V. Khokhlov, Propagation of finite disturbances in a relaxing medium, Akust. Zh., 8(1):107 (1962).

63.   A. L. Polyakova, Finite Disturbances in a Relaxing Medium, Akust. Inst., Moscow (1962).

64.   Z. A. Gol'dberg, Interaction of plane longitudinal and transverse elastic waves, Akust. Zh., 6(3):307 (1960).

65.   A. A. Gedroits, L. K. Zarembo, and V. A. Krasil'nikov, Elastic waves of finite amplitude in solids and lattice anharmonicity, Vest. MGU, Ser. III, 3:92 (1962).

66.   A. A. Gedroits and V. A. Krasil'nikov, Zh. Éksp. Teor. Fiz., 43(5):1592 (1962).

67.   F. R. Rollins, Interaction of ultrasonic waves in solid media, Appl. Phys. Lett., 2(8):147 (1963).

68.   V. V. Shklovskaya-Kordi, Experimental Investigations of Finite-Amplitude Ultrasonic Waves, Akust. Inst., Moscow (1966).

69.   L. A. Pospelov, Propagation of finite amplitude elastic waves, Akust. Zh., 11(3):359 (1965).

70.   R. N. Thurston, Ultrasonic data and the thermodynamics of solids, Proc. IEEE, 53(10):1320 (1965).

71.   S. V. Krivokhizha, D. I. Mash, V. V. Morozov, and I. L. Fabelinskii, Induced Mandel'shtam-Brillouin scattering in a quartz single crystal in the temperature interval 2.1-300°K, ZhÉTF Pis. Red., No. 3, p. 378 (1966).

72.   A. L. Polyakova, Nonlinear effects in a hypersonic wave, ZhÉTF Pis. Red., 4(4):132 (1966).

73.   M. G. Kogan, Energy Losses of the Mechanical Oscillations of Magnetostrictive Transducers and Tools for Ultrasonic Processing, Publ. No. 3, Leningrad, Dom Nauch. Tekh. Propagandy, Leningrad (1962).

74.   P . A . Bezuglyi, V . L . Fil', and A . A . Shevchenko, Nonlinear effects in the
        absorption of ultrasound in superconducting indium, Zh. Éksp. Teor. Fiz.,
        49[6(12)]:1715 (1965).

PART II

ACOUSTIC RADIATION PRESSURE

Z. A. Gol'dberg

Introduction

The acoustic radiation pressure is customarily interpreted as the time-average pressure acting on an object in a sound field. The object in this case is conceived in the broadest sense, i.e., a body in a sound field, an interface between two media, or a single particle of a medium set against the other particles of the same medium.

In the linear approximation, the forces acting on an object in a sound field are periodic functions of the time with a frequency equal to the acoustic frequency. Averaged over the time, they are equal to zero. The linear approximation is adequate, for example, for investigations of the operation of microphones, where the main item of interest is the periodic force acting on the microphone membrane. Nonzero average forces arise as a consequence of second-order effects. The acoustic radiation pressure, a second-order small quantity, is small in comparison with the periodically-varying sound pressure. In a sound field, in which the sound pressure is equal to 10^3 dyn/cm^2 under normal conditions in air, the acoustic radiation pressure in the case of normal incidence of a sound wave on a perfectly reflecting obstacle is on the order of 1 dyn/cm^2. Consequently, the radiation pressure is significant only in very strong sound fields.

The ponderomotive action of a sound field on resonators was observed by Dvořak as early as 1876, and in 1878 Rayleigh laid the theoretical foundation of the effect [1]. Rayleigh later redirected his attention to the problem [2] and derived an equation for the acoustic radiation pressure on a perfectly reflecting solid wall. The Rayleigh equation was confirmed quantitatively by the experiments of Al'tberg [3] and Zernov [4] in the laboratory of

P. N. Lebedev. Starting with the classical studies of Rayleigh, the problem of the radiation pressure has been discussed in the scientific periodical literature until the present day [5-7]. This interest is attributable to the ever-growing use of high-intensity sound fields in ultrasonic engineering for the production of emulsions, the dispersion of solids in liquids, coagulation processes, the degassing of liquids and melts, the cleaning and degreasing of metal parts, the drilling of holes and formation of notches in solids, etc. [8, 9]. Radiation pressure assumes a decisive role in these processes. Furthermore, the intensity of the sound field is determined on the basis of measurements of the ponderomotive action by means of a Rayleigh disk or radiometer.

As is customary, we discern between the Rayleigh and Langevin radiation pressures [8, 10]. Among the problems involving the Rayleigh radiation pressure, we include cases in which interaction between the sound field and the acoustically unperturbed medium is absent; if, on the other hand, interaction is present, as expressed in the influence of the acoustically unperturbed domain on the time-average pressure, we speak of the Langevin radiation pressure. Typical of sound fields with the Rayleigh acoustic radiation pressure are fields in enclosures, when the mass of the medium in which the oscillations are taking place remains constant. The Rayleigh radiation pressure differs mathematically from the Langevin radiation pressure in its dependence on the parameter characterizing the nonlinear properties of the equation of state.

The object of the present part of the book is to describe the theoretical bases of the acoustic radiation pressure and related problems, the momentum and energy of sound waves, and the ponderomotive action of the sound field on obstacles located in it.

Chapter 1

The Initial Equations

Variables

The fundamental equations of acoustics, as the branch of physics concerned with the oscillatory motion of continuous media, are the equations of the mechanics of continuous media. For a liquid or gas these equations may be written in the form [15]

$$\rho\left[\frac{\partial \mathbf{v}}{\partial t} + (\mathbf{v}\nabla)\,\mathbf{v}\right] = -\,\nabla p + \eta\Delta\mathbf{v} + \left(\frac{\eta}{3} + \eta'\right)\nabla\operatorname{div}\mathbf{v}; \qquad (1)$$

$$\frac{\partial \rho}{\partial t} + \operatorname{div}\rho\mathbf{v} = 0; \qquad (2)$$

$$\rho T\left(\frac{\partial s}{\partial t} + \mathbf{v}\,\nabla s\right) = \sigma'_{ik}\frac{\partial v_i}{\partial x_k} + \varkappa\Delta T. \qquad (3)$$

Here ρ, p, and T are the density, pressure, and temperature; $\mathbf{v}$ is the hydrodynamic velocity, η and η' are the shear and dilatational viscosity coefficients, $\varkappa$ is the thermal conductivity coefficient, s is the entropy per unit mass, and σ'_{ik} is the viscous stress tensor. Here and elsewhere we use double-iterated subscripts to indicate summation over the values 1, 2, 3, where $x_1 = x$, $x_2 = y$, $x_3 = z$, $v_1 = v_x$, etc.

In order to make use of these equations, we need to know in addition, of course, the form of some thermodynamic characteristic function for obtaining the equation of state

$$p = p\,(\rho,\; s) \qquad (4)$$

77

and establishing the form of the dependence of s on whatever variables are adopted as the equilibrium thermodynamic coordinates.

If we multiply Eq. (2) by v_i and add it to the i-th component of Eq. (1), we obtain an equation expressing the conservation of the i-th component of the momentum and representing it in the form (see [15], §15)

$$\frac{\partial (\rho v_i)}{\partial t} = - \frac{\partial \Pi_{ik}}{\partial x_k}. \tag{5}$$

Here

$$\Pi_{ik} = p\delta_{ik} + \rho v_i v_k - \sigma'_{ik} \tag{6}$$

is the momentum flux density tensor; the viscous stress tensor σ'_{ik} is equal to

$$\sigma'_{ik} = \eta \left(\frac{\partial v_i}{\partial x_k} + \frac{\partial v_k}{\partial x_i} - \frac{2}{3} \cdot \frac{\partial v_e}{\partial x_e} \delta_{ik} \right) + \eta' \delta_{ik} \frac{\partial v_e}{\partial x_e}; \tag{7}$$

σ_{ik} is a unit tensor with components equal to unity for i = k and zero for i ≠ k. Every component of the tensor Π_{ik} represents the i-th component of the momentum proceeding through a unit area of the surface perpendicular to the x_k axis.

Most problems in acoustics are solved on the basis of the linearized equations (1)-(4) for an ideal fluid:

$$\rho_0 \frac{\partial \mathbf{v}}{\partial t} = - \nabla p,$$

$$\frac{\partial \rho}{\partial t} + \rho_0 \operatorname{div} \mathbf{v} = 0, \tag{8}$$

$$s = \text{const}, \quad p' = p - p_0 = c^2 \rho'.$$

Here ρ_0 and p_0 are the density and pressure of the acoustically unperturbed medium; ρ' and p' are their values in the sound field,[*] and $c = [(\partial p / \partial \rho)_s]^{\frac{1}{2}}$ is the adiabatic velocity of sound. From Eqs. (8) we obtain in customary fashion the wave equation for the velocity potential (see [15], §63)

$$\frac{\partial^2 \varphi}{\partial t^2} - c^2 \Delta \varphi = 0, \tag{9}$$

---

[*]In the ensuing discussion, we use analogous notation for the other variables.

where

$$v = \nabla\varphi, \quad p' = -\rho_0\frac{\partial\varphi}{\partial t}, \quad \rho' = -\frac{\rho_0}{c^2}\frac{\partial\varphi}{\partial t}. \tag{10}$$

The conditions for applicability of Eqs. (8)–(10) consist in smallness of the hydrodynamic velocity relative to the sound propagation velocity, so that the amplitudes of all variables must be small relative to their unperturbed values:

$$v \ll c, \quad \rho' \ll \rho_0, \text{ etc.,} \tag{11}$$

and the wave absorption must be small:

$$\frac{\alpha}{k} \ll 1, \tag{12}$$

where k = ω/c is the wave number

$$\alpha = \frac{\omega^2}{2\rho c^3}\left[\frac{4}{3}\eta + \eta' + \varkappa\left(\frac{1}{c_v} - \frac{1}{c_p}\right)\right] \tag{13}$$

is the sound absorption coefficient; and c_v and c_p are the specific heats under constant volume and pressure, respectively.

Retaining conditions (11) and (12), we now write the initial equations (1)–(3) with greater accuracy than in the linear acoustics of an ideal fluid, also including amplitude-squared terms and linear terms with dissipative coefficients. If the terms of Eqs. (8) are regarded as first-order small terms, it is reasonable to state that we write Eqs. (1)–(3) correct to and including second-order small terms. It can be shown that the second-order terms are smaller than the first-order terms by a factor of v/c or α/k. In the given approximation, Eqs. (1) and (2) keep the same form, and Eqs. (3) and (4) are written

$$\rho_0 T_0\frac{\partial s'}{\partial t} = \varkappa\Delta T', \tag{14}$$

$$p' = c^2\rho' + \frac{1}{2}\left(\frac{\partial c^2}{\partial\rho}\right)_s \rho'^2 + \left(\frac{\partial p}{\partial s}\right)_\rho s'. \tag{15}$$

Inasmuch as we disregard third-order small terms, the transformation of the second-order terms can be accomplished by relations of linear theory for isentropic wave motions. Consequently, from Eq. (14) after the substitution of T' = $(\partial T/\partial p)_s p'$ and the application of relations (10), we find

$$s' = -\frac{\varkappa}{T}\left(\frac{\partial T}{\partial p}\right)_s \text{div}\, v. \tag{16}$$

Inserting (16) into (15) and invoking the thermodynamic relations, we obtain

$$p' = c^2\rho' + \frac{1}{2}\left(\frac{\partial c^2}{\partial \rho}\right)_s \rho'^2 - \varkappa\left(\frac{1}{c_v} - \frac{1}{c_p}\right) \text{div } \mathbf{v}. \qquad (17)$$

Equations (17), (1), and (2) represent the complete system of equations of nonlinear acoustics. The fundamental dimensionless parameter of nonlinear acoustics governing the nature of the radiated motion is the parameter [16, 17]

$$N = \left[1 + \frac{\rho}{c}\left(\frac{\partial c}{\partial \rho}\right)_s\right]\frac{p'_{\max}}{b\omega}, \qquad b = \frac{4}{3}\eta + \eta' + \varkappa\left(\frac{1}{c_v} - \frac{1}{c_p}\right). \qquad (18)$$

The quantity N involves all the fundamental parameters of the problem. The values of $N \ll 1$ and $N > 1$ correspond to the cases of small-amplitude and finite-amplitude waves in a viscous thermally conducting medium. Therefore, from the value of the parameters N, one can estimate the proximity of the motion in question to the first or second mode.

The fundamental equations are somewhat simplified in the case of potential motion in a sound field, as is inherent, for example, in one-dimensional plane, spherical, and cylindrical waves. Substituting (17) into (1) and making use of the formulas of vector analysis for irrotational motion, we reduce Eq. (1) to the form

$$\rho\frac{\partial \mathbf{v}}{\partial t} + \frac{1}{2}\rho\nabla v^2 = -c^2\nabla\rho' - \frac{1}{2}\left(\frac{\partial c^2}{\partial \rho}\right)\nabla\rho'^2 + b\Delta\mathbf{v}. \qquad (19)$$

Affixing Eq. (2) to the latter, we obtain a complete set of equations. Inasmuch as the viscosity and thermal conductivity coefficients in Eq. (19) are solely concentrated in the constant b, they need only enter into all the relations of nonlinear acoustics for potential motions in the form of the parameter b.

We next derive a formula for the time-average pressure, which we will frequently rely on in the future. We rewrite Eq. (19) correct to and including second-order small terms in the form*

$$\rho_0\frac{\partial \mathbf{v}}{\partial t} + \rho'\frac{\partial \mathbf{v}}{\partial t} + \frac{\rho_0}{2}\nabla v^2 = -\nabla p' + \left(\frac{4}{3}\eta + \eta'\right)\nabla\text{div }\mathbf{v}. \qquad (20)$$

---

*In the first approximation, in general, the motion of the fluid in a sound field is potential [see (10)]; hence, rot $\mathbf{v}$ is at least a second-order quantity. This suffices to render Eq. (19) valid in the general case correct to and including second-order small terms.

The term $\rho'(\partial \mathbf{v}/\partial t)$ can be represented by means of the first-order relations (8) as the gradient of a certain function

$$\rho' \, \frac{\partial \mathbf{v}}{\partial t} = - \, \frac{\rho'}{\rho_0} \, \nabla p' = - \, \nabla \left(\frac{c^2 \rho'^2}{2\rho_0} \right).$$

With this in mind, we average Eq. (20) over the time and integrate over space. Since the term $\rho_0 (\partial \mathbf{v}/\partial t)$ vanishes in averaging over a large time interval, we obtain

$$\overline{p'} = \frac{c^2 \overline{\rho'^2}}{2\rho_0} - \frac{\rho_0 \overline{v^2}}{2} + \left(\frac{4}{3} \eta + \eta' \right) \operatorname{div} \overline{\mathbf{v}} + \text{const.} \qquad (21)$$

We note the time-averaging operation by an overhead bar.

§ 2. Initial Equation in Lagrange Variables

By contrast with the Euler variables, the Lagrange variables a, b, and c are associated with definite particles of the medium. The three-dimensional equations of motion in Lagrange variables are too cumbersome and are therefore rarely used. However, one-dimensional problems are often conveniently solved in Lagrange variables. As a matter of fact, in Lagrange variables the one-dimensional problem reduces to the solution of just one equation. It does not contain the nonlinear term $(\mathbf{v}\nabla)\mathbf{v}$ characteristic of the Euler variables. Moreover, in Lagrange variables it is a simple matter to write down the boundary condition for the displacement of the radiating surface. In the final equations, on the other hand, it is relatively easy to go from Lagrange variables to Euler variables. We first present the formulas for transformation from one set of variables to the other, then the fundamental equation for one-dimensional plane motion.

We choose the Lagrange variable a so that, in the acoustically unperturbed medium, it will coincide with the Euler coordinate x_0. We denote a certain quantity, expressed in Lagrange variables, by $L(a, t)$ and the same quantity in Euler variables by $E(x, t)$. The function $L(a, t)$ refers to a fluid particle, its value at time t corresponding to the point in space $a + \xi$, where ξ is the displacement of the particle. Therefore,

$$L \, (a, \ t) = E \, (a + \xi, \ t). \qquad (22)$$

Expanding Eq. (22) in a series on the displacement and retaining only two terms of the expansion, we obtain

$$L(a, t) = E(a, t) + \xi \frac{\partial E}{\partial a}. \tag{23}$$

We can verify that the second term of the series is much smaller than the first, or that

$$\xi \frac{\partial E}{\partial a} \ll E. \tag{24}$$

Inasmuch as $\partial E / \partial a \sim kE$, condition (24) assumes the form $\xi k \ll 1$, which is true in our case, because

$$\xi k \sim \frac{v}{c} \ll 1.$$

It is apparent from Eq. (23) that the disparity between the Lagrange and Euler variables begins with second-order quantities. This permits us in (23) to replace $\partial E / \partial a$ by $\partial L / \partial a$, whereupon we obtain the following equation for transforming from Lagrange variables to Euler variables, correct to and including terms in ξ:

$$E = L - \xi \frac{\partial L}{\partial x}. \tag{25}$$

Thus, applying (25) to the expressions for the hydrodynamic velocity and density,

$$v = \frac{\partial \xi}{\partial t}, \qquad \rho = \rho_0 \left[1 - \frac{\partial \xi}{\partial a} + \left(\frac{\partial \xi}{\partial a} \right)^2 \right], \tag{26}$$

represented in Lagrange variables, we obtain their expressions in Euler variables:

$$v = \frac{\partial \xi}{\partial t} - \xi \frac{\partial^2 \xi}{\partial t \partial x}, \qquad \rho = \rho_0 \left[1 - \frac{\partial \xi}{\partial x} + \left(\frac{\partial \xi}{\partial x} \right)^2 + \xi \frac{\partial^2 \xi}{\partial x^2} \right]. \tag{27}$$

The fundamental equations for the one-dimensional plane flow of an ideal fluid in Lagrange variables have the form (see [15, §2])

$$\rho dx = \rho_0 da; \tag{28}$$

$$\rho_0 \left(\frac{\partial v}{\partial t} \right)_a = \left(\frac{\partial p}{\partial a} \right)_t. \tag{29}$$

The equation of motion of a viscous thermally conducting fluid is obtained by the insertion of p', determined by Eq. (17), into Eq. (29), and the addition of a term to account for the viscosity of the medium, where this term must have the same form as in the Euler variables. If the relations (26)* are also included, the initial equation reduces to the form

$$\frac{\partial^2 \xi}{\partial t^2} - c^2 \frac{\partial^2 \xi}{\partial a^2} - \frac{b}{\rho_0} \frac{\partial^3 \xi}{\partial t \partial a^2} + 2\varepsilon c^2 \frac{\partial \xi}{\partial a} \cdot \frac{\partial^2 \xi}{\partial a^2} = 0. \tag{30}$$

Here

$$\varepsilon = 1 + \frac{\rho_0 \left(\frac{\partial c^2}{\partial \rho}\right)_s}{2c^2}.$$

---

* Apropos, the expression (26) for the density can be deduced from Eq. (28) and the relation $a = x - \xi$.

Chapter 2

Radiation Pressure

§ 1. General

By definition, the acoustic radiation pressure must be numerically equal to the time-average momentum flux per unit area of surface of the obstacle at a given point. This implies a coordinate system in which the given surface element of the obstacle is at rest, otherwise the momentum flux at the given point will not be equal to the momentum flux relative to a moving surface element of the obstacle at that point.

The flux of the i-th component of the momentum proceeding through a unit surface area of the obstacle is determined by the expression $\Pi_{ik}n_k$, where $\mathbf{n}$ is the unit normal vector to the surface. Therefore, the i-th projection of the radiation pressure is equal to

$$P_{ri} = \overline{\Pi'_{ik}n_k} = \overline{p'n_i} + \overline{\rho v_i\,(v_k n_k)} - \sigma'_{ik}n_k. \tag{31}$$

The prime on Π_{ik} and p indicates that the acoustic corrections to the corresponding variables are involved.

The acoustic radiation pressure is a vector, generally having nonzero normal and tangential components to the surface of the obstacle. If the first term on the right-hand side of Eq. (31) only affects the normal component of the radiation pressure, then the second and third terms can contribute to both components [18-23].

In an ideal fluid the radiation pressure vector is

$$\mathbf{P}_r = \overline{p'\mathbf{n}} + \overline{\rho\mathbf{v}\,(\mathbf{v}\mathbf{n})}. \tag{32}$$

On the other hand, in the case of a solid obstacle in a sound field

$$\mathbf{P}_r = \overline{p'} \cdot \mathbf{n}, \tag{33}$$

because on the surface of the solid the normal component of the velocity is

$$(\mathbf{v} \cdot \mathbf{n}) = 0.$$

The term $\rho \mathbf{v}(\mathbf{v} \cdot \mathbf{n})$ must appear in the case of an obstacle with a nonsolid surface, for example, in the case of a sound wave impingent on an interfacial surface between two fluids. The force elicited by this term is caused by the momentum imparted to the obstacle by the moving medium as it permeates the obstacle.

Let us suppose that the sound wave impinges on the obstacle and is completely absorbed by it. Then the surface of the obstacle perpendicular to the direction of the velocity $(\mathbf{n} \| \mathbf{v})$ must be acted upon by the normal component of the radiation pressure

$$P_{rn} = \overline{p'} + \overline{\rho v^2}. \tag{34}$$

In the case of the surface tangent to the velocity $(\mathbf{n} \perp \mathbf{v})$,

$$P_{rn} = \overline{p'}. \tag{35}$$

The tangential components of the radiation pressure in these two cases are equal to zero. If the sound wave impinges at some angle relative to the surface of the obstacle, the latter is acted upon by a radiation pressure with nonzero tangential and normal components, where

$$P_{rt} = \overline{\rho v_t v_n}, \qquad P_{rn} = \overline{p'} + \overline{\rho v_n^2}. \tag{36}$$

Here v_n and v_t are the normal and tangential components of the velocity at the given point.

In a viscous thermally conducting medium, the radiation pressure is determined by Eq. (31). The influence of the viscosity is accounted for in the term $\overline{\sigma'_{ik}} \cdot n_k$. Moreover, the pressure and hydrodynamic velocity now depend on the dissipative coefficients, as they must be found from the equations for a viscous thermally conducting medium. Generally speaking, the friction forces due to the viscous stresses have both tangential and normal components. For example, in the case of an obstacle with a solid surface perpendicular to the x axis

$$P_{rx} = \overline{p'} - \overline{\sigma'_{xx}}, \quad P_{ry} = -\overline{\sigma'_{yx}}, \quad P_{rz} = -\overline{\sigma'_{zx}}; \tag{37}$$

all components of the radiation pressure can have nonzero values. However, if $\bar{v}$ is a second-order small variable, then $\overline{\sigma'_{ik}}$ is a third-order quantity, because the viscosity coefficients introduce another order of smallness. For the same reason the term b div $\bar{v}$ in Eq. (21) is a third-order small term and can be neglected relative to the second-order term $\rho_0 \bar{v}^2 / 2$. Thus, if $|\bar{v}| \sim v(v/c)$, then

$$b \operatorname{div} \bar{v} : \frac{\rho_0 \bar{v}^2}{2} \approx \frac{b\omega}{\rho c^2} \sim \frac{\alpha}{k} \ll 1.$$

Therefore, if we confine ourselves to the second approximation, we can neglect the terms $\overline{\sigma'_{ik} n_k}$ and b div $\bar{v}$ in Eqs. (31) and (21), respectively. It is important to remember that these terms can become significant if in the sound field there is a sufficiently developed "acoustic streaming" effect.

We note, in addition, that for the calculation of the radiation pressure correct to second-order small terms, in general, it is sufficient to have solutions for ρ' and v in the first approximation. This result is implied by Eqs. (21) and (31), by which $\bar{p}'$ and P_r are squared functions of ρ' and v, correct to a constant. But the constant in these formulas can be determined from additional conditions. Thus, in the case of problems involving the Rayleigh radiation pressure, the constant is found from the condition of constant mass in a certain volume; in problems involving the Langevin radiation pressure, it is deduced from the condition of zero perturbations at infinity.

The conditions stipulating the Rayleigh or Langevin radiation pressure have been discussed by various authors [10, 24-28]. Several papers have been devoted to the analysis of just the Langevin [29-32] or Rayleigh [33-37] radiation pressure predominantly in one-dimensional plane sound fields. We will discuss this entire topic in detail below.

The results of experimental studies concerned with direct measurements of the Rayleigh [38-40]* or Langevin [10, 41-44] radiation pressure are generally consistent with the theoretical results.

---

*It has already been remarked [45] that the Langevin radiation pressure was measured by Al'tberg and Zernov. However, the agreement between the experimental results of [4] and the values computed according to the Rayleigh equation is clearly attributable to the small difference between the Langevin and Rayleigh radiation pressures in the case investigated.

A great many theoretical and experimental studies have been devoted to acoustic radiometers and the Rayleigh disk, but this will be taken up in Chaps. 4 and 5.

§ 2.   The Rayleigh Radiation Pressure

The first theoretical papers devoted to the radiation pressure were authored by Rayleigh. For plane standing waves in a layer of an ideal fluid between two stationary planes, Rayleigh [2] derived the following expression for the radiation pressure on one of the planes:

$$P_{rn} = \overline{p'} = \frac{\gamma + 1}{8} \rho_0 v_0^2. \tag{38}$$

Here v_0 is the amplitude of the hydrodynamic velocity in the standing wave, and $\gamma = 1 + (\rho/c^2)(\partial c^2/\partial \rho)_s$ is a parameter characterizing the nonlinear properties of the medium; for an ideal gas γ coincides with the exponent in the adiabatic equation

$$\frac{p}{p_0} = \left(\frac{\rho}{\rho_0}\right)^\gamma,$$

where $\gamma = c_p/c_v$.

Rayleigh derived Eq. (38) on the basis of Eqs. (17) and (21) for an ideal fluid. He used the condition of constant mass in a tube closed off by fixed pistons. A similar, but more general analysis avoiding the constant mass condition in the given part of the field has been given by Mendousse [46].

Dependent on the parameter γ, the radiation pressure is customarily called the Rayleigh acoustic radiation pressure. In cases typified by the Rayleigh radiation pressure, there is no contact between the sound field and the acoustically unperturbed medium in which the radiation pressure is equal to p_0. By measuring the Rayleigh radiation pressure one can, in principle, determine the parameter γ.

Cases are often encountered in practice that depart somewhat from the problem investigated by Rayleigh. In particular, in measurement tubes, one of the pistons is ordinarily fixed, while the other executes harmonic oscillations, thus stimulating oscillations of the medium in the given volume. In this connection we consider the radiation pressure, in a field of plane and spherical standing waves, on one of the pistons under forced oscillation conditions [47]. We regard the fluid in this case as ideal.

We first examine a plane standing wave between two-plane-parallel rigid boundaries, one of which is stationary, the other oscillating. This problem contains as a special case the Rayleigh problem, in which both boundaries are stationary. We carry out the analysis in Lagrange variables, by means of which we readily describe the boundary condition at the oscillating surface.

Taking the time average of the equation of motion written in Lagrange variables (29), and recognizing that $\partial v / \partial t = 0$, by virtue of the finiteness of v, we obtain

$$\overline{p'_L} = \text{const} = A. \tag{39}$$

In Lagrange variables $\overline{p'_L}$ is the radiation pressure on the moving particles of the medium. On a nonmoving particle or on a stationary solid obstacle, the radiation pressure has the same value in Lagrange and Euler variables, namely $\overline{p'_L} = \overline{p'_E}$. The constancy of $\overline{p'_L}$ in Lagrange variables implies, however, that the radiation pressure on the stationary and on the oscillating solid surfaces is identical and equal to a constant A. The value of the constant is found by the equation of state (15), which we represent in the form

$$p' = c_0^2 \rho' + \frac{\gamma - 1}{2} \cdot \frac{c^2}{\rho} \rho'^2. \tag{40}$$

Let the Lagrangian coordinate of the oscillating boundary $a_0 = 0$, and let the coordinate of the stationary boundary $a_l = l$. Averaging Eq. (40) over the time and length l with regard to the result (39), we obtain

$$\overline{P'_L} = A = \frac{c_0^2}{l} \int_0^l \overline{\rho'_L} \, da + \frac{\gamma - 1}{2} \cdot \frac{c^2}{\rho l} \int_0^l \overline{\rho'^2} \, da. \tag{41}$$

By the equation of continuity (28)

$$\rho_0 da = \rho \, dx = (\rho_0 + \rho') \, (da + d\xi)$$

we transform the first integral on the right-hand side of (41) to the form

$$\int_0^l \overline{\rho'_L} \, da = -\int_0^l \overline{\rho' \frac{\partial \xi}{\partial a}} \, da - \rho_0 \int_0^l \overline{d\xi}. \tag{42}$$

Recognizing that in the linear approximation [see (26)] $\rho' = -\rho_0(\partial\xi/\partial a)$, we obtain from expression (41) with regard to Eq. (42)

$$\overline{P'_L} = \frac{\gamma+1}{2l} \cdot \rho c^2 \int_0^l \overline{\left(\frac{\partial\xi}{\partial a}\right)^2} da - \frac{c^2\rho}{l}[\overline{\xi}(l,\,t) - \overline{\xi}(0,\,t)]. \tag{43}$$

For stationary or harmonically oscillating boundaries

$$\overline{\xi}\,(l,\ t) = \overline{\xi}\,(0,\ t) = 0.$$

When both boundaries are fixed, and the displacement field in the linear approximation has the form

$$\xi = \frac{v_0}{\omega_i}\sin ka\cos\omega t,\ \ kl = n\pi,$$

where n is an integer, Eq. (43) reduces to the Rayleigh equation (38).

If one of the boundaries is fixed and the other oscillates harmonically, the sound field can be described in the linear approximation by an equation of the form

$$\xi = -\,\xi_0\,\frac{\sin k\,(l-a)}{\sin kl}\cos\omega t, \tag{44}$$

where ξ_0 is the displacement amplitude of the oscillating boundary at the point $a = 0$. Substituting expression (44) into (43), we obtain an equation for the radiation prssure in Lagrange variables:

$$\overline{P'_L} = \frac{\gamma+1}{8}\rho v_0^2\left(1 + \frac{\sin 2kl}{2kl}\right), \tag{45}$$

where $v_0 = \xi_0\omega/\sin kl$ is the particle velocity amplitude of the medium. The radiation pressure on the stationary and oscillating boundaries is determined by the same expression. For $2kl \gg 1$ or $2kl = (2n+1)\pi$, Eq. (45) transforms to (38). For $kl \ll 1$, as the resonance interval is approached ($kl \to n\pi$), the velocity v_0 and, with it $\overline{p'_L}$, increases without limit.

In Euler variables the pressure $\overline{p'_E}$, according to Eq. (25) and the first-approximation relation $p' = -\rho c^2(\partial\xi/\partial a)$, has the form

$$\overline{P'_E} = \overline{P'_L} + \rho c^2\xi\,\overline{\frac{\partial^2\xi}{\partial x^2}}.$$

Substituting ξ as determined by Eq. (44), we obtain

$$\overline{p'_E} = \overline{p'_L} - \frac{\rho v_0^2}{2} \sin^2 k\,(l-x). \qquad (46)$$

In the absence of $\overline{p'_L}$, the pressure $\overline{p'_E}$ is a function of the given point in space.

We now consider a spherical standing wave between two concentric spheres, with the inner one pulsating harmonically and the outer one fixed. In the linear approximation, this standing wave is described by the formula

$$\xi = \frac{R}{\omega}\left(\frac{k\cos kR}{R} - \frac{\sin kR}{R^2}\right)\sin \omega t, \qquad (47)$$

where R is the Lagrangian spherical coordinate. The boundary condition on the stationary outer sphere of radius R_2, $\xi\,(R_2) = 0$, leads to the relation

$$\tan kR_2 = kR_2.$$

We proceed from the equation of motion, which is written in Lagrange variables under spherical symmetry

$$\frac{\partial p}{\partial R} = -\rho_0\,\frac{R^2}{r^2} \cdot \frac{\partial v}{\partial t},$$

where $r = R + \xi\,(R, t)$ is the Eulerian spherical coordinate. Integrating this equation over R with regard to expression (47), we obtain an equation for the radiation pressure on the stationary outer sphere

$$\overline{p'}\,(R_2) = \frac{\rho B^2 k^2}{4} \cdot \frac{\sin^2 kR_2}{R_0^2} + D, \qquad (48)$$

where D is a constant, which has to be determined by the same method as the constant A for the plane-wave case. An analogous solution is obtained when the inner sphere is fixed and the outer one oscillates.

If $R_1 = 0$, i.e., if the internal sphere is omitted, then

$$\overline{p'}\,(R_2) = \frac{\rho B^2 k^2}{8}\,(3\gamma - 1)\,\frac{\sin^2 kR_2}{R_2^2}, \qquad (49)$$

which exactly agrees with the result of Lucas [48, 49], obtained by the method of adiabatic invariants.

If both spheres are fixed [ξ (R_1) = 0], then

$$\bar{p'}(R_2) = \frac{\rho B^2 k^2}{4} \left\{ \frac{\sin^2 kR_2}{R_2^2} + \frac{3(\gamma - 1)}{2(R_2^3 - R_1^3)} \left[R_2 \sin^2 kR_2 - R_1 \left(1 - \frac{\sin 2kR_1}{2kR_1} \right) \right] \right\} ;$$

(50)

for $kR_1 = n\pi$, where n is an integer greater than zero,

$$\bar{p'}(R_2) = \frac{\rho B^2 k^2}{4} \cdot \frac{\sin^2 kR_2}{R_2^2} + \frac{3\rho B^2}{2(R_2^3 - R_1^3)} \left[\frac{1}{R_1} + \frac{\gamma - 1}{4} k^2 (R_2 \sin^2 kR_2 - R_1) \right].$$

For $R_1 \approx R_2 \gg \lambda$, where λ is the wavelength, Eq. (48) goes over to Eq. (45), as it should.

Inasmuch as the radiation pressure acts on the inner sphere as well, this suggests the possibility of building rather unique "acoustic bearings," in which the central part (of spherical or cylindrical form) would be supported by the radiation pressure. However, the practical realization of this motion runs up against substantial difficulties, as is particularly evident from the work of Seegal [50], who has investigated the problem in detail.

Among the problems associated with the Rayleigh radiation pressure is the one-dimensional problem treated by Airy and later analyzed in detail by Fubini – Chiron [51]. This problem concerns a traveling sound wave generated by the oscillation of a plane perpendicular to an axis a. The initial and boundary conditions of the problem are:

1) For t < 0 and any a > 0 the displacement ξ = 0.

2) For t $\geq$ 0 the displacement of the plane (a = 0) is ξ = $\xi_0(1 - \cos \omega t)$.

We are interested only in the radiation pressure on the radiating plane as determined by Eq. (43).

We first consider the wave in an ideal fluid. To avoid calculations of the displacement field in the second approximation, we carry out space averaging over the length l = ct, i.e., over the entire perturbed domains. Then at the leading front of the wave ξ (l, t) = 0, because the radiation begins with zero displacement

$\xi(0, 0) = 0$, and with zero velocity $v(0, 0) = 0$. By making the time interval small enough, one can regard $\xi(0, t)/l$ as negligibly small. Consequently,

$$\overline{P_L} = \frac{\gamma+1}{2lt}\, \rho c^2 \int\limits_0^l da \int\limits_{a/c}^t \left(\frac{\partial \xi}{\partial a}\right)^2 dt. \tag{51}$$

In the time integral, the lower limit corresponds to the time of arrival of the perturbation at point a, and prior to this time the integrand is zero. Substituting into Eq. (51) the solution of the wave equation for ξ in correspondence with the solution of our problem in the first approximation

$$\xi = \xi_0\, [1 - \cos\,(\omega t - ka)]\mathrm{i},$$

after integration we obtain the familiar result

$$\overline{P_L} = \frac{\gamma+1}{8}\, \rho\xi_0^2\omega^2. \tag{52}$$

Equation (43) can also be used to calculate the dissipative correction for the radiation pressure, because the dissipative terms in the equation of motion and in (17) do not affect the results (39) and (41) used for the derivation of (43). The solution of the linearized equation (30) in this case may be written in the form

$$\xi = \xi_0\left[1 - \cos\omega\left(t - \frac{a}{c}\right)\right](1 - \alpha a). \tag{53}$$

The factor $(1 - \alpha a)$ is the approximate value of the exponential $\exp(-\alpha a)$ at distances $a \ll 1/\gamma$ over which the solution (53) is valid.

Substituting the dissipative correction for the displacement

$$\xi' = -\,\xi_0\alpha a\,[1 - \cos\,(\omega t - ka)]$$

into the result (43), we find that the second-order dissipative correction to the radiation pressure is equal to

$$-\frac{c^2\rho}{l}\,\overline{\xi'(l, t)} = \xi_0\rho c^2\alpha. \tag{54}$$

Adding the results (52) and (54), we obtain an expression for the radiation pressure on a radiating plane in a dissipative medium:

$$\overline{P_L} = \frac{\gamma+1}{8}\, \rho\xi_0^2\omega^2 + \xi_0\rho c^2\alpha. \tag{55}$$

The ratio of the second term on the right-hand side of (55) to the first term is equal to [cf. (18)]

$$\frac{8}{\gamma + 1} \cdot \frac{\alpha}{\xi_0 k^2} \approx \frac{4}{\gamma + 1} \cdot \frac{b\omega}{p'_{max}} = \frac{2}{N} \,.$$ (56)

When this ratio becomes on the order of unity, the dissipative correction begins to affect the radiation pressure significantly.

§ 3.   Langevin Radiation Pressure

Consider a sound wave propagating in an unbounded space and decaying to zero at infinity. For the given sound field, Eq. (21) may be assumed to have the form

$$\overline{p'} = -\frac{\rho \overline{v^2}}{2} + \frac{c^2 \overline{\rho'^2}}{2\rho} \,.$$ (57)

Here, first of all, the term b div $\overline{\mathbf{v}}$ is omitted as a negligibly small quantity if $\overline{\mathbf{v}}$ is second-order small, as mentioned already in the discussion of Eq. (37); second, the constant is chosen equal to zero, in order to meet the condition of zero perturbation at infinity. If the constant did not vanish, we would be required to use Eq. (17), whereupon the equation for $\overline{p}'$ would contain the parameter γ [46]. Generally, in problems where the contact between the sound field and the unperturbed medium permits the constant to be set equal to zero, the parameter γ must be absent in the radiation pressure equations. In such cases it is customary to speak of the Langevin radiation pressure.

We now deduce formulas for the Langevin radiation pressure on an acoustically completely absorbing obstacle, an obstacle having a reflection coefficient of one, and at an interface between two fluids. We investigate these problems by the usual procedure (see, e.g., [10, 15, 31, 43]). Then in §2 of Chap. 3, we return to the same problems in order to show that, from our point of view, certain results need to be modified in the case of plane waves.

1. Let the second wave be completely absorbed by the obstacle. In the field of a plane traveling sound wave (see, e.g., [15, 63]) $\rho' = \rho v/c$, and according to (57) $\overline{p}' = 0$. Therefore, an obstacle that absorbs the plane sound wave is acted upon by the radiation pressure described by Eqs. (34) and (36) with p' = 0. In vector form

this radiation pressure is

$$\mathbf{P}_r = \rho \mathbf{v} \, (\mathbf{vn}).\tag{58}$$

In particular, for $\mathbf{n} \parallel \mathbf{v}$

$$\mathbf{P}_r = \rho \overline{v^2} \mathbf{n} = 2 \overline{E}_\mathrm{K} \mathbf{n},\tag{59}$$

where E_K is the kinetic energy density in the wave.

In a spherical traveling wave the velocity potential is

$$\varphi = \frac{\varphi_0}{r} \cos(\omega t - kr).$$

By Eqs. (10) and (57), we now obtain $\overline{p}' = -\rho \varphi_0^2 / 4r^4$. Consequently, if the spherical wave is absorbed, the radiation pressure vector is described by the complete Eq. (32)

$$\mathbf{P}_r = \overline{p'} \mathbf{n} + \rho \mathbf{v} \, (\mathbf{vn}).\tag{60}$$

2. Let us imagine that a sound beam propagating in an unbounded medium is completely reflected from a stationary solid obstacle. In this event the radiation pressure on the obstacle is completely determined by $\overline{p}'$ [see (33)]. Near the solid wall, standing waves are formed. The interaction of the sound field with the unperturbed medium such that the constant vanishes in Eqs. (21) and (57) is realized through the lateral boundaries of the beam. According to Eqs. (33) and (57), the radiation pressure on the stationary solid obstacle is

$$\mathbf{P}_r = \frac{c^2}{2\rho} \overline{\rho'^2} \mathbf{n}.\tag{61}$$

We determine the radiation pressure on those portions of the obstacle where the sound field can be approximated by a plane standing wave. We draw the x axis perpendicular to the surface of the portion in question, where we affix the coordinate origin. The field of a plane standing wave is characterized in the linear approximation by values

$$v = -v_0 \sin kx \cos \omega t,$$

$$\rho' = \frac{\rho v_0}{c} \cos kx \sin \omega t.\tag{62}$$

Substituting (62) into (57), we obtain near the investigated portion of the obstacle

$$\overline{p}' = \frac{\rho v_0^2}{4} \cos 2kx.\tag{63}$$

Near the surface of the obstacle x = 0, we have p' = $\rho v_0^2/4$, and the radiation pressure, according to Eq. (61), is

$$P_r = \frac{\rho v_0^2}{4}\, n. \tag{64}$$

The numerical value of the radiation pressure on the surface of a solid obstacle parallel to the particle velocity vector in the plane standing wave is determined by Eq. (63). It is directed perpendicular to the surface.

3. We now determine the radiation pressure exerted by a plane sound wave on an interface between two media when the direction of wave propagation is perpendicular to the interfacial surface. The radiation pressure on the interface is equal to the difference between the momentum fluxes through two fixed surfaces on either side of and parallel to the interface. It is important to bear in mind here that a reflected wave propagates along with the incident wave in the first medium, while in the second medium a refracted wave is transmitted. The relationship between the waves is determined by the boundary conditions, which state that the pressures and normal projections of the velocities are equal at the interface:

$$\rho_1 c_1\,(v - v_1) = \rho_2 c_2 v_2, \quad v + v_1 = v_2. \tag{65}$$

Here v, v_1, and v_2 are the velocities of the incident, reflected, and refracted waves, respectively, and the subscripts 1 and 2 refer to the first and second media. In the linear approximation, each wave is still governed by the well-known relations between the traveling-wave variables in linear theory:

$$p' = \rho_1 c_1 v, \quad \rho_1' = -\frac{\rho_1 v_1}{c_1}, \quad p_2' = c_2^2 \rho_2', \text{ etc.} \tag{66}$$

Using relations (34), (57), and (66), we write the equation for the radiation pressure at the interface between two media in the form

$$P_r = P_{r1} - P_{r2} = -\frac{\rho_1 \overline{(v + v_1)^2}}{2} + \frac{c_1^2 \overline{(\rho' + \rho_1')^2}}{2\rho_1} + \rho_1 \overline{(v + v_1)^2} +$$

$$+ \frac{\rho_2 \overline{v_2^2}}{2} - \frac{c_2^2 \overline{\rho_2'^2}}{2\rho_2} - \rho_2 \overline{v_2^2} = \rho_1\,(\overline{v^2} + \overline{v_1^2}) - \rho_2 \overline{v_2^2}. \tag{67}$$

Eliminating $\overline{v_1^2}$ and $\overline{v_2^2}$ from expression (67) by means of relations (65), we obtain

$$P_r = 2\rho_1 \overline{v^2}\left[\frac{\rho_1^2 c_1^2 + \rho_2^2 c_2^2 - 2\rho_1\rho_2 c_1^2}{(\rho_1 c_1 + \rho_2 c_2)^2}\right]. \tag{68}$$

In the case of media invested with identical wave impedances, $\rho_1 c_1 = \rho_2 c_2$,

$$P_r = \rho_1 \overline{v^2} - \rho_2 \overline{v_2^2} = \rho_1 \overline{v^2}\left(1 - \frac{\rho_2}{\rho_1}\right). \tag{69}$$

Note that we have deliberately avoided the customary derivation in energy terms. The reason for this will become apparent in §1 of Chap. 3.

We cite another formula for the radiation pressure on an interface, when the direction of propagation of a plane wave forms an angle θ_1 with respect to the interfacial surface (see, e.g., [15], §65):

$$P_{rn} = \rho_1 \overline{v^2} \sin\theta_1 \cos\theta_1 \left[\cot\theta_1 (1 + R) - \cot\theta_2 (1 - R)\right]. \tag{70}$$
$$P_{rt} = 0.$$

Here the angle of refraction is defined by the relation

$$\theta_2 = \frac{\sin\theta_1}{\sin\theta_2} = \frac{c_1}{c_2}$$

and the reflection coefficient is defined by the relation

$$R = \left(\frac{\rho_2 \tan\theta_2 - \rho_1 \tan\theta_1}{\rho_2 \tan\theta_2 + \rho_1 \tan\theta_1}\right)^2$$

Chapter 3

Problems Associated with
Radiation Pressure

§ 1. The Energy of Sound Waves

The radiation pressure is often expressed in terms of the energy density of the sound field. It is generally assumed that in the time average, the kinetic and potential energy densities are equal in the wave. This means that the time-average density of twice the kinetic energy is replaced by the time-average total energy density. However, as shown by Andreev [52] (see also [53-57]), this approach is impossible in general. In the present section, we discuss the problem in detail for an ideal fluid.

The total energy density E of a sound field is made up of the kinetic energy density E_K and the potential energy density E_p:

$$E = E_{\text{к}} + E_p. \tag{71}$$

In the equations for the average values of the kinetic and potential energies, only the maximal terms are included, i.e., the equations are limited to squared terms. Then

$$E_{\text{к}} = \frac{\rho_0 v^2}{2}.$$

The potential energy of the sound wave per unit mass is equal to the work that can be extracted from that mass in expanding its volume from $1/\rho$ to $1/\rho_0$:

$$A_p = \int_{1/\rho}^{1/\rho_0} p\, d\left(\frac{1}{\rho}\right). \tag{72}$$

99

Substituting into expression (72)

$$p = p_0 + c^2 \rho' \quad \text{and} \quad d\left(\frac{1}{\rho}\right) = -\frac{1}{\rho_0^2}\left(1 - 2\frac{\rho'}{\rho_0}\right)d\rho,$$

after integration we obtain

$$A_P = \frac{1}{\rho_0^2}\left[p_0\rho' + \left(\frac{c^2}{2} - \frac{p_0}{\rho_0}\right)\rho'^2\right]. \tag{73}$$

The potential energy per unit volume is

$$E_P = \rho A_P = \frac{1}{\rho_0}\left(p_0\rho' + \frac{1}{2}c^2\rho'^2\right). \tag{74}$$

Consequently,

$$E = \frac{\rho_0 v^2}{2} + \frac{c^2}{2\rho_0}\rho'^2 + \frac{p_0}{\rho_0}\rho'. \tag{75}$$

In the case of a plane traveling sound wave

$$\frac{\rho_0 \overline{v^2}}{2} = \frac{c^2}{2\rho_0}\overline{\rho'^2}. \tag{76}$$

Therefore, if Eq. (75) did not contain the linear term on ρ', the time-average total energy density would be equal to twice the average value of the kinetic energy. However, in general $\overline{\rho}'$ is not equal to zero. In fact, according to (17), in an ideal fluid

$$\overline{\rho'} = \frac{\overline{p'}}{c^2} - \frac{1}{2c^2}\left(\frac{\partial c^2}{\partial \rho}\right)\overline{\rho'^2}. \tag{77}$$

In the case of a plane traveling wave radiated by a periodically oscillating plane [see (52) and (53)], $\overline{\rho'_L} = \frac{3-\gamma}{8}\rho k^2\xi_0^2$, so that

$$\left(\frac{\overline{E_P}}{\overline{E_K}}\right)_L = \frac{\gamma+3}{2\gamma}, \tag{78}$$

and at a fixed point of space, i.e., in Euler variables, we find by means of relation (25)

$$\left(\frac{\overline{E_P}}{\overline{E_K}}\right)_E = \frac{\gamma-1}{2\gamma}.$$

For air (γ = 1.4), the ratio of the average potential and kinetic energies departs significantly from unity. Indeed, the same is true of other gases and liquids. A similar result is obtained from the investigation of the kinetic and potential energies per unit mass [see (73)] or per unit mass of the unperturbed volume when the potential energy is equal to $\rho_0 A_p$ [58]. In the expression (75) for the energy density, the usual convention is to write only the first two terms, despite the general lack of justification for this. As a result, one obtains equal time-average kinetic and potential energy densities in the traveling plane wave.

The term $(p_0/\rho_0)\rho'$ in expression (74) turns out to be unessential if one considers the acoustic potential energy of the entire fluid. Thus, after integration over the entire volume of the fluid we have

$$\int_V \rho' \, dV = 0, \tag{79}$$

inasmuch as the total quantity of fluid does not change. For this reason the average value of the potential energy over the volume is

$$\langle E_p \rangle = \left\langle \frac{c^2}{2\rho} \rho'^2 \right\rangle. \tag{80}$$

Consequently, the kinetic and potential energies of the traveling plane sound wave, averaged over the entire volume of the fluid, are equal. But if we consider the part of the sound field set off, for example, by a closed surface fixed in space, then $\overline{\int \rho' dV}$ yields the time-average variation of the mass in the isolated volume. Its nonzero value is the reason for the inequality between the potential and kinetic energies in the given volume [46].

We now consider the sound wave energy from the point of view of energy conservation (see [15], § 6):

$$\frac{\partial}{\partial t} \left(\frac{\rho v^2}{2} + \rho \varepsilon \right) = - \operatorname{div} \left[\rho \mathbf{v} \left(\frac{v^2}{2} + w \right) \right], \tag{81}$$

where ε is the internal energy per unit mass and $w = \varepsilon + p/\rho$ is the enthalpy. Adding to Eq. (81) the equation of continuity multiplied by ε_0:

$$- \varepsilon_0 \frac{\partial \rho}{\partial t} = \varepsilon_0 \operatorname{div} \rho v,$$

and stopping with second-approximation terms, we obtain

$$\frac{\partial}{\partial t}\left[\frac{\rho v^2}{2} + \rho\,(\varepsilon - \varepsilon_0)\right] = -\,\mathrm{div}\,[\rho v\,(w - \varepsilon_0)]. \tag{82}$$

Expanding the acoustic potential energy per unit volume $\rho\,(\varepsilon - \varepsilon_0)$ in a power series on ρ', we obtain an expression that coincides with (74):

$$E_\mathrm{p} = \rho\,(\varepsilon - \varepsilon_0) = \frac{p_0}{\rho_0}\,\rho' + \frac{c^2\rho'^2}{2\rho}\,.$$

We represent both sides of Eq. (82) as the sum of two terms:

$$\frac{\partial}{\partial t}\left(\frac{\rho v^2}{2} + \frac{c^2\rho'^2}{2\rho}\right) + \frac{\partial}{\partial t}\left(\frac{p_0}{\rho_0}\,\rho'\right) = -\,\mathrm{div}\,(p'\mathbf{v}) - \mathrm{div}\left(\frac{p_0}{\rho_0}\,\rho\mathbf{v}\right). \tag{83}$$

According to the equation of continuity,

$$\frac{\partial}{\partial t}\left(\frac{p_0}{\rho_0}\,\rho'\right) = -\,\mathrm{div}\left(\frac{p_0}{\rho_0}\,\rho\mathbf{v}\right). \tag{84}$$

Consequently,

$$\frac{\partial}{\partial t}\left(\frac{\rho v^2}{2} + \frac{c^2\rho'^2}{2\rho}\right) = -\,\mathrm{div}\,(p'\mathbf{v}). \tag{85}$$

Therefore, the total sound energy density can be separated into two parts:

$$E_1 = \frac{\rho v^2}{2} + \frac{c^2\rho'^2}{2\rho} \quad \text{and} \quad E_2 = \frac{p_0}{\rho_0}\,\rho',$$

where the law of energy conservation applies to each part. It may be stated also that the energy E_2 does not enter the medium from the sound radiator [55]. Thus, at each instant of time, by virtue of Eq. (79), the integral over the entire fluid volume is equal to

$$\int_V E_2 dV = 0. \tag{86}$$

This means that the sound energy transmitted to the medium by the radiator is contained only in the expression $\int_V E_2 dV$. The energy density E_2, on the other hand, is solely attributable to the nonuniform distribution of the material density in the sound wave. The law of conservation of the energy E_2 reduces to the law of mass conservation.

Consequently, the potential energy associated with the term E_2 is sometimes insignificant, in which case it is permissible to use the conventional expression for the energy [59]. However, the time-average value of E_2 does not vanish. In the time average, therefore, the kinetic and potential energy of a plane sound wave are not equal, and the time-average total energy density is not twice the kinetic energy density.

§ 2. Momentum Flux in Sound Waves

It is well known that the propagation of a light wave neces- sarily involves the transfer of momentum in space. There is a widespread opinion that sound waves, like light waves, also con- sistently transfer momentum in the time average. However, in the propagation of a light wave through a surface perpendicular to the direction of propagation, momentum is always transferred in one direction. In the propagation of a sound wave, on the other hand, the momentum flux during the oscillation period is trans- ferred through such a surface both in the direction of wave prop- agation and in the opposite direction. Therefore, although at every instant of time there is a flux of momentum in one direction or the other, it is not clear whether the momentum flux has a nonzero value in the time average. We will show that plane sound waves propagating in unbounded space do not transfer momentum in the time average.

Consider an irrotational sound field in a viscous thermally conducting medium. Adding to the equation of motion (1) the con- tinuity equation (2) multiplied by $\mathbf{v}$, and recognizing that for irro- tational motion

$$(\mathbf{v}\nabla)\mathbf{v} = \nabla \frac{v^2}{2}, \quad \text{a} \quad \Delta\mathbf{v} = \nabla \operatorname{div} \mathbf{v},$$

we obtain

$$\frac{\partial(\rho\mathbf{v})}{\partial t} + \nabla\left[p' - \left(\frac{4}{3}\eta + \eta'\right)\operatorname{div}\mathbf{v} + \rho v^2\right] + \rho\mathbf{v}\operatorname{div}\mathbf{v} - \frac{\rho}{2}\nabla v^2 = 0. \qquad (87)$$

If

$$\operatorname{rot}\left(\rho\mathbf{v}\operatorname{div}\mathbf{v} - \frac{\rho}{2}\nabla v^2\right) = 0,$$

which is true, for example, in the case of one-dimensional motion, we can introduce a certain function ψ, such that

$$\nabla\psi = \rho v \operatorname{div} \mathbf{v} - \frac{\rho}{2}\nabla v^2. \tag{88}$$

Averaging Eq. (87) over the time with regard for relation (88), we obtain

$$\nabla\left[\overline{p'} - \left(\frac{4}{3}\eta + \eta'\right)\operatorname{div}\overline{\mathbf{v}} + \overline{\rho v^2} + \overline{\psi}\right] = 0,$$

from which it follows that

$$\overline{p'} = -\overline{\rho v^2} + \left(\frac{4}{3}\eta + \eta'\right)\operatorname{div}\overline{\mathbf{v}} - \overline{\psi}. \tag{89}$$

As noted already in the discussion of Eqs. (37), the term $(\frac{4}{3}\eta + \eta')\times$ div $\overline{\mathbf{v}}$ is a third-order small quantity. Therefore, correct to and including the second approximation,

$$\overline{p'} = -\rho_0\overline{v^2} - \overline{\psi}. \tag{90}$$

Consider a plane sound wave propagating in unbounded space and decaying to zero at infinity. According to Eq. (88), ψ = const. Complying with the condition of zero perturbation at infinity, we obtain

$$\overline{\psi} = \operatorname{const} = 0. \tag{91}$$

In the final analysis

$$\overline{p'} = -\overline{\rho v^2}, \tag{92}$$

and the momentum flux density in the direction of wave propagation* is

$$n_i\overline{\Pi'_{xi}} = \overline{\Pi'_{xx}} = \overline{p'} + \overline{\rho v^2} = 0, \tag{93}$$

where $\mathbf{n}(1, 0, 0)$ is the unit vector in the direction of wave propagation. Consequently, in the given approximation, plane sound waves do not transfer momentum in the direction of wave propagation in the time average.

---

*Strictly speaking, for a viscous medium, it is necessary in Eq. (93) to also include the viscous stress tensor. The additional term, however, is negligible for the same reason as the term $(\frac{4}{3}\eta + \eta')$ div $\overline{\mathbf{v}}$ in Eq. (89).

Note that Eq. (57), which is often used in this case, leads to another result:

$$\overline{p'} = 0, \qquad \overline{\Pi'_{xx}} = \overline{\rho v^2}.$$

However, if dissipative terms are included in Eq. (57), we arrive at (93) [60].

Now consider a traveling spherical wave propagating in unbounded space under the condition $kr \gg 1$. In the linear approximation

$$v = -\frac{A}{r} e^{-\alpha r} \cos(\omega t - kr). \tag{94}$$

The momentum flux density, in the direction of propagation of the spherical wave, is determined by the time average component of the momentum flux density tensor:

$$\overline{\Pi'_{rr}} = \overline{p'} + \overline{\rho v^2}.$$

According to relations (90), (88), and (94),

$$\overline{\Pi'_{rr}} = -\overline{\psi} = -2\rho_0 \int_{\infty}^{r} \frac{\overline{v^2}}{r}\, dr = -\rho_0 A^2 \int_{\infty}^{r} \frac{e^{-2\alpha r}}{r^3}\, dr. \tag{95}$$

Integrating twice by parts, we reduce expression (95) to the form

$$\overline{\Pi'_{rr}} = \rho_0 \overline{v^2} \cdot f(2\alpha r), \tag{96}$$

where

$$f(2\alpha r) = 1 - 2\alpha r - (2\alpha r)^2\, \frac{E_i(-2\alpha r)}{e^{-2\alpha r}}$$

$\left[E_i(x) = \int_{-\infty}^{x} \frac{e^t}{t}\, dt \right.$ is the integral exponential function]. Consequently, spherical waves do transfer momentum in the direction of wave propagation in the time average.

Consider an arbitrary sound field in contact with the unperturbed medium. Contact can exist along the lateral surface of a sound beam of finite cross section or, in the case of propagation in unbounded media, in domains that the sound wave is prevented from reaching by absorption or divergence of the sound field. If in this case the sound wave may be regarded in some domain of

space as a plane wave propagating along the x axis, then in that
domain

$$\overline{\Pi'_{xx}} = -\overline{\psi} = \text{const.} \tag{97}$$

The question arises to whether the above constant can be set equal
to zero by satisfying the condition of zero field in the unperturbed
medium. It would seem that this were so, indicating an absence
of momentum flux and, hence, of radiation pressure in plane sound
waves. However, the experimental studies of Al'tberg, Zernov,
and other authors [3, 4, 10, 42, 43] have shown that in the case of
contact between the sound field and the unperturbed medium, plane
standing and traveling waves exert a radiation pressure on an
obstacle. It is possible that the observed nonzero Langevin acoustic radiation pressure is due to a departure of the sound field from
plane, or to the generation of acoustic streaming [61].

We now deduce an equation for the radiation pressure on a
fluid layer bounded by two parallel planes. In the case of a plane
wave, according to Eq. (97), the layer is not affected by radiation
pressure. In the case of a spherical wave, we consider the direction of propagation perpendicular to the planes bounding the segregated volume element. In this direction a unit surface of the
given layer is acted upon by a force equal to the difference between the radiation pressures on either side of the layer:

$$\overline{P_{r2}} - \overline{P_{r1}} = \overline{p'_1} - \overline{p'_2} + \rho_0 (\overline{v_1^2} - \overline{v_2^2}). \tag{98}$$

For $\alpha r \ll 1$, according to (96), we obtain

$$\overline{P_{r2}} - \overline{P_{r1}} = \rho_0 (\overline{v_1^2} - \overline{v_2^2}) = 2 (\overline{E_{k1}} - \overline{E_{\kappa 2}}). \tag{99}$$

In this case the segregated fluid element is acted upon by a radiation pressure equal to the difference between the doubled values
of the kinetic energy densities at the boundaries of the layer.

If the segregated layer represents an interface between two
media, the constant in Eq. (97) can differ on either side of the interfacial surface. This would indicate the presence of radiation
pressure on an interface between two fluids on the side of a plane
sound field. Not to be overlooked, however, is the fact that this
constant could be zero, indicating an absence of radiation pressure at the interface. In either instance, there is no basis for assuming that the radiation pressure in this case is equal to the
difference between the time averages of twice the kinetic energy

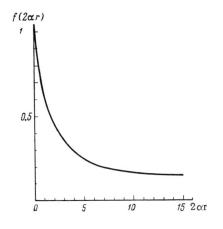

Fig. 1. The function $f(2\alpha r)$ determining the momentum flux density.

densities on either side of the boundary, as is customarily done in correspondence with Eq. (67). We feel that the foregoing calculations should prompt experiments to measure the Langevin radiation pressure in plane sound waves.

It may seem strange that qualitatively different results are obtained for the momentum flux and radiation pressure in plane and spherical waves. As a matter of fact, at large distances from the radiation, a divergent spherical wave is nearly plane. In order to explain this, we analyze the variation of the momentum flux density in a spherical wave during the course of its propagation. It is evident from Eq. (96) that as $\alpha \to 0$ (p' $\to$ 0) we obtain the familiar result for an ideal fluid:

$$\overline{\Pi'_{rr}} = \rho_0 \overline{v^2}. \tag{100}$$

This results corresponds to the case $\alpha r \ll 1$. If $\alpha r \sim 1$, then $\overline{p'} \sim \rho_0 v^2$, i.e., in this case it is required for the calculation of the momentum flux density to take account of dissipative terms in the initial equations. This is especially necessary for $\alpha r > 1$. As $2\alpha r \to \infty$ we find that $f(2\alpha r) \to 0$ (Fig. 1), i.e., in the limit we obtain the result corresponding to plane waves. Consequently, the problem of whether a wave can be assumed plane in a given interval must be solved as a function of the quantity αr.

Radiation Forces Acting
on a Particle in a Sound Field

§ 1. Derivation of the Initial Equation

A particle immersed in a sound field is subjected to the ponderomotive action of the field. The investigation of the corresponding forces has relevant bearing on a number of practical problems. For example, by measuring the force acting in a sound field on an obstacle, one can gain information on the energy of a sound wave. The various devices used for this purpose are discussed in books by Matauschek [12] and Bergmann [8] and in recent papers [62-67]. The ponderomotive forces acting on a particle in a sound field promote coagulation, degassing, and other processes. The acceleration of such processes ties in with the fact that particles suspended in a sound field accumulate near sites where their potential energy is a minimum. The increase in the particle density promotes, for example, acceleration of the coagulation processes, especially since the shorter distances between particles in the sound field causes appreciable interaction forces between the particles [9, 68-72], and these forces affect the coagulation process as well. A detailed review of the problem of acoustic aerosol coagulation and its causes is presented in books by Fuks [73] and Mednikov [9].

In order to gain insight into the factors causing the ponderomotive forces, we consider a body suspended in a sound field. If the density of the body is equal to the density of the surrounding medium, then under the action of sound waves, the body will oscillate

together with the particles of the medium and be acted upon by the
force that would act on the medium in the volume of the body if
the latter were not there. If the densities of the body and surrounding medium are unequal, the body executes motion relative
to the medium, resulting in an additional reaction force on the
body. The force acting on a particle is determined by the momentum acquired by the particle during the scattering of a wave impingent on it. If the body is capable of absorbing sound energy,
there is an additional mechanism for the onset of forces, because
in general, momentum is gained with the absorbed energy of the
particle. In a viscous medium, moreover, each element of the
surface of a body in a sound field is acted upon by the force of
friction.

King has deduced a formula for the force acting on a rigid
sphere [74] and disk [75] in a field of plane traveling or standing
waves in an ideal fluid. King's method was to solve the hydrodynamic equations for the ideal fluid with subsequent calculation of
the forces acting on the obstacle. These calculations were later
repeated by simpler methods [76-82]. We illustrate the method of
direct calculation of the radiation forces in the example of the derivation of a formula for the time-average force acting on a particle in the field of a plane traveling sound wave in an ideal fluid.

The magnitude of the force is equal to the momentum transmitted to the particle by the wave per unit time, or the average
momentum flux through a closed surface surrounding the particle. The surface can be chosen arbitrarily, the only provision
being that it enclose the particle (because in an ideal fluid, the
time-average momentum can only be transmitted to the particle).

Let the particle be situated in the field of a plane traveling
wave propagating along the x axis. We investigate the time-average value of the x projection of the force acting on the particle:

$$F_{rx} = - \oint \overline{\Pi'_{xk}} \, dS_k = \oint P_{rx} \, dS. \qquad (101)$$

The equations for the other projections can be obtained analogously [83]. But if the body is symmetric about the x axis (sphere,
disk), by virtue of symmetry, only the projection of the force on
the x axis will have a nonzero magnitude. Substituting $\overline{\Pi'_{ik}}$ for an

ideal fluid [see (7)] into (101), we obtain

$$F_{rx} = - \oint [\overline{p'n_x} + \overline{\rho v_x (\mathbf{vn})}] \, dS, \tag{102}$$

where $\mathbf{n}$ is the unit outward normal vector to the surface. We represent the velocity $\mathbf{v}$ as the sum of the incident and reflected wave velocities:

$$\mathbf{v} = \mathbf{v_I} + \mathbf{v_R}. \tag{103}$$

We perform the integration in Eq. (102) over a spherical surface far enough from the scattering center to permit the asymptotic expression to be used for the scattered wave. Inasmuch as all terms in Eq. (102) are at most second-order small, for their transformation, we can use the relations of linear theory:

$$p'_I = \rho c \, |\, \mathbf{v_I}\, |, \quad p'_R = \rho c \, |\, \mathbf{v_R}\, |. \tag{104}$$

With relations (103) and (104) we separate out three terms in the radiation pressure P_{rx}:

$$P_{rx} = P_{Ix} + P_{Rx} + P_{IRx}, \tag{105}$$

where

$$P_{Ix} = - \overline{p'_I n_x} - \overline{\rho v_{Ix} (\mathbf{v_I n})};$$

$$P_{Rx} = - \overline{p'_R n_x} - \overline{\rho v_{Rx} (\mathbf{v_R n})};$$

$$P_{IRx} = \rho \, [\overline{(\mathbf{v_I} \, \mathbf{v_R})} - \overline{v_I v_R}] \, n_x - \overline{\rho v_{Ix} (\mathbf{v_R n})} - \overline{\rho v_{Rx} (\mathbf{v_I n})}.$$

Here $v_I = |\, \mathbf{v_I}\, |$; $v_R = |\, \mathbf{v_R}\, |$.

Recognizing that in the absence of the sphere $F_r = 0$, i.e., $\oint P_{nx} \, dS = 0$, and that according to Eqs. (57) and (104), $p'_R = 0$, we reduce F_{rx} to the form

$$F_{rx} = - \rho \oint [\overline{v_R^2} \cos \theta + \overline{v_I v_R} (1 + \cos \theta)] \, dS. \tag{106}$$

Here θ is the angle between the x axis and the scattering direction $\mathbf{n}$. For our ensuing transformations, we make use of the fact that in the given case the time-average energy flux through any closed surface in the fluid is zero:

$$\oint \mathbf{Q} \, dS = 0, \tag{107}$$

where

$$Q = \rho v \left(\frac{v^2}{2} + w \right)$$

is the energy flux density vector [see (81)]. Confining the expression for $\mathbf{Q}$ to second-order terms, we obtain

$$Q = \rho w \mathbf{v}, \tag{108}$$

where $w = w(p)$ need only be expanded in a series to first-order terms:

$$w = w_0 + \left(\frac{\partial w}{\partial p} \right)_s p' = w_0 + \frac{p'}{\rho_0}. \tag{109}$$

With regard for relations (108) and (109), Eq. (107) assumes the form

$$w_0 \oint \rho v dS + \oint p' \mathbf{v} \cdot dS = 0. \tag{110}$$

The first integral is zero in the time average, as in our problem there must be no flow of matter through the closed surface. The second integral is written by means of expressions (103) and (104) in the form

$$\rho c \oint (v_I + v_R) \{ \mathbf{v}_I + \mathbf{v}_R \} \cdot dS = 0, \tag{111}$$

or

$$\oint [\overline{v_R^2} + \overline{v_I v_R}(1 + \cos \Theta)] \, dS = 0. \tag{112}$$

Here we have dropped the following integral, which is equal to zero:

$$\overline{v_I^2} \oint \cos \theta \, dS = 0.$$

Eliminating the term with $\overline{v_I v_R}$ from expressions (106) and (112), we obtain

$$F_{rx} = \rho \oint \overline{v_R^2} (1 - \cos \theta) \, dS. \tag{113}$$

Consequently, the force is determined solely by the momentum transported by the scattered wave. In other fields the interference terms between the incident and reflected waves turn out to be more significant, producing considerably larger forces than in traveling plane waves.

Introducing the effective differential scattering cross section into Eq. (113) (see [15], §76):

$$d\mathfrak{s} = \frac{\overline{v_R^2}}{\overline{v_I^2}}\, dS,$$

we obtain

$$F_{rx} = \rho \overline{v_I^2} \oint (1 - \cos\theta)\, d\mathfrak{s}. \tag{114}$$

In order to continue our calculations, we need to have the scattered wave solution, the determination of which involves considerable computational labor as a rule. At distances much greater than the wavelength and linear dimensions of the particles, the quantity v_R must have the asymptotic form

$$v_R \approx \frac{k f\,(\theta,\,\varphi)}{r} \cos{(kr - \omega t)}, \tag{115}$$

where the scattering amplitude $f(\theta,\,\varphi)$ is a function of the scattering direction. With regard to (115) we obtain

$$\overline{v_I^2}\, d\mathfrak{s} = \overline{v_R^2}\, dS = \frac{k^2}{2}\,|\,f(\theta,\,\varphi)\,|^2\,\frac{dS}{r^2},$$

as well as

$$F_{rx} = \frac{\rho k^2}{2} \oint (1 - \cos\theta)\,|\,f(\theta,\,\varphi)\,|^2\,\frac{dS}{r^2}. \tag{116}$$

§ 2.   Forces Acting on a Spherical Obstacle

The simplest results are obtained in the case when the radius of the sphere is much smaller than the wavelength ($R \ll \lambda$). Thus, if a solid sphere is left fixed in the field of a plane traveling wave, as occurs when the density of the body (sphere) $\rho_b \gg \rho$, then the effective differential cross section is [15]

$$d\mathfrak{s} = \frac{R^2}{9}\,(kR)^4\left(1 - \frac{3\cos\theta}{2}\right)^2\,\frac{dS}{r^2},$$

and the force computed according to (114) is

$$F_{rx} = \frac{11}{18}\,\pi R^2\,(kR)^4 \rho v_0^2, \tag{117}$$

where v_0 is the velocity amplitude in the incident wave.

For a compressible sphere

$$d\sigma = \frac{R^2}{9}\,(kR)^4\left(a_1 - \frac{3}{2}\,a_2\cos\theta\right)^2\frac{dS}{r^2},\tag{118}$$

where

$$a_1 = 1 - \frac{c^2\rho}{c_b^2\rho_b},\qquad a_2 = 2\,\frac{(\rho_b - \rho)}{2\rho_b + \rho};$$

c_b is the velocity of sound in the sphere. Substituting the result (118) into Eq. (114), we obtain

$$F_{rx} = \frac{2}{9}\,\pi R^2\,(kR)^4\rho v_0^2\left(a_1^2 + a_1 a_2 + \frac{3}{4}\,a_2^2\right).\tag{119}$$

In Gor'kov's paper [81], an expression is derived for the force acting on a small particle ($R \ll \lambda$) in an arbitrary field other than a field similar to that of a plane traveling wave:

$$\mathbf{F}_r = -\nabla U,\quad U = 2\pi\rho R^3\left(\frac{\overline{p_I'^2}}{3\rho^2c^2}\,a_1 - \frac{\overline{v_I^2}}{2}\,a_2\right).\tag{120}$$

The result (120) provides a means for the rapid derivation of formulas for the radiation force when the expressions for p_I' and v_I in the linear-acoustical approximation are known.

In the case of a standing wave, when

$$v_I = v_0\cos\omega t\sin kx,\qquad p_I' = -\rho c v_0\sin\omega t\cos kx,$$

Eqs. (120) lead to the force

$$F_{rx} = \pi R^2\,(kR)\,\rho v_0^2\cdot f\left(\frac{\rho_b}{\rho},\,\frac{c_b}{c}\right),$$

$$f\left(\frac{\rho_b}{\rho},\,\frac{c_b}{c}\right) = \left[\frac{\rho_b + \frac{2}{3}(\rho_b - \rho)}{2\rho_b + \rho} - \frac{1}{3}\cdot\frac{c^2\rho}{c_b^2\rho_b}\right].\tag{121}$$

The particles must move toward points where the potential energy U is minimal. In a standing wave, the minimum potential energy for particles with $f > 0$ occurs at the velocity antinodes, whereas at the velocity nodes, where F_{rx} is also zero, the particle equilibrium is unstable. The converse of this result is obtained for particles with $f < 0$. For solid particles ($c_b \to \infty$), $f > 0$, if

$\rho_b > \frac{2}{5}\rho$; and $f < 0$ for $\rho_b < \frac{2}{5}\rho$. We note also that in standing waves, the radiation forces acting on a small particle are much greater than in traveling waves, because the small factor kR is contained in (119) to the fourth power and in (121) to the first power.

In a convergent or divergent spherical wave with velocity potential $\varphi = (\varphi_0/r) \sin \alpha$, where $\alpha = \omega t \pm kr$,

$$v_1 = - \frac{\varphi_0}{r^2} \sin \alpha \pm \frac{k\varphi_0}{r} \cos \alpha, \quad p_1' = - \frac{\omega\varphi_0}{r} \rho \cos \alpha,$$

and the potential energy calculated according to Eq. (120) is

$$U(r) = \pi R^3 \rho k^2 \varphi_0^2 \left[\frac{a_1}{3r^2} - \frac{a_2}{2} \left(\frac{1}{r^2} + \frac{1}{k^2 r^4} \right) \right]. \tag{122}$$

The extremal values of the potential energy occur at points r_0:

$$r_0^2 = \frac{2}{k^2 \left(\frac{2}{3} \cdot \frac{a_1}{a_2} - 1 \right)}.$$

Depending on the values of a_1, a_2, the potential energy minimum occurs at different points. An analysis of expression (122) shows that particles situated at a distance greater than r_0 must be repulsed to infinity by the field; at distances smaller than r_0, repulsion must be replaced by attraction toward the center of the field (see also [84]).*

If $a_1 < \frac{3}{2}a_2 < 0$, then at the distance r_0, the potential energy U is minimized. In other cases the particles must either be drawn to infinity $(a_1 > \frac{3}{2}a_2 < 0)$, or tend toward the center $(a_1 < \frac{3}{2}a_2 > 0)$.

In connection with the problem of the acoustic coagulation of aerosols, the regular motion of a particle due to a sound wave in a viscous medium has been investigated in several studies (see [9], Chap. 2, §6, as well as [7, 86-88]). The starting point in these papers is the equation for the force acting on a sphere immersed in a stationary flow of a viscous fluid, whose velocity far from the sphere is equal to u under the condition $Re = uR\rho/\eta \ll 1$:

$$F = 6\pi R\eta u + \frac{9}{4}\pi\rho R^2 u^2. \tag{123}$$

---

*Concerning the action of a cylindrical field on a sphere, see [85].

Here the first term on the right-hand side represents the Stokes force, and the second term represents the additional term in the Oseen approximation. This equation can also be used for the quasi-stationary regime, when $R^2\rho\omega/\eta \ll 1$ (see [15], § 24).

In the periodic quasi-stationary regime, when the velocity **v** varies harmonically, the time-average Stokes force is equal to zero, because **v** = 0. However, with allowance for 1) the temperature dependence of the viscosity coefficient, 2) the Oseen correction, 3) the nonlinear velocity corrections, which are not equal to zero in the time average, or any combination of these factors (for more details see [9, 34, 76, 89]), the force can have a nonzero value.

In the first case

$$\overline{F} = 6\pi R\overline{\eta\,(T)\,v};\tag{124}$$

in the second case

$$\overline{F} = \frac{9}{4}\,\pi\rho R^2\overline{v\,|\,v\,|};\tag{125}$$

and in the third case

$$\overline{F} = 6\pi R\eta\overline{v}.\tag{126}$$

For a stationary sphere, the quantity v in the indicated equations is the velocity of the fluid at the site of the sphere when the latter is not there. In the case of a moving sphere, it is required to use the difference between that velocity and the velocity of the sphere.

The average forces elicited by viscous friction can turn out to be larger than the radiation forces due to sound scattering [34]. Apropos, such a comparison is not entirely justified, because the equations for the radiation forces in an ideal fluid are valid under the condition $R^2\rho\,\omega/\eta \gg 1$, whereas Eqs. (124)-(126) were obtained under the converse condition, $R^2\rho\omega/\eta \ll 1$; in other words, the indicated equations must be applied in opposing limiting cases.

Yosioka and Kawasima [90] have derived the most general equations for the radiation forces acting on a free-floating compressible sphere of any radius in the field of plane traveling and standing waves. They showed that in a plane standing wave field, spheres whose radii are larger than their resonance values must move toward the velocity antinodes, whereas spheres of radius

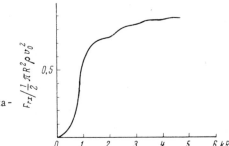

Fig. 2. Radiation forces acting on a stationary sphere.

smaller than the resonance value must move toward the velocity nodes of the standing wave. Spheres having resonance dimensions are not subjected to the action of the radiation forces. These results have been confirmed experimentally [91]. Their results for kR ≪ 1 are described by Eqs. (119) and (121), and for incompressible spheres ($c_b \to \infty$) coincide with the corresponding equations of King [74].

In the case of small particles, rather cumbersome relations are obtained. A graph of the radiation forces acting on a stationary sphere for arbitrary values of kR is shown in Fig. 2; we borrowed the graph from [92]. For values of kR ≫ 1 in a traveling plane wave field, as apparent from the graph,

$$F_{rx} = \frac{1}{2}\,\pi R^2 \rho v_0^2.$$

§3. Forces Acting on a Disk

We wish to consider the radiation forces and torque acting on a disk in a sound field.

Equations for the force acting on a solid unattached disk whose radius is much smaller than the wavelength (R ≪ λ) and whose thickness l ≪ R have been derived by King [75]. If the axis of the disk coincides with the direction of the particle velocity, this radiation force in a traveling wave is equal to

$$F = \frac{8}{27\pi^2}\,S(kR)^4\left(\frac{\delta}{\delta+1}\right)^2 \rho v_1^2; \tag{127}$$

and in a standing wave it is

$$F_r = \frac{2}{3\pi}\,S \cdot kR \cdot \rho v_2^2\left(\frac{\delta}{\delta+1}\right)\left[1 + \frac{\delta}{\delta+1}\cdot\frac{(kR)^2}{5}\right]\sin 2kh. \tag{128}$$

Here $S = \pi R^2$; $\delta = \frac{3}{8} \cdot (m/\rho R^3]$; m is the mass of the disk, h is the distance from the surface of the disk to the velocity node, and v_1, v_2 are the velocity amplitudes in the traveling and standing wave, respectively. For a stationary disk $\delta / (\delta + 1) = 1$.

In the general case, the radiation force equations are rather awesome [93, 94]. Graphs showing the variation of the radiation forces in the domain $kR \sim 1$ for various values of the parameter δ are given in Awatani's paper [93] (see also [94, 95]) both for plane traveling waves (Fig. 3) and for plane standing waves (Fig. 4). In every case the particle velocity is perpendicular to the surface of the disk.

Unlike the sphere, the disk is acted upon in a sound field by a torque as are other bodies incidentally. Rayleigh suggested the utilization of this fact to measure the intensity of sound fields [96]. The corresponding device (Rayleigh disk) proved to be well suited to measurements of the particle velocity in a wave (see, e.g., [62, 97, 99]). If the disk is stationary, where $kR \ll 1$ and $l \ll R$, the average torque acting on the disk is

$$\overline{M}_0 = \frac{4}{3} R^3 \rho \overline{v^2} \sin 2\theta, \qquad (129)$$

where $\overline{v^2}$ is the mean-square particle velocity and θ is the angle between the disk axis and the direction of the particle velocity in the unperturbed sound field [100].

Several papers have already been published in which various factors affecting the torque have been investigated [101-106]. Rasmussen [106] has analyzed in detail the corrections to the

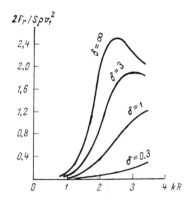

Fig. 3. Radiation forces in the field of plane traveling waves.

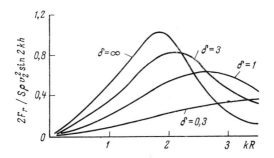

Fig. 4. Radiation forces in the field of plane standing
waves.

König equation (129) for wave diffraction by the disk and the motion of the latter. If the torque is represented in the form

$$\overline{M} = H\overline{M}_0,$$

then for the inclusion of wave diffraction in the case $kR < \frac{1}{2}$ and
$\theta = 45°$,

$$H = 1 + \frac{1}{5}(kR)^2 + 0\,(kR)^4;$$

for $kR \sim 1$ and $\theta = 45°$, the values of H calculated by Kawai [104]
are presented below:

kR	1.0	1.4	1.5	2.0	2.5	3.0	3.5	4.0
H	1.22	1.39	1.42	1.175	0.596	0.405	0.323	0.079

Rasmussen [106] has also obtained an equation for a correction
coefficient H to account for the motion of the disk in the case
$kR \ll 1$:

$$H = \frac{(1 - \rho/\rho_b)^2}{1 + \frac{4}{3\pi}\frac{\rho}{\rho_b} \cdot \frac{2R}{l}} \cdot$$

In the article it is explained why the above equation diverges from
the results of Wood [102] and King [101], who also disagree with
each other.

Below we give the values of H obtained by Kawai for $kR \sim 1$
and $\theta = 45°$ [104] (the second row corresponds to a mica disk in
air, the third to a brass disk in water):

kR	1.5	2.0	2.5	3.0	3.5	4.0
$\rho/\rho_b = 4.63 \cdot 10^{-4}$	1.014	0.009	0.982	1.007	1.019	1.20
$\rho/\rho_b = 1.15 \cdot 10^{-4}$	0.15	0.18	0.33	0.42	0.47	1.33

Radiation Pressure Effects
and Their Applications

The radiation pressure is manifested in the ponderomotive action of the sound field on an obstacle. By measuring the radiation forces, one can acquire information on their related variables, the most important of which is the energy density of the sound field. Moreover, it may be essential to know the radiation forces for the analysis of certain effects that occur in a sound field. In this section, without elaborating in detail, we discuss the possible applications of effects induced by the radiation pressure.

§ 1. The Fountain Effect

The most striking ponderomotive action of a sound field is seen in the fountain effect, which is fairly easily realized. When a sound beam is incident on an interface between two media, a swelling of the interfacial surface takes place, developing ultimately into a fountain as the intensity is increased. Hertz and Mende [10] observed the swelling of an interface after the incidence of a sound beam on the boundary between two liquids having identical wave impedances. It turned out, in accordance with Eq. (69), that the direction of yield of the interface was independent of the direction of propagation of the sound beam, remaining parallel at all times with the velocity of the liquid characterized by the smaller sound energy density. In general, the dimensions of the domelike swell of the surface depend on the intensity of the sound beam. Kornfel'd and Triers [107] established a relation between the height of

the swell and the acoustic radiation pressure impingent on the free
surface of the liquid from below. As a result, they were able on the
basis of microscope measurements of the height of the well to
find the value of the radiation pressure and, hence, the intensity
of the wave incident on the liquid surface (see also [108]).

§ 2.   Absolute Measurement

of the Field Intensity

The fact that the radiation pressure is proportional to the
sound energy density is utilized in experimental practice for absolute measurements of field intensity by means of radiometers,
i.e., devices for the measurement of radiation forces. Several
different types of radiometers have been described in the scientific literature. The radiometers consist of a receiving section
in the form of a plane, cone, sphere, or other figure, which is
placed in the investigated domain of the field during the measurements, and a device for determining the magnitude of the radiation forces acting on the receiver. Sometimes the whole system
is placed in a chamber with sound-absorbing walls, into which the
transmitted sound beam is directed. The design features of the
radiometers are dictated by the required accuracy, irrespective
of whether the beam intensity is measured in the average over the
cross section or at a point, whether the measurements are performed under laboratory or field conditions, etc. We shall not
describe the actual structure and experimental capabilities of
the various radiometers and Rayleigh disks. Pertinent descriptive
material may be found in adequate detail in volume 2 of the present series (Part VII), as well as in books by Matauschek [12] and
Bergmann [8]. We shall not give a detailed comparison of the results of radiometer measurements of the radiation forces with
theory, inasmuch as the experimental and theoretical are almost
always found to agree; instead, we shall rest content with a few
relevant comments.

Radiometers containing a receiving element, whose linear
dimensions are much larger than the wavelength, make it possible
to measure the space average (for example, the average over the
cross section of the sound beam) of the field intensity. In radiometers of this type, the receiver generally spans the entire cross
section of the beam [12, 66, 67]. If, on the other hand, the re-

ceiver is placed inside the sound field, then, according to the theory, the radiation pressure is independent of the receiver dimensions [97, 109]. The field intensity in such situations must be calculated on the basis of the equations given in Chap. 2.* These equations are simplest in the case of the Langevin radiation pressure. According to Eq. (59), the momentum flux density in the direction of wave propagation is equal to $\overline{2E_K}$. Consequently, the radiation pressure on a perfectly absorbing receiver is equal to $\overline{2E_K}$ and acts in the direction of wave propagation, independently of the angle of incidence. Accordingly, the normal and tangential components of the radiation pressure are equal to

$$P_{rn} = \overline{2E}_\kappa \cos \alpha,$$
$$P_{rt} = 2\overline{E}_\kappa \sin \alpha, \tag{130}$$

where α is the angle between the direction of propagation and the normal to the receiving surface. In the case of a perfectly reflecting receiver, according to Eqs. (36) and (64), the normal component of the radiation pressure is doubled, the tangential component vanishing altogether. This dependence of the normal and tangential components of the radiation pressure on the angle of incidence for an acoustically perfectly absorbing or perfectly reflecting obstacle has been corroborated experimentally by Herrey [43]. This also confirms the vector character of the radiation pressure. Results consistent with the theory have also been obtained in measurements of the Rayleigh radiation pressure [38-40].

For local measurements of field intensity, radiometers are used in which the receiving section is much smaller or on the same order as the wavelength. The various types of radiometers differ both in their receiving section (of spherical [84, 65], disk [95], or other configuration), and in their devices for determination of the radiation forces. In some cases a torsional, e.g., balance system is used [110], while in others there are devices incorporated in which the receiver is held in place by the exact compensation of the radiation forces, as for example by forces of electromagnetic origin [111]. The analytical formulas for the corresponding radiometers are given in Chap. 4.

---

*Also within the scope of Chap. 2 are cases when the radiometer receiving element, regardless of its dimensions, is embedded in one of the walls confining the sound field [4, 38].

In the comparison of the theoretical data for the radiation pressure in plane traveling waves with the experimental data, it is important to keep in mind that real fields can depart considerably from the ideal plane traveling-wave field. Yosioka, Kawasima, and Hirano [112] have investigated theoretically and experimentally the influence of spherical divergence of a wave and the contribution of possible standing waves. The experiments were performed in water with a radiometer incorporating spheres of various sizes and materials as the receiving element. It was shown that the theory exhibits good agreement with experiment only at distances

$$r \gg \frac{R^2}{\lambda},$$

where R is the radius of the radiating source and λ is the wavelength. In the event that reflected waves are present, the radiation forces acting on the sphere must be calculated according to the equation

$$F_r = (A^2 - B^2) F_{rtw} + ABF_{rsw}. \tag{131}$$

Here A and B are the respective amplitudes of the traveling wave and reflected wave propagating oppositely to the incident wave. F_{rtw} is the radiation force due solely to a traveling wave of unit amplitude, and F_{rsw} is the radiation pressure in the standing wave formed by a direct and a reverse wave of unit amplitude at a distance from the velocity node equal to the distance from the sphere to the point of minimum velocity. The reduction observed by the authors in the radiation forces as the source is approached at a distance on the order of R^2/λ, in conflict with the theory, is possibly attributable to the singularity that we have indicated in nearly plane fields [see Chap. 3, §2; cf. (97) and (100)]. Similar anomalies have also been observed in [97]. As we have pointed out, there is no basis for regarding the momentum flux density of plane waves as equal to twice the kinetic energy density, nor the radiation pressure, for example, on an obstacle that completely absorbs a plane wave incident upon it as equal to twice the average kinetic energy density of the field. The radiation pressure on such an obstacle must be determined, in our opinion, by Eq. (96). Here the value of r must be assumed equal to the distance from a virtual point source whose field in the given domain would coincide with the actual field. The final judgment as to the validity of the equations for the radiation forces induced in almost-plane sound fields will require an experiment in which the radiation forces and energy density of the acoustic field are measured by two independent techniques. This will also necessitate a monitoring of the structure of the field in order to be certain to what extent the radiated field departs from a plane-wave field. We note, of course, the many papers in which a radiometer has been used to measure field intensities, as opposed to the very few papers in which the radiation pressure equations have actually been tested.

For the measurement of field intensity with a radiometer, acoustic streaming has a detrimental effect. In the case of a one-dimensional plane sound field, however, the momentum flux density in the direction of wave propagation is constant [see (97)]. Even in the presence of acoustic streaming, this result prevails in domains where the motion may be regarded as one-dimensional and planar. Thus, Borgnis [32] was in a position to conclude that the radiation pressure on a perfectly absorbing obstacle in this case is independent of the distance between the sound source and obstacle and is equal to $2\overline{E_K}$. The latter result has been confirmed experimentally [42]. By contrast with the foregoing studies, we postulate that there is no basis for assuming that the constant in Eq. (97)* is equal to the field intensity divided by the velocity of sound (cf. [12]).

§ 3. Nonlinear Properties of a Medium

In the equations for the Rayleigh acoustic radiation pressure, the proportionality factor between the radiation pressure and the kinetic energy density in a wave depends on the parameter γ characterizing the nonlinear properties of the medium [see (38)]. Consequently, in the simultaneous measurement of the Rayleigh radiation pressure and kinetic energy density, the corresponding equations permit one to calculate the parameter γ. In [38, 39] the values of γ for various gases have been obtained as the result of an independent measurement of the field intensity and isotropic part of the Rayleigh radiation pressure in standing waves. The isotropic part of the Rayleigh radiation pressure is interpreted as the constant in Eq. (21). Its value in a standing wave can be obtained by setting p', determined by (21) at a node of the standing wave, equal to its value according to Eq. (38):

$$\frac{c^2 \overline{\rho'^2}}{2\rho_0} + \text{const} = \frac{\gamma + 1}{8} \rho v_0^2.$$

---

*See also the comment immediately following this equation.

Making use of (62), we obtain

$$\text{const} = \frac{\gamma - 1}{8}\, \rho v_n^2.$$

However, the above method of measuring γ has not yet gained as wide acceptance as other methods [14, 113].

§ 4. Acoustic Streaming Velocity and the Coefficient of Sound Absorption

Piercy and Lamb [114], regarding the radiation pressure as the cause of acoustic streaming, obtained for one specific problem a simple formula relating the acoustic streaming velocity to the coefficient of sound absorption. This enabled them on the basis of a measurement of the acoustic streaming velocity to estimate the sound absorption coefficient. Their experimental arrangement (see also [115, 116]) is shown in Fig. 5. During the transmission of sound through the chamber, the liquid in the bleeder tube is acted upon by a radiation pressure equal to the time-average value of the pressure $\bar{p}$. If ultrasound is absorbed, the pressure

$$\bar{p} = -\rho \overline{v^2} + \text{const},$$

where

$$v = v_0 e^{-\alpha z} \sin(\omega t - kz),$$

will differ at the ends of the bleeder tube [see (90)], and

$$\Delta \bar{p} = \bar{p}_2 - \bar{p}_1 = \frac{\rho v_0^2}{2}\left(e^{2\alpha z_1} - e^{-2\alpha z_2}\right), \tag{132}$$

where v_0 is the particle velocity amplitude at the radiator, and z_1, z_2 are the distances from the radiator to the nearest and farthest ends of the bleeder tube. As a result, Poiseuille flow is set up in the bleeder with a velocity

$$u = \frac{\Delta \bar{p}\,(R^2 - r^2)}{4 \eta l}. \tag{133}$$

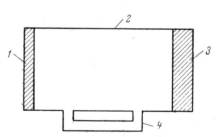

Fig. 5. Diagram of Piercy and Lamb's experiment. 1) Ultrasonic radiator; 2) measurement chamber; 3) ultrasonic absorber; 4) bleeder tube.

Here r is the distance from the bleeder tube axis to the given point and l is the length of the bleeder. This made it possible to relate the flow velocity to the absorption coefficient and thereupon to compute the latter on the basis of a measurement of the former.

Another possible application of the effects induced by the radiation pressure is the idea mentioned previously of fabricating bearings using the radiation pressure for support. Some of the difficulties encountered in the implementation of this idea are indicated in [50].

Herrey [43] has experimentally investigated the radiation forces acting on the receiving disk of a radiometer as a function of the disk material and angle of incidence. He showed how the dependence of the components of the radiation force vector on the angle of incidence [see (130)] could be measured to ascertain the dependence of the average coefficients of absorption, reflection, and transmission of various materials.

Knowledge of the radiation forces is also required in order to calculate the motion of particles in a sound field [7, 9, 90, 91].

References

1. Rayleigh (J. W. Strutt), The Theory of Sound, Vol. 2, New York (1965).
2. Lord Rayleigh, On the momentum and pressure of gaseous vibrations, and on the connection with the virial theorem, Phil. Mag., 10:364 (1905); 3:336 (1902).
3. V. Al'tberg, On the pressure of sound waves and on the absolute measurement of sound intensity, Zh. Russk. Fiz.-Khim. Obshch., Chast' Fiz., 34(4):459 (1903); Ann. Phys., 11:405 (1903).
4. V. D. Zernov, Comparison of methods for the absolute measurement of sound intensity, Zh. Russk. Fiz.-Khim. Obshch., Chast' Fiz., 38(7):410 (1906); Ann. Phys., 21:131 (1906).
5. R. Lucas, La pression de radiation en physique et particulierement en acoustique [Radiation pressure in physics and in particular in acoustics], Fifth Internat. Congr. Acoustics, Liège, Vol. 2, p. 163 (1965).
6. W. E. Smith, Radiation pressure forces in terms of impedance, admittance, and scattering matrices, J. Acoust. Soc. Am., 37(5):932 (1965).
7. A. S. Denisov, D. B. Dianov, A. A. Podol'skii, and V. I. Turubarov, Drift of an aerosol particle in a sound wave distorted by the presence of the second harmonic, Akust. Zh., 12(1):31 (1966).
8. L. Bergmann, Ultraschall und seine Anwendung in Wissenschaft und Technik, Edwards, Ann Arbor, Michigan.
9. E. P. Mednikov, Acoustic Coagulation and Precipitation of Aerosols, Izd. AN SSSR (1963).
10. G. Hertz and H. Mende, Der Schallstrahlungsdruck in Flüssigkeiten [Acoustic radiation pressure in fluids], Z. Phys., 114:354 (1939).

11.   E. Skudzryk, Die Grundlagen der Akustik [Fundamentals of Acoustics], Vienna
       (1954).

12.   I. Matauschek, Einfuhrung in die Ultraschalltechnik [Introduction to Ultra-
       sonic Engineering], Berlin (1961).

13.   L. D. Rozenberg, V. F. Kazantsev, L. O. Makarov, and D. F. Yakhimovich,
       Ultrasonic Cutting, Izd. AN SSSR (1962).

14.   L. K. Zarembo and V. A. Krasil'nikov, Introduction to Nonlinear Acoustics
       (High-Intensity Sonic and Ultrasonic Waves), Izd. "Nauka" (1966).

15.   L. D. Landau and E. M. Lifshits, Mechanics of Continuous Media, Moscow
       (1954).

16.   Z. Gol'dberg, Second-approximation acoustic equations and the propagation
       of finite-amplitude plane waves, Akust. Zh., 2(3):325 (1956).

17.   D. T. Blackstock, Thermoviscous attenuation of plane, periodic, finite-am-
       plitude sound waves, J. Acoust. Soc. Am., 36(3):534 (1964).

18.   L. Brillouin, Sur les tensions de radiation [On the radiation pressure forces],
       Ann. Phys., 4(10):528 (1925).

19.   L. Brillouin, Les pressions de radiation et leur aspect tensorial [Radiation
       pressures and their tensorial aspect], J. Phys. Radium, 17(5):379 (1956).

20.   F. E. Borgnis, Acoustic radiation pressure of plane compressional waves, Rev.
       Mod. Phys., 25(3):653 (1953).

21.   F. E. Borgnis, Über die Bewegungsgleichung und den Impulssatz in viskosen
       und kompressiblen Medien [Equation of motion and momentum principle
       in viscous and compressible media], Acustica, 4(4):407 (1954).

22.   O. K. Mawardi, Sur la pression de radiation en acoustique [On the acoustic
       radiation pressure], J. Phys. Radium, 17(5):384 (1956).

23.   J. Mercier, De la pression de radiation dans les fluides [Radiation pressure
       in fluids], J. Phys. Radium, 17(5):401 (1956).

24.   C. Schaefer, Zur Theorie des Schallstrahlungsdruckes [On the theory of the
       acoustic radiation pressure], Ann. Phys. (5), 35(6):473 (1939).

25.   F. Bopp, Energetische Betrachtungen zum Schallstrahlungsdruck [Energy
       analysis of the acoustic radiation pressure], Ann. Phys., (5), 38:495 (1940).

26.   R. T. Beyer, Radiation pressure in a sound wave, Am. J. Phys., 18(1):25(1950).

27.   G. Richter, Zur Frage der Schallstrahlungsdruckes [On the acoustic radia-
       tion pressure], Z. Phys., 115:97 (1940).

28.   E. Karaskiewicz, Radiation pressure of an acoustical plane wave, Bull. Soc.
       Amis Sci. Lettres de Poznan Ser. B, 14:73 (1958); Postepy Akustyki, 8(1):79
       (1957).

29.   P. Biquard, Les ondes ultras-sonores (II) [Ultrasonic waves (II)], Rev. Acoust.,
       1:315 (1932).

30.   P. Biquard, Les ondes ultras-sonores (II), [Ultrasonic Waves (II)] Rev. Acoust.,
       1:315 (1932).

31.   F. E. Borgnis, Acoustic radiation pressure of plane-compressional waves at
       oblique incidence, J. Acoust. Soc. Am., 24(5):468 (1952).

32.   F. E. Borgnis, On the forces due to acoustic wave motion in a viscous medium
       and their use in the measurement of acoustic intensity, J. Acoust. Soc. Am.,
       25(3):546 (1953).

33.   A. A. Éikhenval'd, Large-amplitude sound waves, Usp. Fiz. Nauk, 14(5):552 (1934).

34.   P. J. Westervelt, The mean pressure and velocity in a plane acoustic wave in a gas, J. Acoust. Soc. Am., 22(3):319 (1950).

35.   R. Lucas, Les tensions de radiation en acoustique [Acoustic radiation pressure forces], J. Phys. Radium, 17(5):395-399 (1956).

36.   D. T. Blackstock, Normal reflection of finite amplitude plane waves from a rigid wall, Proc. Third Internat. Congr. Acoustics, Stuttgart, Vol. 1, p. 309 (1959).

37.   J. Mendousse, Acoustic radiation pressure, Compt. Rend., 208:1977 (1938).

38.   M. Mathiot, Étude experimentale du terme isotrope de la tension de radiation acoustique dans un gaz [Experimental study of the isotropic term in the acoustic radiation pressure in a gas], Compt. Rend., 255(1):64 (1962).

39.   M. Mathiot, Étude experimentale du terme isotrope de la tension de radiation acoustique dans un gaz [Experimental study of the isotropic term in the acoustic radiation pressure in a gas], Ann. Phys., 1:235 (1966).

40.   R. V. Dombrovskii, Report to the Colloquium at the Acoustics Institute of the Academy of Sciences of the USSR (1966).

41.   C. Florisson, Procédé d'étalonnage d'une sonde acoustique au moyen du pendule absolu de pression de radiation [Procedure for the absolute calibration of an acoustic probe by means of a pendulum from the radiation pressure], J. Phys. Radium, 17:411 (1956).

42.   W. G. Cady and C. E. Gittings, On the measurement of power radiated from an acoustic source, J. Acoust. Soc. Am., 25(5):892 (1953).

43.   E. M. J. Herrey, Experimental studies on acoustic radiation pressure, J. Acoust. Soc. Am., 27(5):891 (1955).

44.   V. Gavreau, Pression de radiation sonore d'apres la théorie cinétique des gaz [Acoustic radiation pressure in terms of the kinetic theory of gases], J. Phys. Radium, 17(10):899 (1956).

45.   E. J. Post, Radiation pressure and dispersion, J. Acoust. Soc. Am., 25(1):55 (1953).

46.   J. S. Mendousse, On the theory of acoustic radiation pressure, Proc. Am. Acad. Arts and Sci., 78:148 (1950).

47.   Z. A. Gol'dberg and K. A. Naugol'nykh, The Rayleigh sound pressure, Akust. Zh., 9(1):28 (1963).

48.   R. Lucas, Sur les tensions de radiation des ondes acoustices [Radiation pressure forces of sound waves], Nuovo Cimento (9), 7(2):236 (1950).

49.   R. Lucas, Sur les pressions des radiation des ondes spheriques [Radiation pressure of spherical waves], Compt. Rend., 230:2004 (1950).

50.   M. J. Seegal, Acoustic radiation pressure bearing, J. Acoust. Soc. Am., 33(5):566 (1961).

51.   E. Fubini-Chiron, Anomalie nella propagazione di onde acustiche di grande ampiezza [Anomalies in the propagation of large-amplitude sound waves], Alta Frequenza, 4(5):530 (1935).

52.   N. N. Andreev, Über die Energieausdrucke in der Akustik [Energy expressions in acoustics], J. Phys. (USSR), 2:305 (1940).

53.   A. Schoch, Zur Frage nach dem Impuls einer Schallwelle [On the momentum
       of sound wave], Z. Naturforsch., 7a:273 (1952).

54.   J. Markham, Second order acoustic field; relation between energy and in-
       tensity, Phys. Rev., 89:972 (1953).

55.   N. N. Andreev, Certain second-order quantities in acoustics, Akust. Zh.,
       1(1):3 (1955).

56.   N. N. Andreev, Einige Fragen der nichtlinearen Akustik [Some problems in
       nonlinear acoustics], Proc. Third Internat. Congr. Acoustics, Stuttgart (1959),
       Vol. 1, p. 304 (1961).

57.   J. Fazanowicz, Ped i energia ciagu falowego, Postepy Akustyki, 8(1):181
       (1957).

58.   J. Markham, Second order acoustic field; energy relations, Phys. Rev.,
       86(5):712 (1952).

59.   A. Schoch, Remarks on the concept of acoustic energy, Acustica, 3(3):181
       (1953).

60.   Z. Gol'dberg, On momentum flux in sound waves, Preprints Fifth Internat.
       Congr. Acoust., Liège, Vol. K44 (1965).

61.   L. D. Rozenberg and L. O. Makarov, Causes of the swelling of the surface of
       a liquid under the influence of ultrasound, Dokl. Akad. Nauk SSSR, 114(2):275
       (1957).

62.   A. N. Golenkov and I. G. Rusakov, Optimum Rayleigh disks for the measure-
       ment of sound intensity in water, Trudy Inst. Komiteta Standartov, Mer i
       Izmeritel'nykh Priborov pri Sovete Ministrov SSSR, No. 45(105), p. 63 (1960).

63.   J. A. Newell, A radiation pressure balance for the absolute measurement of
       ultrasonic power, Phys. Med. Biol., 8(2):215 (1963).

64.   J. Cabrielli and G. Jenretti, Torsion balance for radiation pressure measure-
       ments, Acustica, 13(3):175 (1963).

65.   A. T. Kosolapov, Application of King's formula for measurement of ultra-
       sonic intensity, Uch. Zap. Mordovsk. Univ., Ser. Fiz. Nauk (Saransk), No. 36,
       p. 112 (1964).

66.   G. Kossoff, Balance technique for the measurement of very low ultrasonic
       power outputs, J. Acoust. Soc. Am., 38(8):880 (1965).

67.   O. E. Tsok, Balance for the measurement of ultrasonic intensity, Izmeritel',
       Tekh., 7:42 (1965).

68.   W. Dörr, Anziehende und abstossende Kräfte zwischen Kugeln im Schallfeld
       [Attractive and repulsive forces between spheres in a sound field], Acustica,
       5(3):163 (1955).

69.   V. F. Kazantsev, Motion of gas bubbles in a liquid under the action of Bjerknes
       forces arising in a sound field, Dokl. Akad. Nauk SSSR, 129(1):64 (1959).

70.   T. F. W. Embleton, Mutual interaction between two spheres in a plane sound
       field, J. Acoust. Soc. Am., 34(11):1714 (1962).

71.   V. I. Timoshenko, Aggregation of aerosol particles in a sound field under
       the conditions of Stokes law flow, Akust. Zh., 11(2):222 (1965).

72.   N. L. Shirokova and O. K. Éknadiosyants, Interaction of aerosol particles in
       an acoustic field, Akust. Zh., 11(3):409 (1965).

73.   N. A. Fuks, Mechanics of Aerosols, Izd. AN SSSR (1955).

74. L. V. King, On the acoustic radiation pressure on spheres, Proc. Roy. Soc.,
     A147(861):212 (1934).

75. L. V. King, On the acoustic radiation pressure on circular discs; inertia and
     diffraction corrections, Proc. Roy. Soc., A153(878) (1935).

76. P. J. Westervelt, The theory of steady forces caused by sound waves, J. Acoust.
     Soc. Am., 23(3):312 (1951).

77. J. Awatani, Studies on acoustic radiation pressure: I. General considerations,
     J. Acoust. Soc. Am., 27(2):278 (1955).

78. J. Awatani, Note on acoustic radiation pressure, J. Acoust. Soc. Am., 29(3):392
     (1957).

79. A. Johansen, Force agissant sur une sphere suspendue dans un champ sonore
     [Force acting on a sphere suspended in a sound field], J. Phys. Radium, 17(5):400
     (1956).

80. H. Olsen, H. Wergeland, and P. J. Westervelt, Acoustic radiation force, J.
     Acoust. Soc. Am., 30(7):633 (1958).

81. L. P. Gor'kov, Forces acting on a small particle in a sound field in an ideal
     fluid, Dokl. Akad. Nauk SSSR, 140(1):88 (1961).

82. W. E. Smith, Average radiation-pressure forces produced by sound fields,
     Australian J. Phys., 17(3):389 (1964).

83. P. J. Westervelt, Acoustic radiation pressure, J. Acoust. Soc. Am., 29(1):26
     (1957).

84. T. F. W. Embleton, Mean force on a sphere in a spherical sound field, J.
     Acoust. Soc. Am., 26(1):40 (1954).

85. T. F. W. Embleton, The radiation force on a spherical obstacle in a cyl-
     indrical sound field, Can. J. Phys., 34(3):276 (1956).

86. A. S. Denisov, A. A. Podol'skii, and V. I. Turubarov, Entrainment of aerosol
     particles in an acoustic field at Reynolds numbers R ≤ 1, Akust. Zh., 11(1):115
     (1965).

87. A. A. Podol'skii and V. I. Turubarov, Dependence of the degree of slip past
     aerosol particles on the amplitude of the sound field at Reynolds numbers
     0.5 ≤ R ≤ 1, Trudy LIAR, No. 45, p. 60 (1965).

88. A. A. Podol'skii and V. I. Turubarov, Drift of aerosol particles in a sound
     field under asymmetric distortion of the acoustic waveform, Kolloidn. Zh.,
     27(3):425 (1965).

89. I. N. Kanevskii, Steady forces arising in a sound field, Akust. Zh., 7(1):3
     (1961).

90. K. Yosioka and Y. Kawasima, Acoustic radiation pressure on a compressible
     sphere, Acustica, 5(3):167 (1955).

91. K. Yosioka, Y. Kawasima, and H. Hirano, Acoustic radiation pressure on
     bubbles and their logarithmic decrement, Acustica, 5(3):173 (1955).

92. H. Olsen, W. Romberg, and H. Wegeland, Radiation force on bodies in a sound
     field, J. Acoust. Soc. Am., 30(1):69 (1958).

93. J. Awatani, Studies on acoustic radiation pressure: II. Radiation pressure
     on a circular disk, J. Acoust. Soc. Am., 27(2):282 (1955).

94. H. H. Jensen and K. Saermark, On the theory of the Rayleigh disk and the
     sound pressure radiometer, Acustica, 8(2):79 (1958).

95. K. Budal, E. Hoy, and H. Olsen, Measurements of acoustic radiation force, J. Acoust. Soc. Am., 31(11):1536 (1959).

96. Lord Rayleigh, On an instrument capable of measuring the intensity of aerial vibrations, Phil. Mag., 14:186 (1882).

97. J. Hartmann and T. Mortensen, A comparison of the Rayleigh disk and the acoustic radiometer methods for the measurement of sound-wave energy, Phil. Mag., 39(292):377 (1948).

98. W. West, The accuracy of measurements by Rayleigh disk, Proc. Phys. Soc., B62(355):437 (1949).

99. A. Kösters, Über Schallschnellemessungen in Flüssigkeiten mit der Rayleightschen Scheibe [Sound velocity measurements in fluids with the Rayleigh disk], Akust. Beih., 3:AB171 (1952).

100. W. König, Hydrodynamisch-akustische Untersuchungen [Hydrodynamic-acoustical investigations] (III), Ann. Phys., 43:43 (1891).

101. V. King, On the theory of the inertia and diffraction corrections for the Rayleigh disc, Proc. Roy. Soc., A153:878 (1935).

102. A. B. Wood, Theory of the Rayleigh disc, Proc. Phys. Soc., 47(262):779 (1935).

103. J. Awatani, Anomalous behavior of Rayleigh disk for high-frequency waves, J. Acoust. Soc. Am., 28(2):297 (1956).

104. N. Kawai, Sci. Rep. Tohoku Univ., Ser. 1, 35:210 (1951).

105. G. Maidanic, Torques due to acoustical radiation pressure, J. Acoust. Soc. Am., 30(7):620 (1958).

106. C. G. Rasmussen, An experimental investigation of the diffraction correction for a Rayleigh disc, Acustica, 14(3):148 (1964).

107. M. Kornfel'd and V. I. Triers, Swelling of the surface of a liquid under the influence of ultrasound, Zh. Tekh. Fiz., 26(12):2778 (1956).

108. V. V. Bogorodskii, E. D. Pigulevskii, and V. G. Prokhorov, Method for the Measurement of Ultrasonic Intensity in Liquids, USSR Patent, Class 42d, 1/01, No. 120927 (1959).

109. K. Negiski and O. Nomoto, Experiment on acoustic radiation pressure, J. Acoust. Soc. Japan, 15(4):224 (1959).

110. I. T. Sokolov, Application of the mathematical theory of King to radiometric measurements of sound pressures in a liquid, Zh. Tekh. Fiz., 15(4/5):223 (1945).

111. G. P. Motulevich, I. L. Fabelinskii, and L. N. Shteingauz, An absolute acoustic microradiometer, Dokl. Akad. Nauk SSSR, 70(1):29 (1950).

112. K. Yosioka, Y. Kawasima, and H. Hirano, On the absolute measurement of ultrasound intensity by radiation force on a solid sphere, Mem. Inst. Sci. and Indust. Res. Osaka Univ., 21:13 (1964).

113. A. B. Coppens, R. T. Beyer, M. B. Seiden, J. Donohue, F. Guepin, R. H. Holdson, and C. Townsend, Parameter of nonlinearity in fluids, J. Acoust. Soc. Am., 38(5):797 (1965).

114. J. E. Piercy and J. Lamb, Acoustic streaming in liquids, Proc. Roy. Soc., A226(1164):43 (1954).

115.   D. N. Hall and J. Lamb, Measurement of ultrasonic absorption in liquids by
        the observations of acoustic streaming, Proc. Phys. Soc., 73(471):354 (1959).
116.   K. P. Nikonov and B. B. Kudryavtsev, Measurement of ultrasonic absorption
        in a liquid by the streaming method, in: Application of Ultrasonics to the
        Investigation of Matter, No. 16, Izd. MOPI (1962), p. 183.

PART III

ACOUSTIC STREAMING

L. K. Zarembo

Introduction

Sound at high intensity levels in gases and liquids is accompanied by stationary (time-independent) flows known as acoustic streaming (other terms encountered in the literature are "acoustic wind" or "quartz wind"). These flows occur either in a free non-uniform sound field or (particularly) near various types of obstacles immersed in a sound field or near oscillating bodies. They are always of a rotational character. Their velocity increases with the sound intensity, but, even at the highest intensities currently available, the velocity remains smaller than the particle velocity in the sound wave.

Stationary vortex flows over an oscillating membrane were apparently first observed by Faraday as early as 1831. The first theory of stationary streaming was formulated by Rayleigh at the end of the last century; he showed that the streaming is elicited by constant forces in the sound field, but their magnitude can only be determined by the inclusion of higher than the first linear approximation. Subsequently there were many papers devoted to acoustic streaming, both theoretical and experimental. Especially relevant here is the boundary streaming near the surface of bodies in a sound field. This streaming, which perturbs the boundary layer, does much to explain a number of observed effects involving the acceleration of transport processes under the influence of sound, e.g., the heat transfer of heated bodies, the variation of concentration in the cleaning of contaminated surfaces, the atomization of liquid fuel droplets in vibratory combustion, etc. The action of sound or noise is a very effective method for the perturbation of boundary layers. As we shall learn presently,

137

under certain conditions sound effects not only reduce the thickness of the boundary layer but also promote turbulence in it.

At the present time, three types of acoustic streaming are
known and have been rather thoroughly investigated.

The first type is streaming in a viscous boundary layer near
obstacles in a sound field. The theory of this type was first given
by Schlichting, who showed that under the influence of sound, the
stationary flows in a boundary layer have a rotational (vortex)
character. The scale of the boundary vortices, as a rule, is determined by the thickness of the acoustic boundary layer, and their
dimensions are much smaller than the wavelength. This is small-
scale streaming.

The second type is streaming outside the boundary layer and
also has a vortex character. The vortex scale in this case is considerably larger than the scale of the boundary layer vortices.
One form of this type of streaming has been investigated by Rayleigh, namely two-dimensional flow between two planes (or in a
cylindrical tube) under the action of a standing wave; the vortices
in this case have a scale equal to the acoustic wavelength.

The third type is streaming in a free nonuniform sound field
in which the inhomogeneity scale of the sound field is much larger
than the acoustic wavelength. We refer to this type of streaming
henceforth as Eckart streaming, after the one who first solved the
problem of the flows generated in a confined volume by a well-
collimated sound beam. The vortex scale of this streaming is determined by the volume and greatly exceeds the acoustic wavelength.
Thus, Eckart streaming, unlike the other two types, is a large-
scale phenomenon.

This classification of streaming in terms of its scale is in
some measure arbitrary. First of all, different types of streaming can occur simultaneously under experimental conditions and,
second, different types (in the above classification) of streaming
can have the same dependence, say, on the viscosity and other
parameters of the medium and sound field.

For all three types of streaming, the viscous forces govern
the conditions of stabilization of the streaming velocity.

The onset of streaming indicates a nonzero time-average mass
flow. It is particularly important to stress this point, because we

know that, in powerful sound fields, such time-average parameters as the pressure, density, and velocity can differ from the corresponding parameters for the unperturbed medium. The occurrence of a constant velocity component (if the medium was at rest prior to the initiation of sound) is by no means sufficient cause for the onset of streaming; this constant velocity component can be offset by a time-constant variation of the density, so that the time-average mass flow is still zero. A constant velocity component arises, for example, in the solution in Euler coordinates of the problem of finite oscillations of an infinite piston in a nondissipative medium; however, as shown by the analysis of this solution, there is no time-average mass flow, a result that is entirely reasonable, considering that the onset of a mass flow for a nonpermeable piston would violate the mass conservation principle. Nevertheless, the latter example does not imply that acoustic streaming cannot occur under definite conditions in an inhomogeneous sound field in a nondissipative medium. So far this problem has scarcely been investigated.

The velocity of stationary acoustic streaming is smaller than the particle velocity amplitude in a sound wave. In the event the streaming velocity is considerably smaller than the particle velocity amplitude, it is convenient to refer to the streaming as slow. Only slow streaming has been investigated by and large, either theoretically or experimentally. In the case of fast streaming, where the flow velocity is of the same order of magnitude as the particle velocity amplitude, research has only begun, and very few papers have been written on this streaming.

As implied by a number of solutions for slow streaming, the ratio of the flow velocity to the particle velocity amplitude is on the order of $M\Phi$, where $M = v_0/c_0$ is the acoustic Mach number (v_0 is the particle velocity amplitude, and c_0 is the velocity of sound in the unperturbed medium) and Φ is a geometric factor proportional to the dimensionless streaming scale. Even at the highest available sound levels, the Mach number does not exceed $\sim 5 \cdot 10^{-2}$, and ordinarily (especially in liquids), even for very high sound intensities, $M \leq 10^{-3}$ to 10^{-4}. It is evident from the condition $M\Phi \ll 1$, which is true of slow streaming, that, as M is increased, the streaming will cease to be slow, first in the case of large-scale (Eckart) streaming, then for Rayleigh streaming, and finally for small-scale Schlichting streaming in a boundary layer. This has also been borne out experimentally; Eckart streaming departs from the theory of relatively low sound intensities, while, in a boundary layer (or, in any case, at distances from the boundary less than the thickness of the acoustic boundary layer), Schlichting streaming remains slow up to the highest acoustic Mach numbers.

As for the acoustic Reynolds numbers

$$\text{Re} = \frac{\rho_0 v_0 \lambda}{2\pi b}$$

(where ρ_0 is the density of the unperturbed medium, λ is the acoustic wavelength, $b = 2\rho_0 c_0^3 \alpha_0 / \omega^2$,* α_0 is the small-amplitude sound absorption coefficient, and ω is the circular frequency), as will become apparent shortly, the slow streaming condition is satisfied for $\text{Re} \ll 1$ in the case of large-scale streaming. The theory of small-scale boundary-layer streaming is applicable even for $\text{Re} > 1$.

At present there are many articles in the periodical literature on the various types of acoustic streaming. Also, there very recently appeared the first reasonably exhaustive survey of the theoretical and experimental work to date on acoustic streaming [1].

---

*For a medium characterized by a shear viscosity η, dilatational viscosity η', and thermal conductivity $\varkappa$

$$b = \left(\frac{4}{3}\eta + \eta'\right) + \varkappa\left(\frac{1}{c_v} - \frac{1}{c_p}\right);$$

here c_v and c_p are the specific heats at constant volume and constant pressure.

Theory of Stationary Streaming in a Sound Field

The theory of acoustic streaming is based on the hydrodynamic equations for a viscous compressible fluid. In Euler coordinates the equation of continuity has the form*

$$\frac{\partial \rho}{\partial t} + \nabla (\rho v) = 0, \tag{1}$$

where ρ is the density of the medium and $\mathbf{v}$ is the velocity vector.

The equation of motion of a viscous compressible fluid has the form

$$\rho \left[\frac{\partial \mathbf{v}}{\partial t} + (\mathbf{v}\nabla)\,\mathbf{v} \right] = \rho \left\{ \frac{\partial \mathbf{v}}{\partial t} + \frac{1}{2}\nabla v^2 - [\mathbf{v}\times(\nabla\times\mathbf{v})] \right\} = \mathbf{F}. \tag{2}$$

Here

$$\mathbf{F} = -\nabla p + \eta\Delta\mathbf{v} + \left(\eta' + \frac{\eta}{3}\right)\nabla\nabla\mathbf{v} = -\nabla p + \left(\frac{4}{3}\eta + \eta'\right)\nabla\nabla\mathbf{v} - \eta\nabla\times(\nabla\times\mathbf{v}) \tag{3}$$

(p is the pressure).

Equation (2) can be transformed with the aid of (1) to the form

$$\frac{\partial (\rho v)}{\partial t} + \mathbf{v}\nabla(\rho v) + \rho(\mathbf{v}\nabla)\,\mathbf{v} = \mathbf{F}. \tag{4}$$

---

*In the ensuing discussion we use the Hamiltonian operator $\Delta\varphi = \operatorname{grad}\varphi$; $\nabla\mathbf{a} = \operatorname{div}\mathbf{a}$; $\nabla\times\mathbf{a} = \operatorname{rot}\mathbf{a}$, and the Laplace operator $\Delta = \nabla^2 = \partial^2/\partial x^2 + \partial^2/\partial y^2 + \partial^2/\partial z^2$.

The force acting on a closed volume that is nonmoving relative to fixed space (Euler coordinates) is determined by the momentum flux through the surface S of that volume:*

$$Q_i = -\oint_S \Pi_{ik} dn_k,\qquad(5)$$

where n_k is the component of the unit vector normal to the surface and

$$\Pi_{ik} = p\delta_{ik} + \rho v_i v_k + \eta\left(\frac{\partial v_i}{\partial x_k} + \frac{\partial v_k}{\partial x_i} - \frac{2}{3}\delta_{ik}\frac{\partial v_l}{\partial x_l}\right) + \eta'\delta_{ik}\frac{\partial v_l}{\partial x_l}\qquad(6)$$

is the momentum flux density tensor. The force **Q**, unlike the volume force **F** of (3), is a surface force.

§ 1.   General Theory of Stationary Acoustic Streaming in the Second Approximation (Slow Streaming)

We represent the density, pressure, and velocity in series form:

$$\rho = \rho_0 + \rho' + \rho'' + \ldots$$
$$p = p_0 + p' + p'' + \ldots\qquad(7)$$
$$\mathbf{v} = \mathbf{v}' + \mathbf{v}'' + \ldots,$$

where the number of primes indicates the order of smallness of the corresponding variable. In the expansion (7) the first-order small variables depend periodically on the time and represent the sound field in the linear approximation. The second-order small variables can have, in addition to terms periodically dependent on the time at twice the sound frequency, constant components as well.

Substituting (7) into (1) and (2) and grouping terms of like order of smallness, we obtain the first-approximation equations

$$\rho_0\frac{\partial \mathbf{v}'}{\partial t} = -\nabla p' + \left(\frac{4}{3}\eta + \eta'\right)\nabla\nabla\mathbf{v}' - \eta\nabla\times(\nabla\times\mathbf{v}'),\qquad(8)$$
$$\frac{\partial p'}{\partial t} + \rho_0\nabla\mathbf{v}' = 0$$

---

*Here and elsewhere summation is carried out over the double subscripts.

and second-approximation equations

$$\rho_0 \frac{\partial \mathbf{v}''}{\partial t} + \eta \nabla \times (\nabla \times \mathbf{v}'') - \left(\frac{4}{3}\eta + \eta'\right)\nabla\nabla\mathbf{v}'' = -\nabla p'' - \mathbf{F}''; \qquad (9)$$

$$\frac{\partial \rho''}{\partial t} + \nabla(\rho'\mathbf{v}') + \rho_0\nabla\mathbf{v}'' = 0, \qquad (10)$$

where

$$\mathbf{F}'' = \frac{\partial(\rho'\mathbf{v}')}{\partial t} + \rho_0\mathbf{v}'\nabla\mathbf{v}' + \rho_0(\mathbf{v}'\nabla)\mathbf{v}'. \qquad (11)$$

Equation (9) can also be represented in another form. We take the curl of (9):

$$\frac{\partial \Omega}{\partial t} - \nu\Delta\Omega = -\nabla \times \mathbf{f}'', \qquad (12)$$

where $\Omega = \nabla \times \mathbf{v}''$, $\mathbf{f}'' = \mathbf{F}''/\rho_0$, $\nu = \eta/\rho_0$. Equation (12) is the expression for forced velocity vortex diffusion in the second approximation. In the special case when the rapidly time-varying (at twice the sound frequency) part $\mathbf{v}''$ is a potential term, Eq. (12) describes the buildup process of acoustic streaming.

It is not too difficult to obtain higher than second-approximation expressions. For the theory of slow streaming, however, the first- and second-approximation equations are sufficient.

Excluding the buildup process from further discussion, we average (9) and (10) over the time:

$$\eta\overline{\nabla \times (\nabla \times \mathbf{v}'')} - \left(\frac{4}{3}\eta + \eta'\right)\overline{\nabla\nabla\mathbf{v}''} = -\overline{\nabla p''} - \overline{\mathbf{F}''}; \qquad (13)$$

$$\overline{\nabla\mathbf{v}''} = -\frac{1}{\rho_0}\overline{\nabla(\rho'\mathbf{v}')}, \qquad (14)$$

where

$$\overline{\mathbf{F}''} = \rho_0\overline{[\mathbf{v}'\nabla\mathbf{v}' + (\mathbf{v}'\nabla)\mathbf{v}']}. \qquad (15)$$

Equations (13) and (14) are the initial expressions for the theory of slow acoustic streaming. They are the equations of motion of a viscous fluid under the influence of external volume forces, i.e., the time-average pressure gradient and time-average volume force $\mathbf{F}$ determined according to (15) from the solutions of the first-approximation equations (8).

Applying the curl operator to (13), we represent it as the Poisson equation

$$\overline{\Delta\Omega} = \frac{1}{\nu}\overline{\nabla \times \mathbf{f}''}. \tag{16}$$

The function $(1/\nu)\overline{(\nabla \times \mathbf{f}'')}$ is the density of the vortex sources. Using the familiar relations of vector analysis, we write $\nabla \times \mathbf{f}''$ in expanded form:

$$\nabla \times \mathbf{f}'' = \nabla v' \cdot \nabla \times v' + \nabla\nabla\, v' \times v' + \nabla \times (v'\nabla)\cdot v' =$$
$$= \nabla v' \cdot \nabla \times v' + \nabla\nabla\, v' \times v' - \nabla \times [v' \times (\nabla \times v')]. \tag{17}$$

Depending on the characteristics of the first-approximation velocity field, Eq. (16) can be simplified somewhat. In the event the field may be regarded as irrotational, i.e., $\nabla \times v' = 0$, Eqs. (16) and (17) imply

$$\overline{\Delta\Omega} = \frac{1}{\nu}\overline{[\nabla\nabla v' \times v']}. \tag{18}$$

In another special case, when the first-approximation velocity field may be treated as solenoidal, i.e., when $\nabla v' = 0$,

$$\overline{\Delta\Omega} = \frac{1}{\nu}\overline{\nabla \times (v'\nabla)v'} = -\frac{1}{\nu}\overline{\nabla \times [v' \times (\nabla \times v')]}. \tag{19}$$

In a homogeneous free sound field $\nabla \times \overrightarrow{v'} = 0$ and Eq. (18) may be used. In the terminology of [2], this case involves v o l u m e s o u r c e s   o f   s t a t i o n a r y   v o r t i c e s . In an inhomogeneous sound field, say, near the boundary of a sound beam or near obstacles placed in the sound field $\nabla \times \mathbf{v}' \neq 0$; according to [2], these domains are s u r f a c e   s o u r c e s   o f   s t a t i o n a r y   v o r - t i c e s .   The condition $\nabla v' = 0$ can be approximately fulfilled, for example, if the dimensions of the obstacle are much smaller than the standing wavelength.

Equation (16) is greatly simplified in the case of plane flow, when the first-approximation sound field is representable in the form

$$\mathbf{v}' = \mathbf{i} \, v'_1 (x, y, t) + \mathbf{j} \, v'_2 (x, y, t),$$

where $\mathbf{i}$ and $\mathbf{j}$ are unit vectors along the respective axes Ox and Oy. In this case (16) has the form [3]

$$\overline{\Delta\Omega}_z = \frac{\partial^2 (\overline{v'_1 v'_2})}{\partial x^2} + \frac{\partial^2 (\overline{v'^2_2})}{\partial x \, \partial y} - \frac{\partial^2 (\overline{v'^2_1})}{\partial x \, \partial y} - \frac{\partial^2 (\overline{v'_1 v'_2})}{\partial y^2}, \tag{20}$$

where v'_1 and v'_2 are the projections of the first-approximation velocity vector on the axes Ox and Oy, and Ω_z is the z component of the vector $\boldsymbol{\Omega}$. Introducing for the streaming velocity a flow function ψ such that

$$\overline{v''_1} = \frac{\partial \psi}{\partial y}, \qquad \overline{v''_2} = -\frac{\partial \psi}{\partial x},$$

from (20) we obtain

$$\nabla^4 \psi = \frac{\partial^2 (\overline{v'_1 v'_2})}{\partial x^2} + \frac{\partial^2 (\overline{v'^2_2})}{\partial x \, \partial y} - \frac{\partial^2 (\overline{v'^2_1})}{\partial x \, \partial y} - \frac{\partial^2 (\overline{v'_1 v'_2})}{\partial y^2}. \tag{20a}$$

Next we consider the constraints imposed on the acoustic amplitudes when the method of successive approximations is used. The expansion (7) presupposes convergence of the series, which requires that each ensuing term be much smaller than the one preceding it at any point of space where the solution is applied. However, an approximate estimation of the terms in (13) shows that the quantity $n\bar{\mathbf{v}}''/L''^2$ (where L'' is the stationary streaming scale) is regarded as a variable of the same order as $\rho_0 v_0^2 / L'$ (where v_0 is the amplitude of $\mathbf{v}'$, and L' is the scale of the sound field in the first approximation). It follows from the condition $\bar{\mathbf{v}}''/v_0 \ll 1$ that the acoustic Reynolds numbers must satisfy the condition

$$\mathrm{Re} \ll \frac{\lambda L'}{L''^2}. \tag{21}$$

Consequently, for large-scale ($L'' \gg \lambda$) streaming, the theoretical results are applicable for small values of Re. As for small-scale streaming, when L' and L'' are both of the order $\delta = (2\nu/\omega)^{\frac{1}{2}}$, according to this approximate estimation, Re is limited to $\ll (k\delta)^{-1}$ and can have large values.

We next delve somewhat into the kinematic characteristic of
the constant component $\overline{\mathbf{v}}''$ of the velocity. In Euler coordinates
time average is not carried out for the velocity of a given fluid
particle, but for the velocities of different fluid particles passing
through a given point of fixed space during the averaging period.
In some problems of nonlinear acoustics, the solution in Euler
coordinates has a constant velocity component, which, however,
does not involve transfer of the medium. The transfer of the medium is determined by the mass flow $\rho\mathbf{v}$. Let us determine, correct
to second-order small terms, the time-average mass flow through
a certain surface S:

$$q = \int_S (\rho_0 \overline{\mathbf{v}}'' + \overline{\rho'\mathbf{v}'})\, d\mathbf{s} = \rho_0 \int_S \mathbf{U}\, d\mathbf{s}, \qquad (22)$$

where the velocity

$$\mathbf{U} = \overline{\mathbf{v}}'' + \frac{\overline{\rho'\mathbf{v}'}}{\rho_0} \qquad (23)$$

represents the equivalent flow velocity of an incompressible fluid;
it follows from (14) that $\nabla\mathbf{U} = 0$ [3]. We can show, correct to second-order small terms, that $\mathbf{U}$ is equal to the time-average velocity in Lagrange coordinates. The transformation from the
Euler velocity to the Lagrange velocity has the following form in
the given approximation

$$v_L = v_E + \boldsymbol{\xi}\nabla v_E, \qquad (24)$$

where $\mathbf{v}_L$ and $\mathbf{v}_E$ are the velocities in Lagrange and Euler coordinates and $\boldsymbol{\xi}$ is the displacement vector. In the second approximation, after the time averaging of (24), we obtain

$$\overline{v_L''} = \overline{v_E''} + \overline{\boldsymbol{\xi}'\nabla v_E'}. \qquad (25)$$

Multiplying the first-approximation equation of continuity (8) by ξ'
and averaging over the time, we obtain

$$\overline{\boldsymbol{\xi}'\nabla v_E'} = \frac{1}{\rho_0}\overline{\rho_E' v_E'}. \qquad (26)$$

Comparing (23) with (25) and (26), we see that the velocity $\mathbf{U}$ is
equal in this approximation to the time-average velocity $\overline{\mathbf{v}}_L''$ in
Lagrange coordinates.

§2. Eckart Streaming

The flow generated under the influence of a well-collimated traveling-wave beam was first investigated by Eckart [4]. Assuming that the first-approximation velocity field is irrotational (potential), one can solve the problem on the basis of (18). It is essential to explore the permissibility of representing a confined sound beam as a potential field. It is readily seen that for a plane wave, $\nabla \times \mathbf{v}'$ depends significantly on the distribution of the particle velocity over the cross section of the beam. In the case of a well-collimated sound beam with a sharp boundary,* $\nabla \times \mathbf{v}'$ goes to infinity at the boundary and is equal to zero over the rest of the space; for a smooth distribution of the particle velocity over the beam cross section, $\nabla \times \mathbf{v}'$ has a nonzero value over the entire domain occupied by the field. In the interval of frequencies on the order of a few megacycles for a source of ultrasound with dimensions much larger than the wavelength, clearly, it may be assumed that the volume of the vortex domain at the boundary of the sound beam is small in comparison with the volume occupied by the sound field, and these surface vortex sources introduce a considerably smaller contribution than the volume sources to the stationary streaming. Eckart's theory, within the limits of its applicability, as will be seen below, is in fully satisfactory agreement with the experimental results.

For the determination of the stationary streaming velocity, we need to find the solution of Eq. (18); for $p' = c_0^2 \rho'$, the solution can also be represented in the form [3]

$$\overline{\Delta \Omega} = \frac{b}{\eta \rho_0^2} \overline{\left(\nabla \frac{\partial \rho'}{\partial t} \times \nabla \rho' \right)}. \tag{27}$$

The particle velocity is directed along the z axis in cylindrical coordinates (r, φ, z) and has the form

$$v' = U(r) e^{-\alpha_0 z} \sin(\omega t - kz), \tag{28}$$

where $\alpha_0 = b\omega^2/2\rho_0 c_0^3$ is the sound absorption coefficient, b = $\frac{4}{3} \eta + \eta'$, and the function U(r) gives the distribution of the par-

---

*Note that such a sound field does not satisfy the wave equation.

ticle velocity over the wave front. Substituting (28) into (27), we can find the solution for stationary streaming. In [4] this solution has been found under the following conditions: 1) The stationary streaming has only a z component on a given interval, 2) it is permissible in expression (28) to assume $e^{-\alpha_0 z} \simeq 1$; 3) sound propagates in a cylindrical tube whose end opposite the sound source is terminated in an absorber, so that a traveling wave propagates in the tube; the mass flow of fluid through all cross sections of the tube is zero. Under these conditions the streaming velocity has the form

$$\overline{v_z''} = \frac{bk^2}{4\eta c_0} \left\{ \int_r^{r_0} sU^2(s) \ln \frac{r_0}{s} ds + \ln \frac{r_0}{r} \int_0^r sU^2(s) ds + \right.$$
$$\left. + \frac{r^2 - r_0^2}{r_0^4} \int_0^{r'} (sr_0^2 - s^3) U^2(s) ds \right\};$$ (29)

here r_0 is the radius of the tube. The solution assumes a simple form for a uniform distribution of the particle velocity over the beam cross section:

$$U(r) = v_0, \ r < r_1,$$
$$U(r) = 0, \ r_1 < r \leqslant r_0;$$ (30)

here r_1 is the radius of the sound beam. Now the transverse distribution of the streaming velocity has the form

$$\overline{v_z''} = \begin{cases} U_0 \left\{ \frac{1}{2} \left(1 - \frac{x^2}{y^2} \right) - \left(1 - \frac{1}{2} y^2 \right)(1 - x^2) - \ln y \right\}, \ 0 \leqslant x \leqslant y, \\ - U_0 \left\{ \left(1 - \frac{1}{2} y^2 \right)(1 - x^2) + \ln x \right\}, \ y \leqslant x \leqslant 1, \end{cases}$$ (31)

where $x = r/r_0$, $y = r_1/r_0$, and

$$U_0 = \frac{b}{4\eta c_0} v_0^2 (kr_1)^2.$$ (32)

A typical streaming velocity distribution is shown in Fig. 1. In the sound beam the flow is directed away from the sound source, and in the periphery of the tube, it is directed toward the source. The streaming velocity is a maximum on the axis of the sound beam

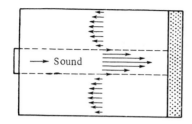

Fig. 1. Diagram of Eckart streaming. The tube is terminated in a
totally sound-absorbing wall.

and depends on r_0/r_1. For $r_0/r_1 = 1$, according to (31), the velocity
reverts to zero*; under actual conditions, due to the influence of
both the boundary layer and, usually, other inhomogeneities of the
sound beam, the streaming velocity merely diminishes under these
conditions without vanishing completely. The maximum velocity
increases slowly with r_0/r_1, maintaining the same order U_0 under
typical experimental conditions. For small values of y, the radius
of stationary flow is somewhat larger than the radius of the sound
beam.

It follows from (32) that the ratio of the streaming velocity to
the particle velocity amplitude in a medium with a small dilatational viscosity is of the order $M(kr_1)^2$. The scale of the streaming
is considerably larger than the wavelength, and, according to (21),
the solution is only applicable for acoustic Reynolds numbers much
smaller than unity. Streaming similar to Eckart streaming but induced by a sawtooth wave of large intensity [5] will be considered
below.

It follows from (32) that the streaming velocity is proportional
to the sound intensity, the frequency squared, and the ratio of the
dilatational to the shear viscosity.† The latter consideration sug-

---

*Note that Eckart streaming is streaming "outside the boundary layer" and if the
dilatational viscosity of the medium is zero, the streaming velocity is independent
of the viscosity.

†This result is attributable to the fact, as already mentioned, that the thin vortex
layer at the boundary of the sound beam has been disregarded in the solution. When
the tube radius is equal to the radius of the sound beam, in order to satisfy the
boundary conditions, it must be required that the component of the velocity on the
surface of the tube vanish. The effect of the walls is felt at a distance $\sim \delta =
(2\nu/\omega)^{1/2}$; in this boundary layer there occur constant forces, which also elicit
streaming in this case.

gested [6] the measurement of the Eckart streaming velocity as a
new method for measuring the dilatational viscosity.* However,
it is seen at once from (32) that the velocity is proportional to the
sound absorption coefficient α_0 and time-average kinetic energy
density $\overline{E}_K = \rho_0 v_0^2 / 4$:

$$U_0 = \frac{2\alpha_0 \overline{E}_K}{\eta} r_1^2. \tag{33}$$

It subsequently became clear that Eckart streaming is caused
by all types of losses in the sound wave (thermal conduction losses,
scattering losses, etc.), whereas deceleration of the flow occurs
only under the influence of viscous forces.

Equation (33) implies that this type of streaming can be interpreted as streaming under the influence of the radiation pressure
gradient, which is proportional to $2\alpha_0 \overline{E}_K$ [7]. In [8], relying on the
law of the conservation of momentum, it has been shown that the
sum of the radiation pressure and dynamic pressure of the resulting stationary streaming is independent of the distance from
the sound source. This result cannot be accepted as one of sufficient rigor, but, as shown by experiments [9, 10] performed at
megacycle frequencies, the total streaming plus radiation pressure is in fact independent of the distance from the sound source
within the experimental error limits. We note that this fact affords a fairly simple means for determining the sound energy flux
through the surface of a source from a measurement of the total
pressure far from the source, because in this plane the streaming velocity is hypothetically equal to zero. At the present time,
the determination of the intensity (or amplitude) directly on the
surface of a radiator at frequencies on the order of several megacycles presents a rather complex experimental problem. The
force acting on an absorbing receiving element of a radiometer in
water as a function of the distance at 15 Mc is shown in Fig. 2
(over a path of 30 cm ~99% of the radiated sound energy has been
absorbed) [9]. To a distance of ~12 cm, laminar flow was observed, and at greater distances the flow became irregular and

---

*At the present time, the only method available for measuring the dilatational vis-
cosity, of course, is to determine it from the results of measurements of the sound
absorption.

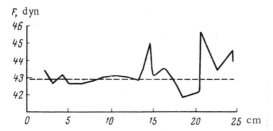

Fig. 2. Force acting on an absorbing radiometer versus the distance from the sound source. The radiometer is acted upon both by the dynamic flow pressure and by the acoustic radiation pressure.

diffuse; in this interval, as apparent from Fig. 2, the deviation of the force from a constant value increased, but remain smaller than 1%.

The solution (31) applies to a well-collimated sound beam. In [11] a solution has been found for a problem similar to the Eckart case when the divergence of the sound beam is taken into account. The first approximation in this case may be represented in the form

$$\rho' := A_0 \frac{\sin (kr - \omega t)}{r} e^{-\alpha_0 r} \cdot \frac{2J_1 (ka \sin \theta)}{ka \sin \theta} , \tag{34}$$

where $A_0 = -\pi v_0 a^2 \rho_0 / c_0 \lambda$, and J_1 is a first-order Bessel function. For $r \gg a$ expression (34), of course, represents the sound field produced by a piston of radius a embedded in an infinite plane baffle and vibrating with an amplitude v_0. Introducing a flow function such that

$$v_r = \frac{1}{r} \frac{\partial \psi}{\partial \theta} + \frac{\psi}{r} \cot \theta, \ v_\theta = - \frac{\partial \psi}{\partial r} - \frac{\psi}{r} ,$$

and substituting (34) into (27), we obtain

$$\nabla^4 \psi = - A_0^2 \frac{bk\omega}{2\eta\rho_0^2} \frac{e^{-2\alpha_0 r}}{r^3} \frac{d}{d\theta} \left[\frac{2J_1 (ka \sin \theta)}{ka \sin \theta} \right]^2. \tag{35}$$

The solution of Eq. (35) has been found in [11] for $e^{-2\alpha_0 r} \simeq 1$ in the domain $\theta \leq \theta_0$, $r_0 < r < r_1$, and for small values of θ_0 has the form

$$\psi = -\frac{A_0^2 bk\omega}{2\eta\rho_0^2}\frac{r}{\theta}\int_0^\theta \theta \left\{\int_0^\theta \frac{1}{\theta}\left[\int_0^\theta \theta\left[\frac{2J_1(ka\theta)}{ka\theta}\right]^2 d\theta\right]d\theta\right\}d\theta + \frac{C_1\theta^3}{16} + \frac{C_2\theta}{2}. \quad (36)$$

The constants C_1 and C_2 are chosen to make the velocities v_r and v_θ vanish for $\theta = \theta_0$. This solution is only applicable for small angles θ_0. Inasmuch as the sound field (34) is nonuniform, i.e., as the amplitude ρ' is a maximum on the axis $\theta = 0$ and decreases with increasing θ, the flow velocity, as shown in Fig. 3, is directed away from the sound source along the axis of the sound beam and toward the source in the periphery of the beam. This solution has been refined in [12], in which the streaming velocity is also determined for the sound field (34), but without the restriction to small angles θ_0.

It can be shown that a sound source in the form of a pulsating sphere (zero-order spherical source) cannot generate streaming. In fact, in this case the particle velocity in spherical coordinates has the form $v_r' = U(r)\sin(\omega t - kr)$; $v_\theta' = 0$; $v_\varphi' = 0$. Inasmuch as $\nabla \times \mathbf{v}' = 0$, it is permissible to use Eqs. (18) or (27). The right-hand side of the equation vanishes. Since $\overline{\mathbf{v}}''$ in this case could have had only a radial component, which vanishes on the surface of the sphere, we find $\overline{\mathbf{v}}'' \equiv 0$. This result is not unexpected, because the onset of streaming in this case would violate the condition of continuity. For the oscillations of a spherical source of other than zero order, of course, streaming is possible.

In [13] the streaming that occurs in the intersection of two plane traveling waves of equal frequency at arbitrary angles is analyzed. The streaming in this case can only have its velocity directed along the bisector of the angle of intersection, and the magnitude of the velocity varies by a cosine law in the perpendic-

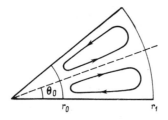

Fig. 3. Diagram of Eckart streaming produced by a divergent sound beam.

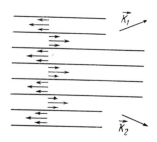

Fig. 4. Streaming under the influence of two intersecting plane waves with wave vectors k_1 and k_2.

ular direction. This creates "layered" streaming, as shown in Fig. 4. The maximum velocity in each layer is $\sim b/2\sqrt{2}\,\eta \cdot (v_0^2/c_0)$.

Because Eckart streaming is a large-scale phenomenon, the theory based on Eqs. (13) and (14) is only applicable for acoustic Reynolds numbers less than unity. In liquids having a relatively low viscosity (0.01 P), this imposes severe limitations on the sound pressure amplitude in the frequency range of several megacycles. It has been shown in a number of experimental studies (see Part II) that, with an increase in the sound intensity, a departure from theory is observed; thus, for example, the streaming velocity is no longer proportional to the intensity. This was originally ascribed to the onset of turbulence. It now appears reasonably certain that this departure is due to the inapplicability of the theory in the domain Re ≳ 1. In this domain, as we know, nonlinear effects such as distortion of the traveling wave profile and the attendant increase in wave absorption begin to play an important part. For Re ≳ 1 a sine wave gradually transforms into a sawtooth at a certain distance from the source. The streaming velocity, as the experimental results indicate, no longer satisfies the slow streaming condition.

The problem of the acoustic streaming elicited by a sawtooth wave has been solved in [14]; the flow velocity in this case was assumed small relative to the particle velocity amplitude of the wave, i.e., the solution is also limited to slow streaming.

The theory of fast acoustic streaming has only recently begun to develop. One constructive approach to the determination of the streaming velocity for a sawtooth wave is indicated in [5]. We shall briefly consider the methods that can be used to determine the streaming velocity in the case when streaming cannot be regarded as slow. The method of successive approximations, naturally, is inapplicable in this case. The parameters of the sound field ρ, $\mathbf{v}$, and p may be written in the form [15, 5]

$$\rho = \rho_0(x, y, z) + \rho_A(x, y, z, t),$$
$$p = p_0^{\cdot}(x, y, z) + p_A(x, y, z, t),$$
$$\mathbf{v} = \mathbf{v}_0(x, y, z) + \mathbf{v}_A(x, y, z, t). \tag{37}$$

This division of the parameters is made so that, for example, the velocity $\mathbf{v}_0$ will represent the time-average value of the velocity $\mathbf{v}$:

$$\mathbf{v}_0 = \frac{1}{(t_2 - t_1)} \int_{t_1}^{t_2} \mathbf{v}\, dt, \tag{38}$$

where $t_2 - t_1$ is a certain time interval equal to a multiple of the period of the sound wave. It follows from (38) that the average of the variables carrying the subscript A over this time interval is zero. Inserting (37) into Eq. (2), represented in the form

$$\frac{\partial \mathbf{v}}{\partial t} + (\mathbf{v}\nabla)\,\mathbf{v} = -\frac{\nabla \rho}{\rho} + \frac{\eta}{\rho}\Delta\mathbf{v} + \frac{\left(\eta' + \dfrac{\eta}{3}\right)}{\rho}\nabla\nabla\mathbf{v}, \tag{2a}$$

assuming that $\rho_A \ll \rho_0$, and taking the time average,* we obtain an equation in the form [15, 5]

$$(\mathbf{v}_0\nabla)\,\mathbf{v}_0 - \frac{\eta}{\rho_0^{\cdot}}\Delta\mathbf{v}_0 - \frac{\left(\eta' + \dfrac{\eta}{3}\right)}{\rho_0}\nabla\nabla\,\mathbf{v}_0 = -\frac{\nabla p_0}{\rho_0} + \mathbf{F}, \tag{39}$$

where

$$\mathbf{F} = -\overline{(\mathbf{v}_A\nabla)\,\mathbf{v}_A} + \frac{1}{\rho_0^2}\left[\overline{\rho_A \nabla p_A} - \overline{\eta\rho_A \Delta\mathbf{v}_A} - \left(\eta' + \frac{\eta}{3}\right)\overline{\rho_A \nabla\nabla\,\mathbf{v}_A}\right]. \tag{40}$$

The time average of the equation of continuity (1) gives

---

*It is assumed in taking the time average that the averaging operation commutes with the operation of differentiation on the coordinates, an assumption that is valid, of course, if the functions and their coordinate derivatives are continuously dependent on the coordinates and time.

$$\nabla \left[\rho_0 \left(\mathbf{v}_0 + \frac{\overline{p_A \, \mathbf{v}_A}}{\rho_0} \right) \right] = 0. \tag{41}$$

Equations (39) and (41) are the initial equations for determining the velocity of fast streaming, as their derivation did not rely on the assumption of a small flow velocity relative to the particle velocity of the sound wave. A logical assumption made in the derivation of the equations is $\rho_A / \rho_0 \ll 1$. It was remarked earlier that the time-average values of the density, pressure, and velocity in strong sound fields can differ from the values of these variables in the unperturbed medium [16]. Consequently, in these equations, in general, ρ_0 and p_0 are not equal to the density and pressure in the absence of sound. The deviation of these variables from their unperturbed values, at any rate, for problems that have been solved in nonlinear acoustics, is of the order M^2. Under the condition $M \ll 1$, the variables ρ_0 and p_0 may be assumed equal to their unperturbed values; then the condition $\rho_A \ll \rho_0$ also corresponds to $M \ll 1$, and Eqs. (39) and (41) can be somewhat simplified. It follows from (41), in particular, that

$$\nabla \, \mathbf{v}_0 = 0, \tag{42}$$

and from (39) that

$$(\mathbf{v}_0 \nabla) \, \mathbf{v}_0 - \frac{\eta}{\rho_0} \, \Delta \mathbf{v}_0 = \mathbf{F}, \tag{43}$$

where $\mathbf{F}$ can be expressed by means of $\rho_A = c_0^{-2} p_A$ only as a function of p_A and $\mathbf{v}_A$. The solution of (42) and (43) requires knowledge of the solution of the problem of sound propagation in the approximation in which (42) and (43) were originally deduced. In [5] a solution has been found for a problem similar to the Eckart case when v_A has the form

$$v_A = U(r) \frac{b\omega}{8 \rho_0 c_0} \sum_{n=1}^{\infty} \frac{\sin n(\omega t - kz)}{\sinh(\beta + \alpha_0 z)}, \tag{44}$$

where

$$U(r) = 1, \ r \leqslant r_1;$$
$$U(r) = 0, \ r_1 < r \leqslant r_0$$

for small values of β and still smaller values of $\alpha_0 z$

$$\beta \simeq \sinh^{-1}\left(\frac{\pi}{2\varepsilon\,\mathrm{Re}}\right); \quad \varepsilon = \frac{1}{2}\frac{\rho_0}{c_0^2}\left(\frac{\partial c^2}{\partial\rho}\right)_S + 1.$$

For $r \leq r_1$ the quantity v_A from (44) represents the solution obtained in [17] for a wave whose profile is considered the "most stable." This profile, of course, is close to a sawtooth. The solution of Eq. (43), in this case, has the form (31), where it is necessary to substitute for U_0

$$U_0 = \frac{b^2\omega^2k^2 r_1^2}{4\rho_0^2 c_0^3 \varepsilon^2} \sum_{n=1}^{\infty} \frac{n^2}{\sinh^2 n\beta} \simeq 10^{-4}\,|\,v_A\,|\,\varepsilon\,\mathrm{Re}\,\mathrm{M}\,(kr_1)^2. \tag{45}$$

Here $|\,v_A\,|$ is the amplitude of v_A. The condition U_0/c_0 implies that (45) is applicable up to very large numbers Re; long before this, clearly, it can happen that the method of deducing (43) from (2a) becomes inapplicable due to the formation of first-order discontinuities in the wave (see the footnote on p. 154). Unlike the velocity of slow Eckart streaming, which varies as the amplitude squared (of the particle velocity or sound pressure), the velocity of fast streaming in a sawtooth wave is $\sim v_0^3$. This bears on the fact that the coefficient of absorption of the sawtooth $\alpha = \alpha_0 \varepsilon \mathrm{Re}$, and (45) can be reduced to the form (33), where instead of the small-amplitude absorption coefficient α_0, we have the sawtooth absorption coefficient α. Thus, fast streaming of this type may also be interpreted as streaming brought on by dissipative losses, which are now already dependent on the wave amplitude.

§3. Streaming Outside the Boundary Layer

The Eckart type of streaming, analyzed in the preceding section, is also essentially streaming outside the boundary layer because the processes in the thin vortex layer at the boundary of the sound beam have been disregarded. In this section we consider certain problems related to the onset of streaming when the sound field is confined by rigid walls. An example of this type of streaming is the Rayleigh variety [18], i.e., two-dimensional streaming elicited by a standing wave between two planes. The oscillations in this standing wave are directed along the axis Ox, and far from the walls have the form $v_1'(x, t) = v_0 \cos kx \cos \omega t$. If one of the planes has the coordinate $y = 0$ and the other plane has the coordinate $y = 2y_1$, then the x and y components of the velocity from (8) for $0 \leq y \leq y_1$ have the form

$$v_1 = v_0 \cos kx \, [- \cos \omega t + e^{-\mu} \cos (\omega t - \mu)]; \qquad (46)$$

$$v_2' = v_0 \frac{k\delta}{\sqrt{2}} \sin kx \left[\left(1 - \frac{\mu}{\mu_1}\right) \cos\left(\omega t - \frac{\pi}{4}\right) - e^{-\mu} \cos\left(\omega t - \frac{\pi}{4} - \mu\right)\right], \; (47)$$

where $\delta = (2v/\omega)^{1/2}$; $\mu = y/\delta$; $\mu_1 = y_1/\delta$.

The motion is symmetric about the plane $y = y_1$; by the substitution $y = 2y_1 - y'$ we obtain the velocity for $y_1 \leq y' \leq 2y_1$. It was assumed in the derivation of (46) and (47) that $k\delta \ll 1$ and $y_1 \gg \lambda$. Rayleigh obtained an approximate solution for the flow in the case when the first-approximation velocity field may be regarded as nearly solenoidal, so that Eq. (19) is applicable. The solution has the form

$$\overline{v_1''} = - \frac{v_0^2}{4c_0} \sin 2kx \left\{\frac{1}{2} e^{-2\mu} + e^{-\mu} \cos \mu + 2e^{-\mu} \sin \mu + \frac{3}{4} - \frac{9}{4}\left(1 - \frac{\mu}{\mu_1}\right)^2\right\};$$

$$\overline{v_2''} = - \frac{v_0^2}{8c_0} k\delta \cos 2 kx \left\{\frac{1}{2} e^{-2\mu} + 3e^{-\mu} \cos \mu + \right. \qquad (48)$$

$$\left. + e^{-\mu} \sin \mu + \frac{3}{2} \mu_1 \left[\left(1 - \frac{\mu}{\mu_1}\right) - \left(1 - \frac{\mu}{\mu_1}\right)^3\right]\right\}.$$

It only approximately satisfies the boundary conditions and, hence, fails to give the true value of the streaming velocity in the boundary layer (as $\mu \to 0$).

Far from the boundary, where the terms $e^{-\mu}$ become negligible, (48) implies

$$\overline{v_1''} = - \frac{3v_0^2}{16c_0} \sin 2kx \left[1 - 3\left(1 - \frac{\mu}{\mu_1}\right)^2\right];$$

$$\overline{v_2''} = - \frac{3v_0^2}{16c_0} ky_1 \cos 2kx \left[\left(1 - \frac{\mu}{\mu_1}\right) - \left(1 - \frac{\mu}{\mu_1}\right)^3\right]. \qquad (49)$$

The behavior of the steaming according to (49) is illustrated in Fig. 5. The streaming pattern comprises vortices spaced at a distance $\lambda/4$ along the axis Ox. The axes of the vortices are found at points with coordinates $x = [(2n - 1)\lambda]/8$, $y = 0.423y_1$ and $x = [(2n - 1)\lambda]/8$, $y = 1.577y_1$, where n is an integer. According to (49), the velocity is independent of the viscosity, although, as will become apparent from the ensuing discussion (see §4), the formation of the vortices is due to interaction with the boundary-layer vortices.

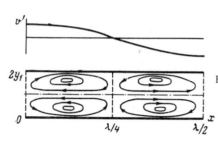

Fig. 5. Diagram of Rayleigh streaming between two planes.

The Rayleigh solution was refined in [3], in which Eq. (20) was
solved for the same problem, i.e., no assumptions were made with
regard to the solenoidal character of the first-approximation velocity. The condition $k\delta \ll 1$ permits one to assume that the derivatives with respect to y in (20) are much larger than the deri-
vatives with respect to x, while $v_2' \ll v_1'$ and $\overline{v}_2'' \ll \overline{v}_1''$. This makes
it possible to simplify Eq. (20) considerably and to find a solution
which still does not qualitatively alter the nature of the vortex flow.

The problem of the acoustic streaming produced by a standing
wave in a narrow cylindrical tube (Kundt tube) for a solenoidal
first-approximation velocity field (which is equivalent to saying
that the domain is smaller than the sound wavelength and $kR \ll 1$,
where R is the radius of the tube) has been treated in [19]. For
the first approximation the Stokes –Kirchhoff solution is adopted

$$v_z' = jv_0 e^{j\omega t} \cos kz \, [1 - e^{-(1+j)\varkappa}];$$

$$v_r' = jv_0 \frac{k\delta}{1+j} e^{j\omega t} \sin kz \, [1 - e^{(1+j)\varkappa}], \tag{50}$$

where $\varkappa = (R - r)/\delta$ and v_0 is the velocity amplitude on the tube
axis. The second approximation is found under the condition that
the thickness of the boundary layer is small relative to the tube
radius. Substituting (50) into (19), we obtain the following relations
for the stationary streaming velocity:

$$\overline{v}_z'' = -\frac{3}{8} \frac{v_0^2}{c_0} \sin 2kz \left\{ 1 - 2\left(\frac{r}{R}\right)^2 + \frac{e^{-\varkappa}}{3} \left[\left(2 + 3\frac{\delta}{r}\right)\cos\varkappa + \right. \right.$$

$$\left. \left. + \left(4 + \frac{\delta}{r}\right)\sin\varkappa + \left(1 + \frac{\delta}{2r}\right)e^{-\varkappa}\right]\right\};$$

$$\overline{v}_r'' = \frac{3}{8} \frac{v_0^2}{c_0} kr \cos 2kz \left\{ 1 - \left(\frac{r}{R}\right)^2 + \frac{2\delta}{3r} e^{-\varkappa} \left[3\cos\varkappa + \sin\varkappa + \frac{1}{2}e^{-\varkappa}\right]\right\}. \tag{51}$$

This solution, like the Rayleigh solution, approximately satisfies the boundary conditions as, for r = R, we have $\overline{v}_z'' \sim \delta v^2{}_0/Rc_0$ and $\overline{v}_r'' \sim k\delta v_0{}^2/c_0$. Near the tube axis ($\varkappa \ll 1$), Eq. (51) implies the parabolic velocity distribution typical of Poiseuille flow

$$\overline{v''_z} = -\frac{3}{8}\frac{v_0^2}{c_0}\sin 2kz\left[1 - 2\left(\frac{r}{R}\right)^2\right];$$

$$\overline{v''_r} = \frac{3}{8}\frac{v_0^2}{c_0}kr\cos 2kz\left[1 - \left(\frac{r}{R}\right)^2\right].$$

(52)

The streaming velocity does not depend on the viscosity of the medium and is proportional to v_0^2/c_0. The streaming pattern has a spatial period of $\lambda/2$; for r = 0.707R, the z component of the streaming velocity goes to zero. The nature of the streaming is similar to the one between two planes.

We point out certain characteristic features of the streaming outside the boundary layer at a large distance from the boundary. As in the case of Eckart streaming, for slow streaming outside the boundary layer, the ratio of the vortex velocity to the particle velocity in the sound wave is $\sim M\Phi$, where Φ is a dimensionless variable depending on the geometry of the sound field, the frequency, and the viscosity. Far from the boundary, as indicated by (49) and (52), the streaming velocity is almost independent of the viscosity of the medium. * The streaming scale $L'' \sim \lambda$, and it follows from (21) that far from the boundary layer the solution is applicable for acoustic Reynolds numbers smaller than unity. For large Re, as indicated by the experimental results [20], the nature of the streaming stays roughly the same, but the vortex velocity is appreciably larger than implied by (49).

§ 4. Acoustic Streaming in the Boundary Layer

In the theory of boundary-layer acoustic streaming, the viscous wavelength $\lambda' = 2\pi\delta$ assumes great significance, where $\delta = (2\nu/\omega)^{1/2}$ is the thickness of the acoustic boundary layer. These quantities, of course, determine the distances over which are

---

* By contrast with Eckart streaming, the solutions presented in this section were obtained without regard for the dilatational viscosity.

propagated the perturbations elicited by the viscous forces (shear or viscosity waves in the case of the shear oscillations of a plane in a viscous medium or perturbations of the amplitude of a sound wave propagating near a surface). The ratio of the viscous wavelength to the acoustic wavelength may be written in the form

$$\frac{\lambda'}{\lambda} = k\delta = \left(\frac{2\omega\eta}{\rho_0 c_0^2}\right)^{1/2} = \left(\frac{3}{2}\frac{M}{Re}\right)^{1/2}. \tag{53}$$

Consequently, the condition $\lambda' \ll \lambda$ is equivalent to the stipulation of small viscous stresses $\sim\omega\eta$ in comparison with the stresses produced by the volume elasticity of the medium $\rho_0 c_0^2$, to small sound absorption per wavelength, or, finally, to acoustic Reynolds numbers $Re \gg M$. The condition $\lambda' \ll \lambda$ holds for practically all the cases treated in the present chapter.

Let us consider for a moment the qualitative characteristics of the processes in an acoustic boundary layer. The velocity on a perfectly rigid surface must vanish; hence, the velocity gradient is large in the boundary layer. This means that the momentum of the sound wave varies sharply in the boundary layer and the forces that produce streaming are large. These forces for a sufficiently thin boundary layer greatly exceed the forces produced in a free sound field by absorption.

The streaming generated near obstacles may be analyzed on the basis of the boundary-layer equations. In the derivation of the equations for a plane boundary layer it is assumed, of course, that the velocity gradients in the direction of the normal to the boundary are far greater than the gradients in the direction of the boundary. Inasmuch as the first of these gradients is determined in the case of interest by the viscous wavelength, while the second is determined by the acoustic wavelength, the condition for applicability of the boundary-layer equations is as follows: $\lambda \gg \lambda'$. The Prandtl equations for the boundary layer near the plane $y = 0$ have the form (see, e.g., [21])

$$\frac{\partial v_1}{\partial t} + v_1\frac{\partial v_1}{\partial x} + v_2\frac{\partial v_1}{\partial y} - \nu\frac{\partial^2 v_1}{\partial y^2} = \frac{\partial U}{\partial t} + U\frac{\partial U}{\partial x},$$
$$\frac{\partial v_1}{\partial x} + \frac{\partial v_2}{\partial y} = 0, \tag{54}$$

where v_1 and v_2 are the x and y components of the velocity and U(x, t) is the known flow velocity far from the boundary. In these equations no allowance is made for the dilatational viscosity or thermal conductivity. Using the perturbation method, we represent the velocity components in the form

$$v_i = v_i' + v_i'' + \cdots \quad (i = 1,2).$$ (55)

Now the boundary-layer equations are written:

in the first approximation

$$\frac{\partial v_1'}{\partial t} - \nu \frac{\partial^2 v_1'}{\partial y^2} = \frac{\partial U}{\partial t};$$

$$\frac{\partial v_1'}{\partial x} + \frac{\partial v_2'}{\partial y} = 0;$$ (56)

in the second approximation

$$\frac{\partial v_1''}{\partial t} - \nu \frac{\partial^2 v_1''}{\partial y^2} = U \frac{\partial U}{\partial x} - v_1' \frac{\partial v_1'}{\partial x} - v_2' \frac{\partial v_1'}{\partial y};$$

$$\frac{\partial v_1''}{\partial x} + \frac{\partial v_2''}{\partial y} = 0.$$ (57)

The boundary conditions for the solution of the problem for the half-space y > 0 are $v_1|_{y=0} = v_2|_{y=0} = 0$, and $v_1 \to U(x, t)$ as $y \to \infty$.

The solution of these equations is simplified by the introduction of the flow function $\psi = \psi' + \psi'' + \cdots$ such that

$$v_1' = \frac{\partial \psi'}{\partial y}; \quad v_2' = -\frac{\partial \psi'}{\partial x}; \quad v_1'' = \frac{\partial \psi''}{\partial y}; \quad v_2'' = -\frac{\partial \psi''}{\partial x}.$$ (58)

In Schlichting's paper [22, 23] these equations have been solved for the standing wave U(x, t) = U_0(x) cos ωt. The solution of Eqs. (56) has the form

$$\psi' = \frac{U_0(x) \delta}{2} \operatorname{Re} \{\sigma'(\mu) e^{j\omega t}\},$$ (59)

where

$$\sigma'(\mu) = 2\mu - 1 + j + (1 - j) e^{-(1+j)\mu}$$

and the dimensionless distance $\mu = y/\delta$ is incorporated. From

(59) the components of the first-approximation particle velocity vector have the form

$$v_1' = U_0(x)\{\cos \omega t - e^{-\mu}\cos(\omega t - \mu)\};$$
$$v_2' = -\frac{dU_0}{dx}\frac{\delta}{\sqrt{2}}\left\{\frac{(2\mu - 1)}{\sqrt{2}}\cos \omega t - \frac{1}{\sqrt{2}}\sin \omega t + e^{-\mu}\cos\left(\omega t - \mu - \frac{\pi}{4}\right)\right\}. \quad (60)$$

The solution of Eqs. (57) in the second approximation is

$$\psi'' = \frac{\delta}{2\omega}U_0\frac{dU_0}{dx}\operatorname{Re}\{\sigma_1''(\mu)e^{2i\omega t} + \sigma_2''(\mu)\}, \quad (61)$$

where

$$\sigma_1''(\mu) = \frac{1+i}{2\sqrt{2}}e^{-(1+j)\sqrt{2}\mu} + j\mu e^{-(1+j)\mu} - \frac{1+i}{2\sqrt{2}}; \quad (62)$$

$$\sigma_2''(\mu) = -\frac{1}{4}e^{-2\mu} - 3e^{-\mu}\cos\mu - e^{-\mu}(2+\mu)\sin\mu + \frac{3}{2}\mu + \frac{13}{4}. \quad (63)$$

The form of the function $\sigma_2''(\mu)$ is shown in Fig. 6.

As apparent from (61), in the second approximation there exists in the acoustic boundary layer, besides the double-frequency oscillations, a constant velocity component. The streaming

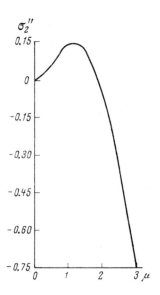

Fig. 6. The function $\sigma_2''(\mu)$.

velocity has the following components:

$$\overline{v''_1} = \frac{1}{2\omega}\, U_0 \frac{dU_0}{dx} \frac{d\sigma''_2}{d\mu}\,;$$

$$\overline{v''_2} = -\frac{\delta}{2\omega}\left[\left(\frac{dU_0}{dx}\right)^2 + U_0 \frac{d^2U_0}{dx^2}\right]\sigma''_2. \tag{64}$$

In the case of a space-periodic wave $U_0(x) = -v_0 \cos kx$, we obtain for the streaming velocity components

$$\overline{v''_1} = -\frac{v_0^2}{4c_0}\sin 2\,kx\left\{\frac{1}{2}\,e^{-2\mu} + (1-\mu)\,e^{-\mu}\cos\mu + (4+\mu)\,e^{-\mu}\sin\mu - \frac{3}{2}\right\};$$

$$\tag{65}$$

$$\overline{v''_2} = -\frac{v_0^2}{4c_0}\,k\delta\cos 2\,kx\left\{\frac{1}{4}\,e^{-2\mu} + 3\,e^{-\mu}\cos\mu + e^{-\mu}(2+\mu)\sin\mu + \frac{3}{2}\,\mu - \frac{13}{4}\right\}.$$

Near the boundary (for $\mu \ll 1$), correct to terms $\sim\mu^2$, we obtain from (65)

$$\overline{v''_1} = -\frac{v_0^2}{4c_0}\,(\mu - \mu^2)\sin 2\,kx;$$

$$\tag{66}$$

$$\overline{v''_2} = \frac{v_0^2}{4c_0}\,k\delta\mu^2\cos 2\,kx.$$

Let us note some of the characteristic features of this streaming: 1) the flow velocity at points where $U_0 = 0$ or $dU_0/dx = 0$ has only a normal component with respect to the boundary; this means that, at particle velocity nodes or antinodes, the flow velocity is either directed toward the boundary or away from it, 2) inasmuch as the sign of the derivative $d\sigma''_2/d\mu$ changes for $\mu \sim 1$ (Fig. 6), (64) indicates that in this domain the sign of the x component of the flow velocity changes, 3) the y component of the velocity normal to the boundary, according to (64), changes sign for $\mu \simeq 1.9$ (for arbitrary x). This determines the dimensions of the boundary-layer vortices.

Henceforth, we shall refer to the quantity Δ, i.e., the distance from the boundary to the position at which the normal component of the flow velocity vector with respect to the boundary vanishes, as the boundary-layer vortex thickness. For Schlichting streaming $\Delta \simeq 1.9\delta$.

Schlichting streaming comprises a series of vortices near a surface, their dimensions equal to $\lambda/4 \times 1.9\delta$. The nature of the

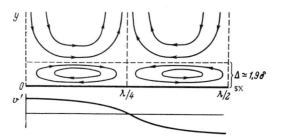

Fig. 7. Schlichting streaming near a plane boundary.

streaming is illustrated in Fig. 7. The transfer of the medium near the boundary surface proceeds from particle velocity nodes to antinodes; on the outer part of the boundary layer, transfer occurs in the opposite direction. It is important to note here that the direction of rotation of the boundary-layer vortices is opposite to the rotation of the vortices outside the boundary layer (see Fig. 5).

The results of the solution of the problem of an acoustic boundary layer near a plane surface can be used to determine the streaming over curved surfaces whose local radius of curvature is considerably greater than the viscous wavelength [24]. This approach has been used to investigate stationary streaming in the boundary layer of a circular cylinder situated at a velocity antinode of a standing wave, so that its axis is perpendicular to the direction of oscillation [22, 25]. The solution (64) can be used here, where $U_0(x) = 2v_0 \sin(x/a) = 2v_0 \sin \varphi$ (φ is the angular coordinate, $\varphi = 0$ corresponds to a branch point of the flow, $\mu = r/\delta$ and $r = 0$ on the surface of the cylinder, and a is the radius of the cylinder). Then the Schlichting solution for the stationary streaming velocity near the cylinder has the form

$$\overline{v''_\varphi} = \frac{v_0^2}{c_0} \frac{1}{ka} \sin 2\varphi \cdot \frac{d\overline{\sigma''_2}}{d\mu};$$

$$\overline{v''_r} = -\frac{2v_0^2}{c_0} \frac{k\delta}{(ka)^2} \cos 2\varphi \cdot \overline{\sigma''_2}.$$

(64a)

The character of the streaming is illustrated in Fig. 8.* In each quadrant there are two vortices, one in a relatively thin boundary

---

*Figure 8 is to be viewed as a qualitative illustration of the flow pattern near a cylinder, because for $a/\delta = 7.1$, the experimental boundary-layer vortex thickness is approximately three times as great.

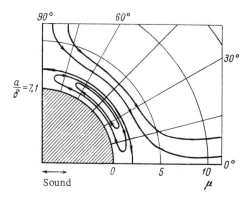

Fig. 8. Schlichting flow near a cylinder.

layer and the other outside the boundary layer. The angular velocity of the boundary-layer vortices is opposite to the angular
velocity of the external vortices. In the case of flow past a cylinder in an unbounded medium, the center of rotation of the external vortex is at infinity, whereas in a bounded medium it is
situated at a finite distance from the cylinder surface. Henceforth, we define the boundary layer vortex thickness Δ as the distance from the surface of the cylinder to the first zero of the radial component of the velocity (for arbitrary φ). As implied by
(66) and Fig. 6, $\Delta = 1.9\delta$ and is independent of the ratio a/δ.
The experimental results (see Chap. 2, § 2) demonstrate that the
thickness of the boundary-layer vortices near a cylinder is consistent with Schlichting's theory only for $a/\delta \gtrsim 30\text{-}40$; for
$a/\delta < 5$ the discrepancy between theory and experiment can be
greater than one order of magnitude. This is fairly logical in that,
as already mentioned, the application of the results obtained for a
boundary layer over a plane surface in order to determine the
streaming near bodies having a radius of curvature a is possible
if $a/\delta \gg 1$. Although this has not been confirmed experimentally,
Schlichting's theory cannot give valid quantitative results for the
streaming velocity at large distances from the surface of the cylinder relative to δ.

A solution to the problem of stationary streaming near a circular cylinder [not on the basis of the boundary-layer equations,
but of the solution for an incompressible medium, followed by the
first-approximation equations (8) and then the second-approxima-

tion equations (16)] has been given in [26, 27]. It follows from the resulting solution that the form of the streaming is not altered qualitatively. However, it is too cumbersome to reproduce here. A particular implication of this solution is the fact that the boundary-layer vortex thickness for small values of a/δ is larger than the Schlichting value; for large a/δ the boundary-layer vortex thickness approaches the Schlichting value. An analysis of the third-approximation equations shows that this approximation does not bear any contribution to the stationary streaming. The solution has also been analyzed in [28], in which, among other things, the flow functions are determined in a Lagrange coordinate system.

In [29] the case of streaming near a rigid sphere has been investigated. In the equatorial plane of the sphere, the streaming is similar to its behavior to streaming near a cylinder, except that the axis of the boundary-layer vortices is slightly displaced. In correspondence with (64a), for a cylinder the vortex axis is situated in the planes $\varphi = 45, 135, 225$, and $315°$, while for a sphere it is in the planes $\varphi = 54°43'$, $125°17'$, $234°43'$, and $305°17'$. The boundary-layer vortex thickness a for large values of a/δ, where Δ is the radius of the sphere (theoretically for $a/\delta > 100$) tends to the Schlichting thickness $\Delta = 1.9\delta$. For $a/\delta \gtrsim 60-70$ the experimental value of Δ is consistent with the Schlichting value.

Near a plane boundary, of course, the Schlichting solution loses its validity when the acoustic wavelength becomes comparable with the viscous wavelength. According to (53), this can happen either in the case of exceedingly viscous liquids or for liquids of medium viscosity at very high frequencies $\omega \sim \rho_0 c_0^2/\eta$, corresponding to the interval of shear viscosity relaxation.

It is not altogether clear at the present time what are the sound intensity limits for applicability of the theory of stationary streaming near a boundary on the basis of the solution of the boundary-layer equations. As already mentioned, the use of these equations is admissible for acoustic Reynolds numbers greater than the Mach number; this does not limit the sound amplitude in any way. The application of the method of successive approximations requires only convergence of the series (55) in the domain of the sound field where the theory is to be applied, i.e., in the latter domain the constant flow velocity must be much smaller than the first-approximation velocity amplitude. Near the boundary both the first-approximation and the second-approximation velocity tend to zero. The method of successive approximation is applicable only when the second-approximation velocity tends to zero while remaining at all times significantly smaller than the first-approximation velocity. It is readily seen from expression (60) that, correct to terms of order μ, the first-approximation amplitude over a plane boundary is $\sim U_0(x)\mu/\sqrt{2}$, whereas the stationary streaming velocity according to (66) is $\sim U_0(x)M\mu/2$, where M = v_0/c_0 is the acoustic Mach number. Consequently, for small M the conditions set forth above are satisfied. However, this sets only a lower limit on the acoustic Reynolds numbers (they must be larger than the Mach number). Clearly there are no upper bounds on the acoustic Reynolds numbers in the theory of boundary-layer streaming if the particle displacement amplitude in the wave is much smaller than the characteristic dimensions of the body. For a displacement amplitude comparable with the dimensions of the body (as shown in [25], in which the velocity near a circular cylinder has been determined with regard for fourth-order small terms), the nature of the resulting streaming can differ qualitatively, as well as quantitatively. The streaming pattern in a standing wave near a cylinder of radius a is shown in Fig. 9 for a ratio of the displacement amplitude ξ_0 to the radius a equal to 0.1; the streamlines have been calculated with the inclusion of fourth-order terms. All other conditions are the same as in Fig. 8. Near the plane $\varphi = 75$-$80°$ the external vortex approaches the sur-

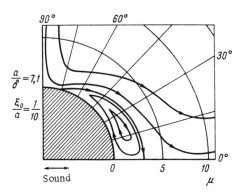

Fig. 9. Schlichting streaming near a cylinder with the inclusion of fourth-order small terms [25].

face of the cylinder, whereas the boundary-layer vortices are displaced nearer to the polar regions (φ = 0 and 180°) and change their form. This variation in the character of the vortices with increasing sound intensity has been observed experimentally in its qualitative aspects [30].

So far we have considered boundary-layer streaming in the case when a fixed obstacle is placed in a sound wave. The streaming produced in a medium near oscillating bodies has been investigated in [31]. Near oscillating bodies the streaming velocity in Lagrange coordinates is invariant under coordinate transformations, such that in the new system the surface of the body is fixed, the surrounding medium executing oscillations. This theorem, which we identify henceforth as the Westervelt theorem, is valid for incompressible acoustic streaming [which always occurs for slow streaming in Lagrange coordinates in correspondence with (23) and (14)] if the surface of the body has sufficiently good geometry and the characteristic dimensions of the body are much smaller than the acoustic wavelength, but greater than the displacement amplitude (of either the surface of the body or surrounding medium). This theorem has been corroborated experimentally in the example of streaming near circular cylinders [28, 32].

We have already discussed the case of streaming produced in a boundary layer by a standing wave. In a standing wave small-scale vortices occur near surfaces; the character of their distribution have a cellular structure. The dimensions of the vortex cells over a plane surface are of the order $\sim \lambda \times \delta$; near obstacles whose dimensions d are smaller than the acoustic wavelength, but larger than the thickness δ of the acoustic boundary layer, the characteristic vortex dimension is $\sim d \times \delta$. For small values of d/δ, the chactacteristic vortex dimension is $> d \times \delta$. For a nonmonochromatic wave, even though the streaming-inducing forces are nonlinear, the time-average forces may be represented as the sum of the forces of each of the Fourier component. It is clear that the thickness of the boundary vortex layer is determined by the lowest frequency of the nonmonochromatic wave. In the boundary layer in this case, there are vortices of different scales, so that the boundary-layer streaming is considerably "mixed up" and may be regarded as turbulent; mass transfer takes place be-

tween the streaming in the boundary layer and the vortex streaming outside that layer.

The only solutions to the problem of boundary-layer streaming produced by a traveling wave are found in [1]. In this case,
let us say near a plane surface, under definite conditions large-
scale vortices can occur, because the time-constant force inducing
streaming near the boundary is considerably greater than the force
far from it. The flow near the boundary has the same direction
as the propagation of sound, while far from the boundary it is in
the opposite direction.

Near solid surfaces both the normal and the tangential components of the stationary streaming velocity vanish. The situation
is different in the case of microstreaming near air bubbles in a
liquid; under the influence of the viscous forces, the gas in the
bubble is entrained by the stationary streaming of the liquid, and
the tangential component of the velocity cannot be assumed equal
to zero at the surface of the bubble. An estimate of the streaming
velocity near bubbles has been made in [24]. The problem of microstreaming near air bubbles suspended in a liquid in the field
of a plane sound wave has also been solved [33]. It was assumed
that the bubble radius a satisfied the condition $\lambda > a > \delta$, and
that the normal component of the stationary streaming velocity
vanished at the bubble surface, while the tangential component is
continuous. In this case the flow function for stationary streaming
in the boundary layer of the bubble assumes the form

$$\psi'' = \frac{2\,v_0^2\delta}{c_0 ka}\sin 2\varphi\left[\frac{5}{16} - \frac{e^{-2\mu}}{16} - \frac{e^{-\mu}}{4}(\sin\mu + \cos\mu)\right]\left[1 + \frac{\Gamma}{(ka)^2}\right], \quad (67)$$

where $\Gamma = [(f_r^2/f^2 - 1)^2 + \sigma^2]^{-1}$; f_r is the resonance frequency of
the bubble, f is the acoustic frequency, and σ is the damping
decrement of the bubble. At the bubble surface, the tangential
component of the velocity has the form

$$\bar{v}_\varphi'' = \frac{v_0^2}{4\,c_0}\frac{1}{ka}\sin 2\varphi\cdot\left[1 + \frac{\Gamma}{(ka)^2}\right]. \quad (68)$$

The agreement between the experimentally observed velocity and
expression (68) is significantly better than according to the estimates given in [24].

Experimental Investigation
of Acoustic Streaming

§ 1. Methods for the Observation of
Streaming and Measurement of
Its Velocity

The ratio of the velocity of stationary streaming to the particle velocity for slow streaming is on the order of magnitude of the acoustic Mach numbers. Experimental investigations have already begun on streaming that does not satisfy the conditions of slow streaming; in this case the streaming velocities can have the order of magnitude of the particle velocity. The maximum streaming velocities observed to date have been several meters per second. Normally, however, the investigation of streaming, particularly the Eckart variety, involves the observation and measurement of relatively small velocities ($\lesssim 1$ cm/sec).

Usually aerosol particles (tobacco smoke or MgO smoke) or suspensions in liquids (aluminum or bronze dust) are used for the observation and determination of the velocity of slow streaming. In general, for every medium it is required to select particles that will be suspended and completely entrained by the flow. The complete entrainment of particles by a flow occurs when the particle radius is very small and their density does not differ too much from that of the medium. The photographing of illuminated particles with a definite exposure makes it possible to determine the streaming velocity from the length of the tracks in the photographs. Sometimes, for a more precise determination of the local velocities and in order to preclude the possibility of different tracks becoming superimposed, strobed illumination is used [15, 34]. For the determination of the velocity in fast streaming, high-speed motion picture films have also been used (up to 1500 frames per second) [20]. Since the streaming velocity differs in different regions of the field, side illumination is created by a narrow light beam, for which long-focus cylindrical lenses are usually used; a narrow light beam makes it possible to investigate the velocity distribution in the illuminated plane. High-intensity light sources are used in order to obtain adequate illumination. This method yields the "instantaneous" velocity field in Lagrange coordinates if it is known that the velocity of the suspended particles is equal to the velocity of the liquid or gas.

For the measurement of the streaming velocity in liquids, thermistors have also been used [35]. The flowing through of liquid cools the thermistor, causing a variation of its resistance. The lag in this type of velocity meter, in the opinion of the author of [35], is large enough that the variable component of the sound field, except at infrasonic frequencies, does not affect the readings of the thermistor. The absolute calibration of this velocity meter can be performed, for example, if the thermistor is placed in a constant artificially generated liquid flow of a known velocity. This method of measuring the streaming velocity, like the method based on the velocity of small suspended particles, has the advantage that the acoustic streaming velocity field is only slightly disturbed. Unlike the suspended particle method, the thermistor affords a means for determining the velocity in Euler coordinates.

In [20, 36] the use of a thermoanemometer for the measurement of streaming in air under the influence of high-intensity sound is discussed. This device comprises a finite wire heated by a current and cooled by the flow. It is shown that under ordinary working conditions, in addition to the constant flow velocity, the thermoanemometer also detects the acoustic particle velocity, and inasmuch as the latter is greater than the constant flow velocity, the noise level is exceedingly high. It has been possible to reduce the level by diminishing the current used to heat the anemometer wire. The reduced current lowered the sensitivity of the anemometer, but this did not matter in the measurement of large velocities.

The acoustic streaming velocity can be ascertained in principle in terms of the dynamic pressure of the stationary streaming. Any number of devices can be used as dynamic pressure sensors, such as a lightweight arm (of the radiometer type) with a receiving element attached to it and placed in the sound field, or a tube of the Pitot type. Measurements of this kind are greatly complicated by the fact that, in addition to the dynamic pressure of the flow, the receivers are sensitive to the acoustic radiation pressure, which can be of the same order of magnitude as the dynamic stationary streaming pressure. In [37] it was proposed that the dynamic streaming pressure be determined by means of the receiving element of a radiometer in the form of a frame, over which is stretched a film that is acoustically "transparent" and impermeable to the flow. Under these conditions the radiation pressure no longer acts on the receiving element of the radiometer as long as, of course, the sound absorption in the film is sufficiently small, while the dynamic streaming pressure is 100% active on the element due to the impermeability of the film. The receiving element of a conventional radiometer (totally or partially absorbing or reflecting) or Pitot tube [20] can be protected against flows by a flow-impermeable acoustically transmissive membrane. In this case a quantity proportional to the radiation pressure is measured. The difference between the total pressure and the radiation pressure makes it possible to determine the dynamic constant streaming pressure and, hence, its velocity. In [6] the acoustic radiation pressure and dynamic pressure of Eckart streaming are separated on the basis of the relatively little-investigated fact that the buildup time of the stationary sound field is considerably shorter than the buildup time of stationary acoustic streaming. The propagation of sound produces at once a deflection of the radiometer under the action of the radiation pressure (but of course only in the event the lag time of the radiometer is small) and then a gradual increase of the deflection under the action of the dynamic streaming pressure. This method has elicited a number of doubts [35], mainly in the fact that, first, the streaming buildup process occurs throughout the entire volume, so that the dynamic streaming pressure, although not completely, begins nevertheless to act when the sound is turned on and, second, the lag time of radiometers is generally so large as to be commensurate with the constant streaming build-up time.

A disadvantage of the Pitot tube and particularly of the radio-meter method is the perturbation of the flow pattern. The radio-meter method has been used heretofore only for the investigation of large-scale streaming, namely of the Eckart type.

In the investigation of slow acoustic streaming, it is occas-sionally necessary to maintain a constant temperature over the entire volume investigated, because temperature irregularities, which become especially marked in the case of illumination of the investigated volume with a high-intensity light source, can lead to convective currents, which introduce error into the measure-ment results. This is particularly pronounced in the observation of slow streaming in gases [38].

Fig. 10. Streaming at 5 Mc in benzene. The diameter of the quartz emitter is 2.5 cm; the diameter of the tube containing the benzene is 5 cm.

The suspended particle method mentioned above has one advantage over the other methods described; it enables one to ascertain the streamlines of the acoustic streaming pattern and
thus to gain an overall picture of that pattern. A photograph of
Eckart acoustic streaming generated by an ultrasonic wave at a
frequency of 5 Mc [39] in benzene is shown in Fig. 10; the flow
was visualized by means of fine aluminum particles.

Steady streaming is readily amenable to qualitative observation if, in the case of a gas, smoke is introduced into the sound
field or, in the case of a liquid, dye is used or, as in [40], the onset
of streaming is determined from the change (upon the propagation
of sound) in the convective flow pattern from a heated body. In [41]
the streamlines of Eckart acoustic streaming at an interface be-

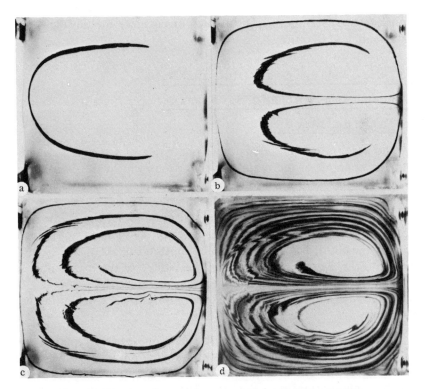

Fig. 11. Visualization of streaming at an interface between two immiscible liquids, glycerin and Vaseline oil. The quartz ultrasound source
is located in the center of the right side of each photograph.

tween two immiscible liquids (glycerin and Vaseline oil) were made
visible by means of a dye dissolved in water. Water has an intermediate density (lower than glycerin and higher than Vaseline oil),
so that a drop of colored water dropped into the Vaseline oil would
arrive at the interface and settle on the glycerin surface. Under
the action of the surface tension forces, the drop spreads over the
surface of the glycerin, forming a spot. With the transmission of
sound, the difference in the flow velocities in glycerin and Vaseline
oil caused the spot to distend further, until at first it assumed the
form of the distribution of the relative flow velocities in the different liquids (Fig. 11a); then, after the dye had entered the streamlines, it "marked" the streamlines (Figs. 11b-11d). This method
relies on the fact that the streaming geometry is independent of the
parameters of the medium (clearly, this is also true of the regions
near the sound source and near an absorber). The method can be
used, not only to investigate the relative streaming velocity for different types of streaming, but also to investigate the buildup time
of stationary streaming.

§ 2. Streaming near Obstacles,
Microstreaming, and Streaming
in Standing Waves

The streaming that occurs near obstacles of various configurations has been observed mainly in situations where the dimensions
of the obstacles are much smaller than the acoustic wavelength.
These observations have been made in standing waves or almost-
standing waves. For the observation of streaming, the Westervelt
theorem (see p. 168) is often used, i.e., the obstacle is not placed
in the sound field, but has oscillatory motion imparted to it. According to the Westervelt theorem, the nature of the streaming and
the magnitude of its velocity in this case must be the same as
though the obstacle were fixed and the medium were executing oscillations.

The experimental setups for observations of this type are
fairly simple. As an illustration we describe an arrangement used
to investigate the streaming near a vibrating cylinder [28], as illustrated schematically in Fig. 12. The audio oscillator AO was
used to drive the dynamic horn 1, which had the rod 2 attached to
its diaphragm. Attached to the rod was the cylinder 3, near which

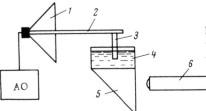

Fig. 12. Diagram of the experimental set-up for the observation of streaming near a vibrating rod.

streaming was observed in the liquid-containing vessel 4. The streaming was observed by means of the totally internally reflecting prism 5 and microscope 6. Fine particles of powdered bronze were added to the liquid, where they were illuminated with a narrow beam from a high-intensity light source in the plane perpendicular to the axis of the cylinder 3 (the light source is not shown).

Notice that even at low acoustic frequencies, in media such as air and water, the thickness of the acoustic boundary layer is very small; at a frequency of 100 cps in water $\delta \simeq 0.6 \cdot 10^{-2}$ cm, and in air $\delta \simeq 2.2 \cdot 10^{-2}$ cm. The observation of the acoustic boundary layer is facilitated somewhat in very viscous liquids; in glycerin at t = 22°C and at 100 cps $\delta \simeq 1.5 \cdot 10^{-1}$ cm. At high frequencies in the sonic range and particularly at ultrasonic frequencies, the thickness of the boundary layer is even smaller. Consequently, all studies of boundary-layer streaming have been carried out at low sonic frequencies.

A great many experimental studies have been concerned with the investigation of stationary vortex streaming near cylindrical rods [22, 27, 28, 30, 38, 42] immersed in a standing sound wave. The principal conditions of these experiments are summarized in Table 1. At the present time it seems unlikely that all of the experimentally observed characteristics of these streaming patterns can be explained fully and without contradiction. For this purpose, in a number of papers certain data are omitted; some of the experimental data are of a qualitative character, or the various parameters of the sound field have not been determined with sufficient accuracy. Perhaps the most valid study is found in [27], in which the streaming of air near a glass rod was investigated. The experimental conditions are given in Table 1.

Based on the sum total of all the experimental evidence, the pattern of formation of vortex streaming near obstacles may

TABLE 1. Principal Conditions of Experiments on Streaming near
Rods or Vibrating Plate Springs

Medium	Fre-quency, cps	Peak dis-placement amplitude, cm	$\delta = (2\nu/\omega)^{1/2}$, cm	Rod diam-eter or plate width, cm	Remarks
Air [30]	50	10^{-2}—10^{-1}	$3 \cdot 10^{-2}$	0.25	Streaming typical of acoustic boundary layer observed under the microscope
Air [38]	89—840	10^{-3}—10^{-2}	$2.3 \cdot 10^{-2}$ — $0.8 \cdot 10^{-2}$	0.0081; 0.2—0.6	Streaming typical of re-gion outside the boundary layer ob-served visually
Water [22]	0.5	0.9	$8 \cdot 10^{-2}$	8	The same
Air [42]	50; 550	$(4—5) \cdot 10^{-2}$	$3 \cdot 10^{-2}$; 10^{-2}	0.16; 0.077	Boundary-layer streaming observed under the microscope on the at-tainment of a certain amplitude, formation of streaming outside the boundary layer
Air [27]	200	$0.6 \cdot 10^{-2}$	$1.5 \cdot 10^{-2}$	0.22	Streaming in and outside the boundary layer ob-served under the micro-scope.
Water and water–glycerin mixture [28]	80—160	$\lesssim 3 \cdot 10^{-2}$	—	0.12—0.4	Boundary-layer observed under the microscope

be qualitatively depicted as follows. When the sound amplitude is
small, clearly-pronounced vortices are formed only in the acoustic
boundary layer. These vortices are seen in Fig. 13 near a cylinder
of radius 0.11 cm at a standing sound wave frequency of 200 cps
[27]. Here there occurs a flow of air toward the cylinder in the
direction of the acoustic oscillations, and a flow away from the
cylinder in the perpendicular direction (linear system of vortices
shown in Fig. 8). Inasmuch as the linear dimensions of these
vortices along the normal to the surface of the obstacle, as a rule,
do not exceed a few thicknesses of the acoustic boundary layer, in
[22, 38, 39], in which streaming was observed visually, the onset
of this type of vortex could not be detected. As for the system of
vortices external to the boundary layer (outer system of vortices
shown in Fig. 8), for relatively small sound amplitudes it is very
indistinct, becoming smeared out by the convective currents. As
the amplitude is increased, the velocity of the boundary-layer vor-

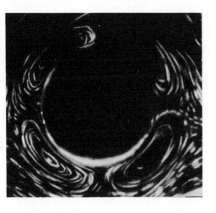

Fig. 13. Stationary vortices in air near a cylindrical rod in a sound field. The frequency is 200 cps; the radius of the glass cylinder is 0.11 cm.

tices increases and begins more and more to form a distinct system of vortices outside the boundary layer. The formation of this vortex system clearly occurs when the displacement amplitude in the sound wave is on the order of the thickness of the acoustic boundary layer. The dimensions of these vortices are determined by the dimensions of the experimental vessel; they can be observed without the microscope. In this sense, there is obviously a possible "threshold" for the onset of macroscopic stationary streaming, as established in [38, 42], i.e., for the onset of a distinct system of vortices on attainment of a definite amplitude. The angular velocity of the vortex system outside the boundary layer has the opposite sense of the vortex system in the boundary layer; in the direction of acoustic oscillation the liquid is transported away from the cylinder and flows toward it in the perpendicular direction (outer vortex system in Fig. 8). Here the velocity of the boundary-layer vortices increases, resulting in a certain reduction in the static pressure in this region and, hence, in a constriction of the boundary-layer vortices, the thickness of the region diminishes, approaching that of the acoustic boundary layer. In addition, a certain sensation of streaming "inversion" is created, as observed in a number of papers [42, 43]. A subsequent increase in the intensity yields a further variation in the nature of the streaming, especially a change in the configuration of the boundary-layer vortices, as shown in Fig. 9; this has also been observed experimentally [30].

Next we consider the dimensions of the vortices near the surface of a cylinder. It was remarked above (see Chap. 1) that the

theory of boundary-layer streaming is applicable when, besides
the condition $\lambda \gg \lambda'$, the local radius of curvature of the obstacle
is considerably greater than the viscous wavelength. For vor-
tices created near a cylinder of radius a, it may be anticipated
that, with a reduction in a/δ, the Schlichting solution will not give
valid results for certain values of a/δ. This has been demonstrated experimentally in the example of the thickness of the
boundary vortex layer [27, 28]. As implied by expressions (64a)
and (63), the distance Δ at which the radial component of the streaming
ing velocity goes to zero (vortex layer thickness) is a function only
of δ; it follows from (63) and Fig. 6 that $\Delta \simeq 1.9\delta$. This is the
vortex layer thickness near the obstacle, according to the theory
of Schlichting. The experimental results obtained in [27, 28] from
the observation of streaming in air and in water-glycerin mixtures
of various concentrations near a circular cylinder are shown in
Fig. 14 (curve 1), which also gives this dependence according to
Schlichting (curve 2). As the figure reveals, the thickness of the
vortex layer is consistent with Schlichting's theory for $a/\delta \gtrsim 30$-
40; for $a/\delta < 5$ the discrepancy between the indicated theory and
experiment is greater than one order of magnitude. The same is

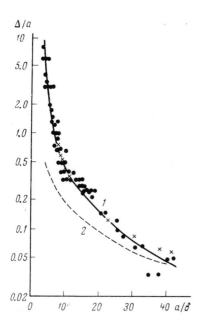

Fig. 14. Dimensionless thickness of the
vortex boundary layer versus the ratio
of the cylinder radius to the thickness
δ of the acoustic boundary layer. 1)
Experimental [27, 28]; 2) Schlichting
theory.

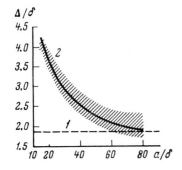

Fig. 15. Dimensionless vortex-layer thickness Δ/δ near the equator of a sphere of radius a versus the ratio a/δ. 1) Schlichting boundary layer, 2) experimental [29]. The scatter of the experimental data is indicated by the hatched region.

true for stationary vortices near a sphere. The experimental results for the ratio Δ/δ in the equatorial plane of a sphere at various values of a/δ are shown in Fig. 15 (curve 2) [29]; the vortices were observed in air at frequencies of 230-1150 cps. Also shown (the horizontal line 1) is the Schlichting value of Δ/δ. For a cylinder, as well as for very tiny spheres, the boundary-layer vortex thickness is greater than the Schlichting value, the latter value being obtained for $a/\delta \gtrsim 60$.

Consider now the streaming produced in a liquid near oscillating gas bubbles [44, 45]. The scale of this kind of streaming is on the order of the bubble radius, and in the literature it is often called microstreaming. The particle velocity of the wall of a gas bubble whose radius is close to the resonance value can greatly exceed the particle velocity in the sound wave. A zero-order radiator (pulsating sphere), as noted in Chap. 1, does not induce streaming. However, if the pulsating bubble is situated, say, near a wall or if it oscillates in a higher than the zeroth mode, streaming can occur near it. This streaming is responsible for a number of effects observed in a field of strong ultrasonic waves; for example, by eliciting a vigorous exchange of liquid near a solid surface, streaming can accelerate the cleaning of contaminated surfaces [45], accelerate the development of a photographic emulsion [46], etc.

Microstreaming has been observed in the vicinity of air bubbles in mixtures of water with glycerin [44]. It has been possible by variation of the concentration of these mixtures to vary the kinematic viscosity of the medium over a wide range (from 0.01 to ~7.5 cm^2/sec). The nature of the microstreaming at constant frequency depended on the kinematic viscosity of the medium and

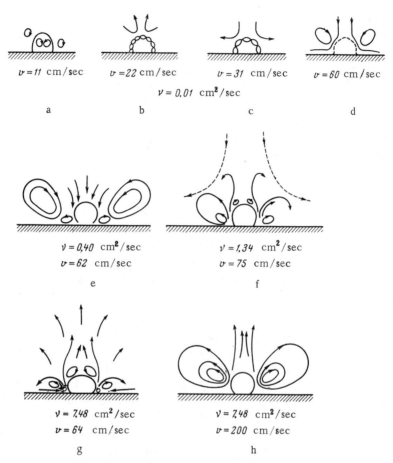

Fig. 16. Microstreaming in the vicinity of oscillating air bubbles; v is the particle velocity of the bubble surface.

on the particle velocity amplitude of the bubble surface. The observations in this study were made on bubbles situated on the surface of a vibrator. The variation of the streaming pattern in a low-viscosity liquid (pure water) with increasing sound amplitude is shown in Fig. 16 (a-d). For velocity amplitudes less than 11 cm/sec of the bubble's surface, there is no clearly marked streaming (Fig. 16a); an increase in the sound intensity produces oscillations in a complex mode and the simultaneous onset of a constant flow of liquid in the upper portion of the bubble (Fig. 16b). With a further increase in the sound intensity, the oscillation mode of the bubble changes suddenly, and simultaneously the direction of flow reverses (Fig. 16c). A greater increase in the intensity is accompanied by a transition of the stable mode into chaotic motion of the bubble surface, the direction of flow remaining unchanged at this time, but vortices forming near the bubble (Fig. 16d).

In medium-viscosity liquids the threshold velocity amplitudes at which streaming occurs turned out to larger than in low-viscosity liquids. Typical microstreaming for this case is shown in Fig. 16e.

The change in the streaming pattern in going from medium-viscosity liquids to high-viscosity liquids is apparent from a comparison of Fig. 16e with Fig. 16f, which illustrates microstreaming for large values of the particle velocity amplitude. As expected, in the high-viscosity liquid the dimensions of the boundary-layer vortices increase, the external vortices extending to a greater distance from the bubble. The streaming about a bubble in a high-viscosity liquid is illustrated in Figs. 16g and 16h. Here an increase in the oscillation amplitude leads to the formation of distinct vortices (see Fig. 16h).

At a greater distance from the bubble, the microstreaming velocity falls off rapidly; in low-viscosity liquids at distances on the order of ten bubble radii, the microstreaming velocity is negligible. For sufficiently intense oscillations the flow entrains even a thin layer of liquid on the surface of the bubble. This has been observed in the motion of fine particles of aluminum dust adsorbed on the bubble surface [44]. At small amplitudes these particles remains at rest. An increase in the amplitude, as already indicated, is accompanied by "inversion" of the flow; with "inversion" of the flow, the layer of adsorbed particles was set in vigorous motion. We note that flow "inversion" set in at bubble surface displacement amplitudes close to the thickness of the acoustic boundary layer.

Microstreaming has also been observed near an interface between two immiscible liquids in [47]. When a sharp needle vibrating at ultrasonic frequency was brought near the liquid interface (water and transformer oil), vortices were created in both liquids. If the needle was brought near the interface, at some distance a surface discontinuity occurred, and the microstreaming of the two liquids merged. This resulted, of course, in the emulsion of one liquid in the other.

The onset of microstreaming in complex biological cell structures has been investigated in [48-50]. As a result of the oscillation of the cell walls under the action of an ultrasonic vibrator in the form of a thin needle, strong micro-eddies occur in each cell near the vibrator, causing intermixing of the intracellular substance. It is interesting that this mixing does not always disrupt the vital activity of the cell. If the jacket of the cell is not disturbed, the cell restores its functions after ultrasonic treatment. Besides the fact that these investigations can yield useful information on certain properties of cells and on the mechanism of the biological action of ultrasound, the vortex velocity can be used to determine certain physical parameters of the intracellular substance (its kinematic viscosity, etc.).

Microstreaming has been observed near a plane surface [51, 52]. A schematic diagram of this streaming, produced in connection with the oscillations of a cylindrical vibrator E near a solid plane surface F, is presented in Fig. 17. The vibrator oscillated at a frequency of 2 kc, and its diameter and distance h were much smaller than the acoustic wavelength in the liquid (the observations were made in mixtures of water with glycerin at various concentrations). The vortices were investigated by means of a stereomicroscope with a magnification of 180. The distances $h \sim 40\text{-}18\delta$. As evident from the figure, near the edge of the vibrator there occurred small vortices C and C', whose dimensions depended both on the viscosity of the liquid (in high-viscosity liquids the diameter of the vortices was large) and on the radius of curvature of the edge of the vibrator E; with a decrease in the radius of curvature, the diameter of the vortices increased. The form of the vortices was independent of the distance between the vibrator and surface. These minute vortices are nothing other than stationary boundary-layer streaming. The diameter of the vortex A' depended on h, where-

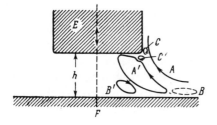

Fig. 17. Microstreaming in a liquid near a vibrator E and fixed solid surface F.

as the diameter of the vortex B' was independent of the radius of curvature of the edge of E and of the liquid viscosity and depended only slightly on h. For a vibrator particle velocity of 2.6 cm/sec, the maximum streaming velocity was on the order of 15 μ/sec, and the streaming velocity exhibited a square-law dependence on the vibrator particle velocity in agreement with the theory. In our present terminology the vortices A, A', B, and B' are v o r t i c e s o u t s i d e t h e b o u n d a r y l a y e r.

Summarizing, we assert that microstreaming occurs in extremely inhomogeneous sound fields. The configuration of the streaming can be very complex and depends both on the geometry of the sound field as well as on the viscosity of the medium and sound frequency. The scales of microstreaming are smaller than the acoustic wavelength. In microstreaming there is clearly always a separation between the streaming in the acoustic boundary layer (analogous to Schlichting streaming) and the one outside the boundary layer.

Streaming in an acoustic waveguide (tube) near a constriction of the waveguide (in the form of a partition with a hole in it) has been observed in [43]. The investigations were carried out at sonic frequencies in tubes containing air, and the sound level was varied over broad limits. The streaming pattern depended strongly on the frequency and amplitude of the sound. At small amplitudes the stationary streaming through the hole was directed away from the sound source. An increase in the sound amplitude, at first, produced vortices near the edges of the hole, then at some critical amplitude caused a reversal in the direction of flow. The critical velocity amplitude, as the experimental results indicate, is $\sim(4\text{-}5)(2\nu\omega)^{1/2}$, i.e., flow inversion, as in the case of streaming near a cylinder of air bubbles in liquids, occurs at a displacement amplitude close to the thickness of the acoustic boundary layer.

A further increase in the sound amplitude induces the superposition of pulsating motion on the constant flow, and the flow through the hole is once again directed away from the sound source, and vortices become periodically detached from the edges of the hole. The rate of transfer of these vortices at a particle velocity amplitude around a few meters per second was on the same order as the particle velocity. The transition to this streaming condition depended not only on the sound amplitude and frequency, but also on the geometric characteristics of the hole (its diameter and the thickness of the partition). Due to vortex formation the impedance of the hole depended on the sound amplitude.

The streaming velocity depends on the sound frequency and viscosity of the medium differently in different spatial domains. The streaming near the surface of an obstacle (Schlichting streaming), according to the theory, has a velocity $v \sim (\omega/\nu)^{\frac{1}{2}}$. Far from the boundary layer (in the region of Rayleigh streaming) $v \sim (a + b\omega^2)^{\frac{1}{2}}$, where a and b are constants that do not depend on the viscosity or frequency. In the intermediate region the streaming velocity depends on the viscosity and frequency in a more complex fashion. At the present time there are no reasonably reliable experimental results on the frequency dependence of the streaming velocity. Streaming produced outside the boundary layer by the oscillations of plates has been observed experimentally [39]. The measurements were performed from low infrasonic to ultrasonic frequencies. Over the entire range the velocity was proportional to the square of the particle velocity. In the low-frequency interval, the streaming velocity was $\sim\omega/\nu$, which does not contradict the foregoing remarks about the frequency dependence of the streaming, because the measurements were performed outside the boundary layer, but not in the Rayleigh streaming domain.

In [38] an experimental test of Rayleigh streaming was carried out, i.e., of the streaming in a standing sound wave in a tube. The results of the study, which was made at sound levels below $\sim$135 dB at a frequency of 600 cps, confirmed Rayleigh's theory not only qualitatively, but quantitatively as well.

The experiments described above, with the exception of those in [43], pertained to the investigation of slow acoustic streaming at not too high sound intensity levels. In [20] streaming analogous to Rayleigh streaming, i.e., in a standing wave, has been investigated in air at sound levels up to 167 dB (particle velocity amplitude up to 17 m/sec). These sound levels were produced by means of a powerful dynamic siren. The standing wave was observed in a circular measurement tube 40 mm in diameter; the constant air flow of the siren was separated from the measurement tube by means of thin plastic membranes. For the measurement of the sonically induced streaming velocity, three methods

were used: visualization of the streaming by means of minute
particles (aluminum dust, crystal powders) in conjunction with
high-speed films, a Pitot tube, and a thermoanemometer with a
reduced filament temperature (from 300 to 30°C) for abatement
of the sensitivity to the variable components of the sound field.
The thermoanemometric measurements were made at pressure
antinodes (and, hence, at velocity nodes) for greater accuracy.
The distribution of the streaming velocity was also determined
from the deflection of thin Kapron filaments inserted into the
measurement tube. The character of the velocity distribution in
a standing wave is consistent with the theoretical distribution
implied by Rayleigh's theory (see Fig. 5). The absolute values of
the streaming velocity, measured by various methods, turned out
to be almost equal to one another, but nearly two orders of magnitude larger than those calculated on the basis of Rayleigh's theory
(in different parts of the tube, the velocity was ~1 to 6 m/sec).
This difference is attributed by the authors, of course, to the inapplicability of Rayleigh's theory at such high intensity levels.
The dependence of the maximum measured streaming velocity on
the sound pressure is shown in Fig. 18, from which it is apparent
that the velocity varies as the square of the sound pressure.

The thickness of the acoustic boundary layer has been measured in the vicinity of a flat plate in air [36]. A plate with dimensions 1×4 cm^2 was oriented in the direction of oscillation

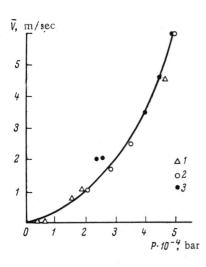

Fig. 18. Maximum stationary streaming
in a standing wave versus the sound pressure amplitude. Methods of velocity measurement: 1) visualization of the streaming by means of air-suspended powders;
2) thermoanemometer; 3) Pitot tube.

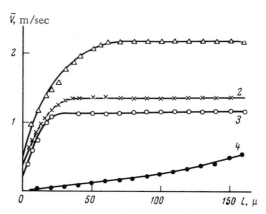

Fig. 19. Stationary streaming velocity near a plate
versus the distance from the plate surface. 1) At
1.2 kc; 2) 2.2 kc; 3) 4 kc; 4) without sound, for the
case of a plate immersed in a flow whose velocity
far from the plate is equal to the particle velocity
amplitude in the standing wave.

in a standing wave generated by a powerful siren. The constant
streaming velocity distribution near the surface of the plate was
determined with a thermoanemometer. The thickness of the
acoustic boundary layer was independent of the sound pressure
amplitude (from 7.6 to 24 $\cdot$ 10^3 bar) and displayed fully satisfactory agreement with the theoretical value $\delta = (2\nu/\omega)^{\frac{1}{2}}$ (at frequencies of 4-1.2 kc). The velocity distribution obtained for a
sound pressure amplitude of 1.2 $\cdot$ 10^4 bar at various frequencies
is illustrated in Fig. 19 (curves 1-3). Also shown in the same
figure is the velocity distribution for a plate in a nonacoustic stationary flow (curve 4) whose velocity far from the plate is equal
to the particle velocity amplitude in the standing wave. Under the
experimental conditions the thickness of the acoustic boundary
layer was approximately two orders of magnitude smaller than
the thickness of the boundary layer for stationary flow past the
plate, thus indicating the possible acceleration of various transport processes in the sound field.

We now discuss the qualitative characteristic of this acceleration. We know that sound acts to accelerate heat-transfer processes in heated bodies and mass-transfer processes (drying in a
sound field, to a certain extent ultrasonic cleaning, the accelera-

tion of atomization of fuels and, as a result, intensification of their burning in the vibration combustion regime, etc.). The mechanism of the acceleration of these processes is sometimes extremely complex, but it is obvious that one of the main contributing factors comprises processes that originate in the boundary layer. The acceleration of slow processes such as heat conduction and diffusion is brought on by the intensification of mass transfer between the boundary layer and the medium outside the boundary layer. Plausible reasons for the acoustic acceleration of mass transfer are, on the one hand, variation of the thickness of the boundary layer in a sound field and, on the other, variation of the streaming pattern in the acoustic boundary layer. The thickness of the boundary layer in stationary flow past a body at a velocity v is of the order $\delta_s \sim (\nu l / v)^{\frac{1}{2}}$, where l is the characteristic dimensions of the obstacle; the acoustic boundary layer near a body whose characteristic dimension is much larger than δ has a thickness $\delta \sim (\nu / \omega)^{\frac{1}{2}}$. In many cases, as evinced in particular by the experimental results of [36], conditions can be created such that $\delta \ll \delta_s$; in this case the concentration or temperature gradients in the sound field will be greater than in nonacoustic stationary flow. Moreover, under certain conditions stationary vortex motion can be generated in the acoustic boundary layer; the boundary layer in an ideal monochromatic standing wave consists of individual vortex cells. This cellular structure becomes less distinct as the amplitude of the traveling wave imposed on the standing wave becomes larger. In the case of nonmonochromatic sound waves, clearly, turbulence can be created in the boundary layer by the sound field. The transition from laminar to turbulent flow in the boundary layer, as we are aware, intensifies the transfer processes.

We should point out one other essential characteristic of the streaming in the acoustic boundary layer. We are concerned with the buildup time of stationary streaming. Inasmuch as Eq. (57), which describes stationary streaming, is a diffusion relation, the time of "vortex diffusion in the acoustic boundary layer" is of the order $\delta^2 / \nu \sim T$, i.e., the stationary streaming is stabilized in a time close to the period of the sound wave. In acoustical applications, where, as a rule, coherent sound sources are involved, such a rapid response on the part of the boundary-layer vortices to external perturbations is of no major importance.

The situation is otherwise in the case of random variable effects
on the boundary layer; since the vortices build up rapidly, such
random effects, obviously, must be very effective in promoting
the transition from a laminar to a turbulent boundary layer.

In concluding this section, it is important to consider a particular proposal for the practical utilization of acoustic streaming.
An acoustic air pump has been proposed [53] and is shown schematically in Fig. 20. The electrodynamic speaker 1, with a membrane 10-12 cm in diameter, is driven off of the 60-cycle ac line,
generating a constant flow in the direction of the tube C; the suction
of air is from the tube A. The pump operates at highest efficiency
when the acoustical system is tuned to resonance, this being done
by variation of the diameter and length of the tube B. According
to the author's report, the pump has an output of 700 liters /min.
The pump creates a uniform flow of air at constant temperature
and is used for the calibration of temperature gauges. The advantage of this type of pump is its absence of rubbing parts, a virtue that renders it useful for low temperature operations.

§ 3. Eckart Streaming

In the preceding section we reviewed the experimental work
on streaming in the vicinity of obstacles (Schlichting streaming)
and in standing sound waves far from boundaries (tentatively
called Rayleigh streaming). In this section we present some of
the experimental results on Eckart streaming, i.e., streaming
created by a well-collimated sound beam in a free sound field.
Unlike the types of streaming considered before, Eckart stream-

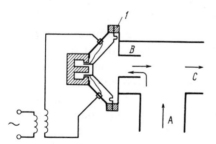

Fig.20. Schematic diagram of the acoustic
air pump.

ing is a large-scale phenomenon. Its vortices (as evinced at least by Figs. 10 and 11) have a characteristic dimension of the same order as the size of the experimental vessel.

Inasmuch as more or less well-collimated sound beams can only be produced in the relatively high ultrasonic range, by far, most of the experimental studies have been conducted in the range of 1 to 10 Mc. The majority of the papers refer to investigations of streaming in liquids, although Eckart streaming has also been observed in gases [54]. The greatest number of studies on Eckart streaming have been carried out, rather than for a direct verification of the theory, for the determination of the ratio of the dilatational to the shear viscosity. The values obtained, which are close to those measured by other acoustical techniques, afford indirect proof of the theory. However, there are papers, [39, 54, 55] in which a direct comparison has been made between the various experimental characteristics and the theory.

For example, streaming in various liquids has been observed [39], beginning at infrasonic frequencies (~4 cps) and ending with frequencies on the order of 10 Mc. It was shown that prior to the onset of turbulence or cavitation the flow velocity is proportional to the sound intensity over the entire range of frequencies. Only the high-frequency part of the experimental results were consistent with Eckart's theory; here the velocity is proportional to the square of the frequency and is almost proportional to the absorption coefficient of the liquid, a result that is qualitatively consistent with the theory.

The transverse distribution of the streaming velocity has been investigated in [54]. It was found that the distribution depends on the homogeneity of the sound field; the best agreement with Eckart's theory was obtained with a Straubel quartz plate as the emitter. In a homogeneous sound field, the streaming velocity distribution adheres quite closely to relation (31). Analogous measurements in gases have been made in [54], with corrections introduced in (31) to account for the inhomogeneity of the sound field; the experimental results are in good agreement with the theoretical.

The results of acoustic streaming studies have been used in a number of papers [6, 11, 15, 35, 56-59] to determine the sound absorption coefficients in liquids. As evident from relations (31) and (33), for the determination of the absorption coefficient by the acoustic streaming method, it is necessary to perform independent measurements of the sound intensity (or acoustic energy density) and streaming velocity. It was mentioned earlier that the radiation pressure injects considerable interference in the determination of the streaming velocity on the basis of the dynamic pressure. The converse is just as true; the dynamic flow pressure introduces errors into the measurement of the radiation pressure by mechanical methods. In measurements of the absorption coefficient by this method, the separation of the dynamic flow pressure and radiation pressure is complicated somewhat further by the fact that conditions must be created so as to conform to the theory. In [6], where the ultrasonic absorption has been determined on the basis of Eckart streaming, the separation of the radiation pressure and dynamic flow pressure was based on the fact that the buildup time of streaming is greater than the deflection time of the radiometer under the action of the radiation pressure; the radiometer has time to be deflected by the radiation pressure (making it possible to determine the acoustic energy density); then the deflection of the radiometer slowly increases under the action of the dynamic flow pressure. This method, however, has its share of drawbacks, which we discussed above.

A somewhat alternative method for determining the sound absorption coefficient has been proposed in [57]. A diagram of the apparatus is shown in Fig. 21. The ultrasonic field (at 1 Mc), generated by the source 1, completely filled the tube containing the test liquid 2; the tube had a bypass capillary duct 3 for backflow. According to relation (31), if the radius of the sound beam

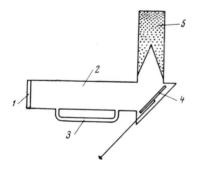

Fig. 21. Diagram of a device used to measure the absorption of ultrasound in liquids in terms of the Eckart streaming velocity.

is equal to the tube radius, the acoustic streaming velocity reverts to zero. Under the experimental conditions, of course, due to the inhomogeneity of the sound field over the tube cross section and the influence of the boundary layer near the walls, and due also to the flow of liquid through the capillary duct 3 in the given apparatus, the transfer of liquid takes place, although its velocity is considerably less than the streaming velocity in the free sound field. The influence of the dynamic flow pressure on the mechanical radiation-pressure sensor 4 was relatively small under these conditions. The sound reflected from receiver 4 was absorbed by the absorber 5. The authors of [58] refrained from an absolute radiometer measurement of the sound field, because the receiving element of the radiometer, as a sound reflector, prevented the creation of a perfect traveling wave (in this investigation the acoustic energy density was determined from the impedance of the emitter in air and in the liquid). According to the Hagen-Poiseuille law, the flow velocity of the liquid in the capillary is equal to

$$v = \frac{\Delta p \cdot r^2}{4 \eta l},$$

where Δp is the pressure difference between the ends of the capillary, whose radius is r and length is l. For small streaming velocities (less than 1 mm/sec in the experiment), the influence of the bends in the tube can be neglected. Since the liquid is incompressible at these velocities, the pressure difference Δp was obtained as a result of the variation of the radiation pressure. Assuming that the radiation pressure $P = \bar{E}$, where $\bar{E}$ is the time-average acoustic energy density, for the velocity of the liquid in the capillary we obtain

$$v = \frac{\bar{E}_0 (1 - e^{-2\alpha_0 l})}{4 \eta l} r^2,$$

where $\bar{E}_0$ is the time-average acoustic energy density at the end of the capillary nearest the sound source, and α_0 is the sound absorption coefficient. The value of $\bar{E}_0$ was determined with receiver 4, and the flow velocity in capillary 3 was determined from the motion of minute aluminum particles suspended in the liquid and observed under the microscope. The device was calibrated beforehand with liquids in which the absorption was known with sufficient accuracy.

TABLE 2. Ratio of the Dilatational to the Shear Viscosity

Liquid	Frequency, Mc	Viscosity ratio η'/η	
		by acoustic streaming	by other acoustical measurement techniques
Water.	5	2.4	2.1
Methyl alcohol.	5	1.3	1.1
Ethyl alcohol	5	3.8	3.6; 2.6; 2.3
Acetone	5	3.1	3.8; 3.1; 2.9
Amyl acetate.	5	9.9	10.2; 12.3
m-Xylene	5	11.0	7.7; 6.5
Carbon tetrachloride	2	28	27; 21; 22

Other modifications of this method for the measurement of ultrasonic absorption are also possible. For example, the absolute value of the sound pressure has been determined from the intensity of light at the first diffraction maximum, while the flow velocity was determined, as always, from the motion of small particles suspended in the liquid [15].

Some experimental data from measurements of the ratio of the dilatational (second) viscosity to the shear viscosity are presented in Table 2 for illustration, along with the results obtained by conventional methods from the absorption of ultrasound.

We need to look into one drawback of sound absorption measurements based on acoustic streaming. Apart from the fact that it is required in these measurements to perform absolute measurements of the sound field, which are characterized, of course, by low accuracy, it is necessary in order to obtain measurable streaming velocities to use rather large sound intensities. In some studies it appears that the acoustic Reynolds numbers were ⩾ 1. The Eckart theory is inapplicable in this domain, and the absorption coefficient in this case is greater than the small-amplitude absorption coefficient, due to nonlinear waveform distortion. The increased absorption causes the streaming velocity to be larger than the Eckart value and, as a result, the experimental ratio of the dilatational to the shear viscosity, or the experimental absorption, determined by this method, is generally somewhat larger than that measured by other methods. Inasmuch as the streaming velocity is $\sim\omega^2$, at low frequencies the sound intensity must be increased in order to obtain velocities appreciable enough to be measured. On the other hand, the Reynolds number is $\sim p\omega^{-1}$, where p is the sound pressure, i.e., in the low-frequency range one is especially prone in streaming velocity measurements of the absorption to fall into the domain of Reynolds numbers greater than unity.

Some authors, ignoring the above consideration, view the method as a promising one for absorption measurements at low frequencies. It has been possible using Eckart streaming to measure the absorption coefficient at a frequency of 130 kc [58], although only for liquids in which the absorption is large $[\alpha_0/f^2 \sim (6-10) \cdot 10^{-14} \sec^2/cm]$. It is not too hopeful that the method will permit the determination of absorption in low-absorption liquids $(\alpha_0/f^2 \sim 10^{-16} \sec^2/cm)$ at frequencies below 10^5 cps. Conceivably, the application of streaming other than the Eckart variety may provide a means for the measurement of absorption at low frequencies.

It has been remarked in several papers [6, 15, 39] that with increasing ultrasonic intensity the Eckart streaming velocity loses its proportionality to the intensity. A more sophisticated study of this effect has been carried out in [34]; a gradual variation in the streaming velocity was observed at a certain point of the sound field as the sound intensity was increased. Figure 22 (curve 1) depicts the streaming velocity in water at a distance of 40 cm from a 1.2-Mc sound source as a function of the sound pressure amplitude near the source. At a sound pressure $p \sim 7$ atm and at a distance of 40 cm a sawtooth is formed. As evident from the figure, the character of the streaming velocity-intensity relation also changes. It was demonstrated in Chap. 1 that the Eckart theory is valid for small acoustic Reynolds numbers, so that departures from that theory are not attributable to turbulence of the acoustic streaming pattern, as postulated in [6, 39], but to waveform distortion and the inapplicability of the theory under the stated conditions.

Also shown in Fig. 22 (curve 2) is the streaming velocity calculated according to (45) for a sawtooth [5]; the assumption of a small streaming velocity relative to the particle velocity am-

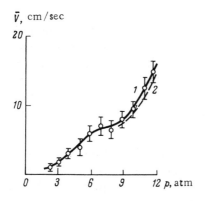

Fig. 22. Streaming velocity due to a sound beam in water. 1) Experimental results [34]; 2) theoretical results [5] (for a sawtooth wave).

plitude was not made in that paper. As **Fig.** 22 indicates, the theoretical results are in good agreement with the experimental.

In conclusion, it is important to note that, as opposed to other aspects of nonlinear acoustics in which nonstationary processes have not been investigated at all, for acoustic streaming an attempt has been made at the experimental assessment of the buildup time of stationary streaming. Inasmuch as the equation describing Eckart streaming is a vortex diffusion equation [4], the order of magnitude of the buildup time (vortex diffusion time) can be estimated from the relation $\tau = L''^2 / \nu$, where L'' is the characteristic dimension of the volume occupied by the streaming and ν is the kinematic viscosity of the medium. For liquids having a relatively low viscosity ($\nu \sim 0.01$ P) and for ultrasonic tanks of conventional dimensions, the buildup time of Eckart streaming as deduced from this relation is on the order of a few minutes. In [40] the buildup of streaming has been observed in water at 200 kc in free space with pulse durations less than 0.02 sec, which contradicts the above approximate estimate. This could be attributed to the relatively large temperature nonuniformity of the medium (the onset of streaming was observed on the basis of the change in the convective current pattern from a heated wire), and, tentatively, to a certain indefiniteness in the interpretation of the buildup time. At any rate, there is no doubt that the buildup time of Schlichting streaming in an acoustic boundary (being close in order of magnitude, as already mentioned, to the wave period) is much smaller than the buildup time of large-scale Eckart streaming.

References

1. W. L. Nyborg, Acoustic streaming, Physical Acoustics (W. P. Mason, ed.), Vol. 2, Part B, Academic Press, New York (1965).
2. H. Medwin and I. Rudnik, Surface and volume sources of vorticity in acoustic fields, J. Acoust. Soc. Am., 25(3):538 (1953).
3. P. J. Westervelt, The theory of steady rotational flow generated by sound fields, J. Acoust. Soc. Am., 25(1):60, 799 (1953).
4. C. Eckart, Vortices and streams caused by sound waves, Phys. Rev., 73(1):68 (1948).
5. Yu. G. Statnikov, Streaming induced by finite-amplitude sound, Akust. Zh., 13(1):146 (1967).
6. E. N. Libermann, Second viscosity of liquids, Phys. Rev., 75(9):1415 (1949).
7. F. E. Fox and K. T. Herzfeld, On the forces producing the ultrasonic wind, Phys. Rev., 78(2):156 (1950).
8. F. E. Borgnis, Theory of acoustic radiation pressure, Rev. Mod. Phys., 25(3):653 (1953).
9. W. Cady and C. Gittings, On the measurement of power radiated from an acoustic source, J. Acoust. Soc. Am., 25(5):892 (1953).
10. E. M. J. Herrey, Experimental studies on acoustic radiation pressure, J. Acoust. Soc. Am., 27(5):891 (1955).
11. I. Johnsen and S. Tjötta, Eine theoretische und experimentalle Untersuchung über den Quartzwind [A theoretical and experimental study of the quartz wind], Acustica, 7(1):7 (1957).
12. S. Tjötta, Steady rotational flow generated by a sound beam, J. Acoust. Soc. Am., 29(4):455 (1957).
13. W. L. Nyborg, Acoustic streaming due to attenuated plane waves, J. Acoust. Soc. Am., 25(1):68 (1953).
14. K. A. Naugol'nykh, On sonically induced streaming, Dokl. Akad. Nauk SSSR, 123(6):1003 (1958).
15. A. I. Ivanovskii, Theoretical and Experimental Investigation of Sonically Induced Streaming, Gidrometeoizdat (1959).
16. P. J. Westervelt, The mean pressure and velocity in a plane acoustic wave in a gas, J. Acoust. Soc. Am., 22(3):319 (1950).
17. R. D. Fay, Plane sound waves of finite amplitude, J. Acoust. Soc. Am., 3(2):222 (1931).
18. Rayleigh (J. W. Strutt), The Theory of Sound, Vol. 2, McGraw-Hill, New York (1948), p. 352.
19. K. Schuster and W. Matz, Über stationare Strömungen in Kundtsche Rohr [On stationary streaming in Kundt tubes], Akust. Z., 5:349 (1940).
20. Yu. Ya. Borisov and Yu. G. Statnikov, Flow currents generated in an acoustic standing wave, Akust. Zh., 11(1):35 (1965).
21. L. D. Landau and E. M. Lifshits, Mechanics of Continuous Media, Gostekhizdat (1954).
22. H. Schlichting, Berechnung ebener periodischer Grenzschichts Strömungen [Calculation of plane periodic boundary-layer streaming], Phys. Z., 33(8):327 (1932).

23. H. Schlichting, Grenzschicht-Theorie [Boundary-Layer Theory], Brann, Karlsruhe (1951).

24. W. L. Nyborg, Acoustic streaming near a boundary, J. Acoust. Soc. Am., 30(4):329 (1958).

25. J. M. Andres and U. Ingard, Acoustic streaming at high Reynolds numbers, J. Acoust. Soc. Am., 25(5):928 (1953).

26. J. M. Andres and U. Ingard, Acoustic streaming at low Reynolds numbers, J. Acoust. Soc. Am., 25(5):932 (1953).

27. J. Holzmark, I. Johnsen, T. Sikkeland, and S. Skavlem, Boundary layer flow near a cylindrical obstacle in an oscillating incompressible fluid, J. Acoust. Soc. Am., 26(1):26 (1954).

28. W. P. Raney, J. C. Corelli, and P. J. Westervelt, Acoustic streaming in the vicinity of a cylinder, J. Acoust. Soc. Am., 26(6):1006 (1954).

29. C. M. A. Lane, Acoustical streaming in the vicinity of a sphere, J. Acoust. Soc. Am., 27(6):1123 (1953).

30. M. Carriere, Analyse ultramicroscopique des vibrations aeriennes [Ultramicroscopic analysis of air vibrations], J. Phys. Radium, 10(5):198 (1929).

31. P. J. Westervelt, Acoustic streaming near a small obstacle, J. Acoust. Soc. Am., 25(6):1123 (1953).

32. S. Skavlem and S. Tjötta, Steady rotational flow of an incompressible viscous fluid enclosed between two coaxial cylinders, J. Acoust. Soc. Am., 27(1):26 (1955).

33. Yu. G. Statnikov, Microstreaming about a gas bubble in a liquid, Akust. Zh., 13(3):464 (1967).

34. E. V. Romanenko, Experimental investigation of acoustic streaming in water, Akust. Zh., 6(1):92 (1960).

35. E. G. Richardson, Acoustic experiment relating to the coefficients of viscosity of various liquids, Proc. Roy. Soc., A226(1164):16 (1954).

36. Yu. Ya. Borisov and Yu. G. Statnikov, Measurement of boundary layer thickness in the presence of a sound field, Akust. Zh., 12(3):372 (1966).

37. G. Spengler, Über den Einfluss des "Quartzwindes" auf Ultraschalleistungmessungen [Influence of the "quartz wind" on ultrasonic power measurements], Naturwissenschaften, 41:59 (1954).

38. E. N. Andrade, On the circulation caused by the vibration of air in a tube, Proc. Roy. Soc., A134(824):445 (1931).

39. A. M. Gabrial and E. G. Richardson, A study of acoustic streaming in liquids over a wide frequency range, Acustica, 5(1):28 (1955).

40. C. L. Darner and E. N. Laid, "Quartz wind" formation time, J. Acoust. Soc. Am., 26(1):104 (1954).

41. L. K. Zarembo and V. V. Shklovskaya-Kordi, Visualization of acoustic streaming at the boundary of two immiscible liquids, Akust. Zh., 3(4):373 (1957).

42. J. D. West, Circulation occurring in acoustic phenomena, Proc. Phys. Soc., B64(378):483 (1951).

43. U. Ingard and S. Labate, Acoustic circulation effects and the nonlinear impedance of orifices, J. Acoust. Soc. Am., 22(2):211 (1950).

44. S. A. Elder, Cavitation microstreaming, J. Acoust. Soc. Am., 31(1):54 (1959).

45. J. Kolb and W. L. Nyborg, Small-scale acoustic streaming in liquids, J. Acoust. Soc. Am., 28(6):1237 (1956).

46. W. L. Nyborg, R. K. Gould, F. J. Jackson, and C. E. Adams, Sonically induced microstreaming applied to a surface reaction, J. Acoust. Soc. Am., 31(6):706 (1959).

47. I. M. Faikin and I. E. Él'piner, Onset of emulsification processes due to microstreaming induced by an ultrasonic field, Akust. Zh., 11(1):126 (1965).

48. I. E. Él'piner, Recent advances in ultrasonic biophysics, Usp. Sovrem. Biol., 61(2):212 (1966).

49. I. E. Él'piner, I. M. Faikin, and O. K. Basurmanova, Intracellular microstreaming induced by ultrasonic waves, Biofizika, 10(5):805 (1965).

50. M. I. Gol'din, I. M. Faikin, and I. E. Él'piner, Microstreaming induced by ultrasonic waves in plant cells containing tobacco mosaic virus injections, Dokl. Akad. Nauk SSSR, 166(5):1221 (1966).

51. F. Y. Jackson and W. L. Nyborg, Microscopic eddying near a vibrating ultrasonic tool tip, J. Appl. Phys., 30(6):949 (1959).

52. F. Y. Jackson, Sonically induced microstreaming near a plane boundary, II, Acoustic streaming field, J. Acoust. Soc. Am., 32(11):1387 (1960).

53. T. M. Dauphinee, Acoustic air pump, Rev. Sci. Instr., 28(6):452 (1957).

54. H. Medwin, Acoustic streaming experiment in gases, J. Acoust. Soc. Am., 26(3):332 (1954).

55. E. W. Samuel and R. S. Shankland, The sound field of a Straubel X-cut crystal, J. Acoust. Soc. Am., 22(5):589 (1950).

56. S. M. Karim and L. Rosenheed, Second coefficient of viscosity of liquids and gases, Rev. Mod. Phys., 24(2):108 (1952).

57. J. E. Piercy and J. Lamb, Acoustic streaming in liquids, Proc. Roy. Soc., A226(1164):43 (1954).

58. D. N. Hall and J. Lamb, Measurement of ultrasonic absorption in liquids by the observation of acoustic streaming, Proc. Phys. Soc., 75:354 (1959).

59. S. M. Karim, Second viscosity coefficient of liquids, J. Acoust. Soc. Am., 25(5):997 (1953).

PART IV

PULSATIONS OF CAVITATION VOIDS

V. A. Akulichev

Introduction

All effects observed in connection with ultrasonic cavitation, such as cavitation erosion, sonoluminescence, cavitation noise, and the initiation of chemical reactions, are related to the existence and characteristic behavior of cavitation voids in an intense ultrasonic wave field. This makes the investigation of the motion of cavitation bubbles or voids one of the central problem areas of ultrasonic cavitation research.

According to current thinking, cavitation voids originate in an ultrasonic wave field at nuclei that are ever-present in water (or other liquid) in the form of solid, vapor, or gas micro-inhomogeneities. In the course of several periods of the ultrasonic field, each cavity that develops from a nucleus becomes infused due to rectified [1] and convective [2] diffusion with a definite quantity of gas dissolved in the water. For a steady field intensity, as well as a definite gas content and water temperature, this quantity remains constant on the average over one period. The quantity of infused gas corresponds to a stationary bubble with a certain equilibrium radius R_0, which could be ascertained experimentally if the ultrasonic field could be suddenly withdrawn and the bubble did not have time to become dissolved. In the present part of the book, we examine the motion of cavitation voids with a definite constant gas content corresponding to a stationary gas bubble of radius R_0, which we henceforth agree to call the initial radius. In other words, diffusion of the gas at the boundary of the bubble is disregarded.

Numerous experiments conducted with high-speed motion picture film [3-6] have shown that cavitation voids, as they expand in the ultrasonic field, preserve a distinct spherical configuration. It is possible that in the final stage of collapse the spherical shape of the cavitation voids is distorted, and it is even possible that the cavitation void will sometimes break up into smaller parts; this was first indicated by Kornfel'd [7]. However, inasmuch as the cavitation voids spend the main part of their incumbency in the ultrasonic field with a spherical shape, we shall assume that they execute only zero-order pulsations. We refer to such voids henceforth as cavitation bubbles.

Under real conditions the cavitation bubble usually exists simultaneously with a set of other bubbles situated at distances smaller than the ultrasonic wavelength. We lack a mathematical description of the motion of such bubbles, because it is vague how to account for their mutual interaction. The problem is significantly simplified if we assume that the interaction is so slight as to be negligible. This is in fact tantamount to the assumption that the given cavitation bubble is an isolated object. In the present study we investigate the differential equations describing the pulsations of such individual cavitation bubbles.

It is postulated that the radius of the cavitation bubble is much smaller than the ultrasonic wavelength (this stipulation is almost always satisfied). It is also assumed that the cavitation bubbles contain, in addition to the gas, vapor whose partial pressure is equal to the equilibrium value for the given temperature of the liquid.

The description of the pulsations of cavitation voids, even when all of the simplifications and constraints indicated above are met, leads to complex nonlinear differential equations, which are not solvable in general form. On the basis of an analysis of the numerical solutions of the equations, we can give certain principles of the pulsations of cavitation bubbles in an ultrasonic field and deduce empirical formulas by which one can estimate the velocity and collapse time of cavitation bubbles. We compare the solution of the equations with the corresponding experimentally described pulsations of actual cavitation bubbles. Moreover, we consider the relation between the pulsations of cavitation bubbles and the radiation of cavitation noise and shock waves.

Chapter 1

Fundamental Equations for the Pulsations of a Cavitation Void

§ 1. Pulsations of a Void in an Incompressible Liquid

Consider an isolated cavitation void of radius R, where R = R(t), executing pulsations in an ideal incompressible liquid. Then for the pressure p and velocity u in the liquid at a point r in space, where $r \geq R$, at time t we have the equation of motion

$$\frac{\partial u}{\partial t} + u \frac{\partial u}{\partial r} = -\frac{1}{\rho} \frac{\partial p}{\partial r} \tag{1}$$

and equation of continuity

$$\frac{\partial}{\partial r}(r^2 u) = 0, \tag{2}$$

where ρ is the density of the liquid. Inasmuch as we are considering irrotational motion of the liquid, we can introduce the velocity potential φ, so that

$$u = -\frac{\partial \varphi}{\partial r}. \tag{3}$$

Now, integrating Eq. (1) from r to ∞, we obtain

$$-\frac{\partial \varphi}{\partial t} + \frac{u^2}{2} + \int_{P_\infty}^{p(r)} \frac{dp}{\rho} = 0, \tag{4}$$

205

since $\varphi = 0$, $u = 0$, and $p(r) = p_\infty$ for $r \to \infty$. It follows from the condition of incompressibility of the liquid that $\rho = \rho_0 = \text{const}$ and

$$\frac{\partial \varphi}{\partial t} - \frac{u^2}{2} + \frac{1}{\rho_0} [P_\infty - p(r)] = 0. \tag{5}$$

From Eq. (2) we find

$$u = \frac{C}{r^2},$$

where the constant C is determined from the stipulation that the velocity is equal to U at the boundary of a sphere of radius R. Then the velocity is

$$u = U \frac{R^2}{r^2}, \tag{6}$$

and the velocity potential is

$$\varphi = U \frac{R^2}{r^2}. \tag{7}$$

Expressing $\partial \varphi / \partial t$ from (7) and substituting it into (5), we obtain [8]

$$\frac{1}{r} \left(\frac{R^2}{2} \frac{dU^2}{dR} + 2 RU^2 \right) - \frac{1}{2} U^2 \frac{R^4}{r^4} + \frac{1}{\rho_0} [P_\infty - p(r)] = 0. \tag{8}$$

If we set $r = R$ and recognize that $U = \partial R / \partial t$, we transform Eq. (8) as follows:

$$R \frac{d^2 R}{dt^2} + \frac{3}{2} \left(\frac{dR}{dt} \right)^2 + \frac{1}{\rho_0} [P_\infty - P(R)] = 0. \tag{9}$$

This equation describes the pulsations of the cavitation void for a pressure P_∞ at infinity and pressure $P(R)$ on the surface of the void. Both the pressure P_∞ and the pressure $P(R)$ can vary according to different laws, making it necessary to investigate diverse cases involving the behavior of cavitation voids.

Solutions of Eq. (9) for the simplest cases of constant pressure at infinity, when $P_\infty = P_0$ (P_0 is the hydrostatic pressure), were first obtained by Lamb [9] and Rayleigh [10]. Assuming that a vacuum exists inside the void [i.e., $P(R) = 0$], the following expression is readily obtained from Eq. (9) for the rate of collapse of the void

$$U^2 = \frac{2}{3}\frac{P_0}{\rho_0}\left(\frac{R_m^3}{R^3} - 1\right),$$ (10)

where R_m is the maximum radius of the void at the beginning of collapse. From this we can integrate to obtain the well-known Rayleigh equation for the collapse time of an empty void in a pressure field P_0:

$$\tau = 0.915\, R_m\left(\frac{\rho_0}{P_0}\right)^{1/2}.$$ (11)

Consider a cavitation void filled with gas and vapor. The variation of the gas pressure in the void is related to the variation of the radius R as follows

$$P_g = P_{g0}\left(\frac{R_0}{R}\right)^{3\gamma} = \left(P_0 + \frac{2\sigma}{R_0}\right)\left(\frac{R_0}{R}\right)^{3\gamma},$$ (12)

where γ is the polytropy index, which determines the state of the gas in the void; $\gamma = 1$ in the case of isothermal pulsations, and $\gamma = \frac{4}{3}$ in the case of adiabatic pulsations. We consider the vapor pressure P_v to be a certain constant determined by the temperature of the liquid, i.e., we assume that the condensation of the vapor and evaporation of the liquid keep pace with the variations of the bubble volume. In this case the following condition must be met at the boundary of the bubble:

$$\left(P_0 + \frac{2\sigma}{R_0}\right)\left(\frac{R_0}{R}\right)^{3\gamma} + P_v = P(R) + \frac{2\sigma}{R},$$

where σ is the surface tension and $2\sigma/R$ determines the associated pressure. From this boundary condition we easily ascertain

$$P(R) = \left(P_0 + \frac{2\sigma}{R_0}\right)\left(\frac{R_0}{R}\right)^{3\gamma} + P_v - \frac{2\sigma}{R}.$$ (13)

In the event the cavitation void is acted upon by an ultrasonic field pressure of amplitude P_m and frequency f, the pressure at infinity may be written in the form

$$P_\infty = P_0 - P_m \sin \omega t,$$ (14)

where $\omega = 2\pi f$.

Substituting (12) and (13) into Eq. (9), we obtain a second-order nonlinear differential equation describing the pulsations of a gaseous cavitation void in an ultrasonic field

$$R \frac{d^2R}{dt^2} + \frac{3}{2} \left(\frac{dR}{dt} \right)^2 +$$
$$+ \frac{1}{\rho_0} \left[P_0 - P_v - P_m \sin \omega t + \frac{2\sigma}{R} - \left(P_0 + \frac{2\sigma}{R_0} \right) \left(\frac{R_0}{R} \right)^{3\gamma} \right] = 0; \qquad (15)$$

Eq. (15) was first deduced and analyzed by Noltingk and Neppiras [11], hence, it is customarily referred to as the Noltingk − Neppiras equation. The numerical solutions obtained for this equation in several papers [11-13] have greatly expanded the notions concerning the possible character of the pulsations of actual cavitation bubbles. A comparison of these solutions with the corresponding experimental results has shown that the Noltingk − Neppiras equation affords a very adequate description of the variation of the radius of a cavitation bubble pulsating in an ultrasonic field. In the final stage of collapse of the cavitation bubble, however, when the rate of motion of its boundary becomes commensurate with the velocity of sound in the liquid, the assumption of an incompressible liquid becomes incorrect. Consequently, the Noltingk − Neppiras equation does not permit a description of the maximum rates of collapse of cavitation bubbles and the related maximum pressures which govern the various phenomena observed in a liquid.

§2. Pulsations of a Void with Regard for Compressibility of the Liquid

For the pulsations of a spherical void in a compressible liquid, we have, along with the equation of motion (1), the following more general equation of continuity

$$\frac{\partial \rho}{\partial t} + \frac{1}{r^2} \frac{\partial}{\partial r} (r^2 \rho u) = 0. \qquad (16)$$

In accordance with (3) we introduce the velocity potential φ and consider the pulsations of the void and concomitant motion of the liquid such that the velocity $u(r)$ for $r \geq R$ is much smaller than the velocity of sound c_0 in the unperturbed liquid, i.e., $u/c_0 \ll 1$, where

$$c_0 = \left(\frac{\partial p}{\partial \rho} \right)^{1/2} \qquad (17)$$

In this approximation we have the following equation for a divergent spherical wave [9]

$$\left(\frac{\partial}{\partial t} + c_0 \frac{\partial}{\partial r}\right)(r\varphi) = 0. \tag{18}$$

Expressing u from (4) and (18), $u = -\partial\varphi/\partial r$, and inserting it into the equation of motion (1), we obtain the expression [14]

$$ru\left(\frac{\partial u}{\partial t}\right) + \frac{r}{\rho}\frac{\partial p}{\partial t} + c_0\frac{u^2}{2} + c_0 \int_{R_\infty}^{p(r)} \frac{\partial p}{\rho} + c_0 ru\frac{\partial u}{\partial r} + \frac{c_0 r}{\rho}\frac{\partial p}{\partial r} = 0. \tag{19}$$

The following relations apply to the pressure P(R) and velocity U on the surface of a void of radius R(t):

$$\frac{\partial P(R)}{dt} = \frac{\partial p}{\partial t} + U\frac{\partial p}{\partial r}, \tag{20}$$

$$\frac{dU}{dt} = \frac{\partial u}{\partial t} + U\frac{\partial u}{\partial r}. \tag{21}$$

The continuity equation (16) may be represented in the form

$$\frac{1}{\rho c_0^2}\frac{\partial p}{\partial t} + \frac{u}{\rho c_0^2}\frac{\partial p}{\partial r} + \frac{\partial u}{\partial r} + \frac{2u}{r} = 0, \tag{22}$$

because p and ρ are related by Eq. (17). On the basis of Eqs. (1) and (22), as well as relations (20) and (21), we can express the partial derivatives [14]. Substituting them into (19) with regard for r = R, we obtain the equation

$$R\left(1 - \frac{2U}{c_0}\right)\frac{d^2R}{dt^2} + \frac{3}{2}\left(1 - \frac{4}{3}\frac{U}{c_0}\right)\left(\frac{dR}{dt}\right)^2 -$$
$$- \int_{P_\infty}^{P(R)}\frac{dp}{\rho} - \frac{R}{\rho}\frac{U}{c_0}\left(1 - \frac{U}{c_0} - \frac{U^2}{c_0^2}\right)\frac{dP(R)}{dR} = 0, \tag{23}$$

where U = dR/dt. In the given approximation $u^2/c_0^2 \ll 1$, and $\rho = \rho_0$, because the variations of ρ are proportional to u^2/c_0^2. Now Eq. (23) reduces to the form

$$R\left(1 - \frac{2U}{c_0}\right)\frac{d^2R}{dt^2} + \frac{3}{2}\left(1 - \frac{4}{3}\frac{U}{c_0}\right)\left(\frac{dR}{dt}\right)^2 + \frac{1}{\rho_0}[P_\infty - P(R)] +$$
$$+ \frac{R}{\rho_0}\frac{U}{c_0}\left(1 - \frac{U}{c_0}\right)\frac{dP(R)}{dR} = 0. \tag{24}$$

In this form the equation for the pulsations of a cavitation void with regard for the compressibility of the liquid in the first approximation has been derived by Herring [15]. Substituting the values of P_∞ and $P(R)$ from (13) and (14) into Eq. (24), we obtain an equation describing the pulsations of a cavitation gas bubble in an ultrasonic field:

$$R\left(1 - 2\frac{U}{c_0}\right)\frac{d^2R}{dt^2} + \frac{3}{2}\left(1 - \frac{4}{3}\frac{U}{c_0}\right)\left(\frac{dR}{dt}\right)^2 + \frac{1}{\rho_0}\left[P_0 - P_v - P_m\sin\omega t + \right.$$
$$\left. + \frac{2\sigma}{R} - \left(P_0 + \frac{2\sigma}{R_0}\right)\left(\frac{R_0}{R}\right)^{3\gamma}\right] + \frac{R}{\rho_0}\frac{U}{c_0}\left(1 - \frac{U}{c_0}\right)\frac{dP(R)}{dR} = 0. \qquad (25)$$

This equation was first obtained by Flynn [12], who also augmented it with a term to account for the viscosity of the liquid in the case of bubble pulsations:

$$R\left(1 - 2\frac{U}{c_0}\right)\frac{d^2R}{dt^2} + \frac{3}{2}\left(1 - \frac{4}{3}\frac{U}{c_0}\right)\left(\frac{dR}{dt}\right)^2 + \frac{1}{\rho_0}\left[P_0 - P_v - P_m\sin\omega t + \frac{2\sigma}{R} + \right.$$
$$\left. + \frac{4\mu U}{R} - \left(P_0 + \frac{2\sigma}{R_0}\right)\left(\frac{R_0}{R}\right)^{3\gamma}\right] + \frac{R}{\rho_0}\frac{U}{c_0}\left(1 - \frac{U}{c_0}\right)\frac{dP(R)}{dR} = 0. \qquad (26)$$

In the ensuing discussion we call (26) the Herring - Flynn equation. By including the compressibility of the liquid (albeit in the first approximation), it provides a means for obtaining more accurate quantitative information on the rate of collapse and minimum radius of the cavitation bubble than is given by the Noltingk-Neppiras equation.

However, for values of U/c_0 close to or in excess of unity, the Herring - Flynn equation can only be used to obtain qualitative pictures of the process of collapse of the cavitation bubble.

Allowance for the compressibility of the liquid in the case of cavitation void pulsations with arbitrary velocities U is possible according to a method set forth by Kirkwood and Bethe [16]. They postulated that in the case of spherical waves of finite amplitude, the quantity $r\varphi$ propagates with a velocity

$$\tilde{c} = c + u, \qquad (27)$$

much as fixed values of the pressure and velocity in a plane Riemannian wave are propagated. Here c is the local value of the sound propagation velocity for arbitrary perturbations and is defined as

$$c = \left(\frac{\partial p}{\partial \rho}\right)_s^{1/2}. \tag{28}$$

Inasmuch as $(r\varphi)$ propagates with the velocity $\tilde{c}$, the quantity $G = r(\partial\varphi/\partial t)$ also propagates with the velocity $\tilde{c}$. From the hydrodynamical equations (1) and (16) we arrive at [16]

$$\frac{\partial\varphi}{\partial t} = h + \frac{u^2}{2}, \tag{29}$$

where h is the specific enthalpy. For isentropic flows, with which we shall be concerned below $h = \int_{P_\infty}^{p} \frac{dp}{\rho}$.

The function $\partial\varphi/\partial t = \Omega$ has been named the kinetic enthalpy. Therefore, for the function

$$G = r\Omega = r\left(h + \frac{u^2}{2}\right) \tag{30}$$

the following equation holds:

$$\left(\frac{\partial}{\partial r} + \tilde{c}\frac{\partial}{\partial r}\right)G = 0. \tag{31}$$

This implies that the function $G(r, t)$ is propagated with the velocity $\tilde{c}$. Consequently, if the value of $G(R, t_R)$ is known on the surface of a radiating sphere of radius R at time t_R, then its value at any point r of space is determined by the equation

$$G(r, t) = G(R, t_R), \tag{32}$$

where

$$t = t_R + \int_{R}^{r} \frac{dr}{\tilde{c}}. \tag{33}$$

Therefore, the analysis of finite-amplitude spherical waves radiated by a cavitation void reduces to the determination of the function $G(R, t_R)$ on the surface of a sphere and the calculation of the time according to Eq. (33).

Consider the pulsation of a cavitation void of radius R. Substituting Eq. (30) into (31), we obtain

$$\tilde{c}\left(h + \frac{u^2}{2}\right) + \frac{r}{\rho}\left(\frac{\partial p}{\partial t} + \tilde{c}\,\frac{\partial p}{\partial r}\right) + ru\left(\frac{\partial u}{\partial t} + \tilde{c}\,\frac{\partial u}{\partial r}\right) = 0. \qquad (34)$$

On the surface of the void of radius R, relations (20) and (21) are still applicable for the pressure P(R) and velocity U in the general case. If we represent the continuity equation in the form

$$\frac{1}{\rho c^2}\frac{\partial p}{\partial t} + \frac{u^2}{\rho c^2}\frac{\partial p}{\partial r} + \frac{\partial u}{\partial r} + \frac{2u}{r} = 0, \qquad (35)$$

then (1), (20), (21), and (35) can be used to express the partial derivatives [16]. Substituting the latter into (34) and putting $r = R$, we obtain

$$R\left(1 - \frac{U}{c}\right)\frac{d^2R}{dt^2} + \frac{3}{2}\left(1 - \frac{U}{3c}\right)\left(\frac{dR}{dt}\right)^2 - \left(1 + \frac{U}{c}\right)H - \frac{U}{c}\left(1 - \frac{U}{c}\right)R\frac{dH}{dR} = 0, \qquad (36)$$

where

$$H = \int_{P_\infty}^{P(R)} \frac{dp}{\rho} \qquad (37)$$

is the free enthalpy on the surface of the sphere.

Inserting into (37) the value of P_∞ and $P(R)$ from (13) and (14) and expressing p from the equation of state for water in the form

$$p = A\left(\frac{\rho}{\rho_0}\right)^n - B, \qquad (38)$$

where A, B, and n are constants (it may be assumed for water that A = 3001 atm, B = 3000 atm, and n = 7), for the free enthalpy on the surface of the sphere we obtain

$$H = \frac{n}{n-1}\frac{A^{1/n}}{\rho_0}\left\{\left[\left(P_0 + \frac{2\sigma}{R_0}\right)\left(\frac{R_0}{R}\right)^{3\gamma} - \frac{2\sigma}{R} + B\right]^{\frac{n-1}{n}} - [P_0 - P_m \sin \omega t + B]^{\frac{n-1}{n}}\right\}. \qquad (39)$$

Equation (36) has been investigated by Cole [16] and Gilmore [17] in application to underwater explosions. When H is expressed by relation (39), Eq. (36) describes the pulsations of a cavitation gas bubble in an ultrasonic field with regard for the compressibility of the liquid. Henceforth, we refer to this as the Kirkwood–Bethe equation.

In Eq. (36) c is the local velocity of sound in the liquid. Substituting (38) into (28) and (37), we obtain

$$c^2 = c_0^2 \left(\frac{\rho}{\rho_0} \right)^{n-1} \tag{40}$$

and

$$H = \int_{\rho_1}^{\rho} \frac{dp}{\rho} = \frac{c_0^2}{n-1} \left[\left(\frac{\rho}{\rho_0} \right)^{n-1} - 1 \right], \tag{41}$$

respectively, where $c_0^2 = An / \rho_0$. From (40) and (41) we obtain

$$c = [c_0^2 + (n-1) H]^{1/2}, \tag{42}$$

i.e., the velocity c is expressed as a function of H. In order to find the solutions of the Kirkwood−Bethe equation, it is required to investigate Eqs. (36), (39), and (42), simultaneously.

Chapter 2

Investigation of the Pulsations of Cavitation Voids

§ 1. Similarity of the Solutions of the Equations for Different Frequencies of the Ultrasonic Field

The general analysis of the equations describing the pulsation of cavitation voids is rendered difficult by the fact that the equations are not solvable in general form; only their numerical solutions can be obtained for concrete special cases characterized by definite values of the ultrasonic frequency and amplitude, as well as by the initial radius of the bubble. Therefore, it is opportune to transform the equations as we have done below, so as to generalize the numerical solutions for different ultrasonic frequencies.

In place of R and t we introduce the new variables a and τ, defined as

$$a = \frac{\omega}{\omega_0} \frac{R}{R_0},\tag{43}$$

$$\tau = \omega t.\tag{44}$$

Here ω_0 is determined according to the formula

$$\omega_0 = \frac{1}{R_0} \left(\frac{P_0 + 2\,\sigma/R_0}{\rho_0}\right)^{1/2},\tag{45}$$

correct to a certain constant equal to $(3\gamma)^{\frac{1}{2}}$, and corresponds to the linear resonance frequency of an equilibrium bubble of radius R_0 [18].

In the new variables (a, τ), the Noltingk–Neppiras equation is reduced at once to the form

$$a\,\frac{d^2a}{d\tau^2} + \frac{3}{2}\left(\frac{da}{d\tau}\right)^2 - \frac{1}{\rho_0\omega_0^2R_0^2}\left[\frac{2\sigma}{R_0}\left(1 - \frac{\omega}{\omega_0 a}\right) + P_{\mathrm{v}} + P_m\sin\tau\right] +$$
$$+ \left(1 - \frac{\omega^{3\gamma}}{\omega_0^{3\gamma}a^{3\gamma}}\right) = 0. \tag{46}$$

We notice that the solution $a(\tau)$ of the above equation for fixed R_0 does not depend explicitly on the ultrasonic frequency if the following condition holds

$$\frac{\omega}{\omega_0 a} \ll 1, \tag{47}$$

which is identical to the condition

$$\frac{R}{R_0} \gg 1. \tag{48}$$

According to [11] and [18], the following estimates are valid for R and R_0:

$$R \sim \frac{1}{\omega}\left(\frac{P_m - P_0}{\rho_0}\right)^{1/2}, \tag{49}$$

$$R_0 \sim \frac{1}{\omega_0}\left(\frac{P_0 + 2\sigma/R_0}{\rho_0}\right)^{1/2}. \tag{50}$$

Consequently, condition (48) transforms to the condition

$$\frac{\omega_0}{\omega}\left(\frac{P_m - P_0}{P_0 + 2\sigma/R_0}\right)^{1/2} \gg 1, \tag{51}$$

which is coarser insofar as the estimate (49) is approximative. Condition (51) is fulfilled in the typical cases

$$\frac{\omega_0}{\omega} > 1; \quad P_m \gg 2P_0 + \frac{2\sigma}{R_0} \tag{52}$$

and

$$\frac{\omega_0}{\omega} \gg 1; \quad P_m \geqslant 2 P_0 + \frac{2\sigma}{R_0}. \tag{53}$$

It is necessary, therefore, that the linear resonance frequency ω_0 of the bubble not be smaller than the ultrasonic frequency and that the pressure amplitude of the ultrasonic field exceeds a certain threshold value determined by the hydrostatic pressure and initial radius of the bubble. When condition (47) is satisfied, the Noltingk–Neppiras equation is greatly simplified, and, according to (46), in the variables (a, τ) assumes the form

$$a \frac{d^2a}{d\tau^2} + \frac{3}{2}\left(\frac{da}{d\tau}\right)^2 - \frac{1}{\rho_0 \omega_0^2 R_0^2}\left[\frac{2\sigma}{R_0} + P_v + P_m \sin \tau\right] + 1 = 0. \tag{54}$$

The absence in this equation of terms bearing an explicit dependence on ω obviates the need to solve it for different ultrasonic frequencies. A solution obtained at some frequency ω_1 models the solution at another frequency ω_2, given the same initial radius R_0 of the bubble. The new variables a and τ in a certain sense represent similarity coefficients.

As an illustration the functions $a(\tau)$, obtained on the basis of numerical solution of Eq. (15) under the initial conditions $R = R_0$ and $U = 0$,* are shown in Fig. 1. The object investigated was a bubble of initial radius $R_0 = 10^{-5}$ cm pulsating adiabatically $(3\gamma = 4)$ in water $(p = 1$ g/cm^3, $\sigma = 75$ dyn/cm) at a pressure $P_0 = 1.0$ atm. The numeral attached to each curve indicates the corresponding pressure amplitude P_m of the ultrasonic field (in atm). The solid curves represent the dependence $a(\tau)$ at frequency $f = 10$ kc, and the dashed curves represent the same at $f = 500$ kc. The condition $\omega_0/\omega \gg 1$ is observed in both cases. The lower part of the figure shows the time variation of the ultrasonic pressure $p_a = -P_m \sin \tau$. The graphs of the solutions reveal that, despite the considerable disparity in the frequencies ω_1 and ω_2, their corre-

---

*All numerical solutions given in the paper were obtained on the Minsk-2 digital computer of the Acoustics Institute of the Academy of Sciences of the USSR. The equations were solved by the Runge–Kutta method with a computational error of 0.01 on a guaranteed accuracy interval $\tau = 6\pi$.

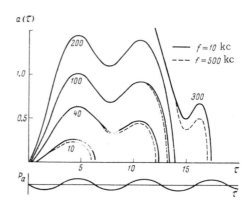

Fig. 1. Solutions of the Noltingk−Neppiras equation.

sponding solutions (for identical values of P_m) in the coordinate system (a, τ) differ insignificantly, not only qualitatively, but quantitatively as well. The difference becomes smaller as the amplitude P_m of the ultrasonic field is increased in correspondence with condition (51).

The functions $a(\tau)$ obtained on the basis of numerical solutions of Eq. (15) for $R_0 = 10^{-4}$ cm are shown in Fig. 2a. It is readily seen that the change in R_0 has not affected the similarity of the solutions of the Noltingk−Neppiras equation for different ultrasonic frequencies. Comparing the numerical solutions in the two figures, we note that with a reduction in the initial bubble radius R_0, all other conditions being equal, the variable $a(\tau)$ also decreases. This is because, according to (43) and (45), a has the following dependence on R_0:

$$a = \omega R \left(\frac{p_0}{P_0 + 2\sigma/R_0} \right)^{1/2}. \tag{55}$$

Inasmuch as R depends very little on the initial radius R_0 [11, 12], when the cavitation pressure threshold is exceeded, particularly when the second inequalities of conditions (52) and (53) hold, the reduction of $a(\tau)$ with R is legitimate.

The above notions regarding the similarity of the solutions in the coordinate system (a, τ) for different frequencies ω of the ultrasonic field prove correct in application to the Herring−Flynn

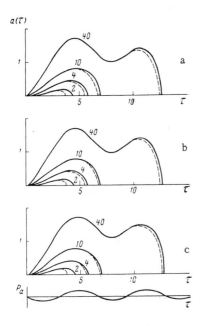

Fig. 2. Solutions of the Noltingk—
Neppiras, Herring—Flynn, and Kirkwood—Bethe equations.

equation. As demonstrated by the numerical solutions of that
equation, for the case of pulsations of a gas bubble in water, the
influence of the viscosity term is small and may be neglected.
If R and t are replaced in Eq. (25) by the variables a and τ in accordance with (43) and (44), then under condition (47) we obtain

$$(1 - 2 M_X) a \frac{d^2a}{d\tau^2} + \frac{3}{2}\left(1 - \frac{4}{3} M_X\right)\left(\frac{da}{d\tau}\right)^2 -$$

$$- \frac{1}{\rho_0 \omega_0^2 R_0^2}\left[\frac{2\sigma}{R_0} + P_v + P_m \sin\tau\right] + 1 = 0, \qquad (56)$$

where $M_X = (\omega_0 R_0 / c_0)(da/d\tau)$. Since this equation also lacks an
explicit dependence on the frequency ω, the above-postulated similarity of the numerical solutions for different ultrasonic frequencies and the possibility of their modeling in the coordinate
system (a, τ) is also valid for the analysis of the Herring—Flynn
equation. This is illustrated by Fig. 2b, which shows the dependences $a(\tau)$ obtained on the basis of numerical solutions of Eq.
(25) for $R_0 = 10^{-4}$ cm. It is quickly seen that, despite the considerable disparity in the ultrasonic frequencies, the numerical
solutions of the Herring—Flynn equation are, with acceptable er-

ror, similar when represented in the coordinate system (a, τ).
The similarity of the solutions for different ultrasonic frequencies
in the coordinate system (a, τ) also extends to the Kirkwood–
Bethe equation. If the new variables a and τ are also introduced
into Eq. (36) in compliance with (43) and (44), then under condition (47) we obtain an equation of the type

$$(1 - M_K) a \frac{d^2 a}{d\tau^2} + \frac{3}{2} \left(1 - \frac{1}{3} M_K\right) \left(\frac{da}{d\tau}\right)^2 - (1 + M_K) \frac{H}{\omega_0^2 R_0^2} = 0, \quad (57)$$

where $M_K = (\omega_0 R_0 / c)(da / d\tau)$; H is determined from (39), (43), and
(45) under condition (47) as follows:

$$H = \frac{n}{n - 1} \frac{A^{1/n}}{\rho_0} \left[B^{\frac{n-1}{n}} - (P_0 - P_m \sin \tau + B)^{\frac{n-1}{n}} \right]. \quad (58)$$

Inasmuch as Eqs. (57) and (58) also lack an explicit dependence
on the frequency ω, the solutions in the coordinate system (a, τ)
must also be similar in the investigation of the Kirkwood–Bethe
equation. This is illustrated in Fig. 2c, which shows the numerical
solutions of Eq. (36) for $R_0 = 10^{-4}$ cm and various amplitudes P_m
of the ultrasonic field.

It ensues from Fig. 2 that the solution of all three equations
for the pulsations of a cavitation bubble are similar in the coordinate system (a, τ) with acceptable accuracy for practical calculations, despite a considerable disparity in the ultrasonic frequencies.

The results illustrated in Fig. 2 indicate that in the coordinate system (a, τ), which correct to constant terms is similar
to the coordinate system (R, t), the solutions of the Herring–Flynn
and Kirkwood–Bethe equations do not differ from the solutions of
the Noltingk–Neppiras equation. This is entirely legitimate, because they should only exhibit a disparity when the velocity U becomes very large in comparison with c_0. Since this is always
manifest in the final stage of collapse, when $R < R_0$, it is impossible to analyze completely the disparity between the solutions of
the equations in the coordinate system (R, t). The disparity is
exhibited by comparison of the rates of collapse U as determined
from the equations; this topic will be discussed below.

The foregoing generalization of the similarity of the numerical solutions of the Noltingk–Neppiras, Herring–Flynn, and Kirkwood–Bethe equations in the coordinate system (a, τ) for different frequencies ω affords the possibility of obtaining not only qualitative, but also quantitative, estimates of the behavior of cavitation bubbles in an ultrasonic field at different frequencies satisfying condition (47) if the solutions at some particular frequency are known.

§2. Structure of the Solutions; Structural Stability

The unavailability of solutions in the general form of the Noltingk–Neppiras, Herring–Flynn, and Kirkwood–Bethe equations prohibits the comprehensive analysis of how the solutions are influenced by such parameters as the ampltude of the ultrasonic field and the radius of the cavitation nuclei. However, it is possible on the basis of the set of numerical solutions, obtained with variation of the fundamental parameters over sufficiently broad limits, to infer some important conclusions regarding the general structure of those solutions.

The results of the preceding section make it possible under certain conditions to relate a numerical solution obtained at one particular frequency with the solution obtained at other ultrasonic frequencies. Our prime concern, therefore, is the investigation of how the solutions of the Noltingk–Neppiras, Herring–Flynn, and Kirkwood–Bethe equations vary with the pressure amplitude of the ultrasonic field.

Consider the numerical solutions of the Noltingk–Neppiras equation (15), which describe the adiabatic pulsations of a gas bubble in water at a hydrostatic pressure $P_0 = 1.0$ atm under typical initial conditions $R = R_0$ and $U = 0$. The bubble is excited by an ultrasonic wave of frequency $f = 500$ kc and pressure amplitude P_m. The dependences of the relative bubble radius R/R_0 on the dimensionless time ωt are shown in Fig. 3 as obtained for bubbles having an initial equilibrium radius $R_0 = 10^{-4}$ cm. The parameter of the illustrated family of curves is the pressure amplitude P_m (in atm). The lower part of the figure shows the time variation of the pressure P_a of the ultrasonic pressure. The analogous dependences for larger equilibrium bubbles, $R_0 = 5 \cdot 10^{-4}$ cm and $R_0 = 10^{-3}$ cm, are presented in Figs. 4 and 5, respectively.

These three cases of the equilibrium radii enable us to enunciate some general notions concerning the structure of the solutions, considering that they represent three distinct typical cases. The natural resonance frequency ω_0 (according to Minnaert [18]) of an equilibrium gas bubble for linear pulsations in the case $R_0 = 10^{-4}$ cm is greater than the ultrasonic frequency ω, hence, we say that the bubble of radius R_0 is "smaller than resonance." In the case $R_0 = 5 \cdot 10^{-4}$ cm at a frequency of 500 kc, the pulsating bubble is a "resonance bubble," because $\omega_0 \simeq \omega$. Finally, for $R_0 = 10^{-3}$ cm we find the third typical case, when the bubble is "larger than resonance" $(\omega_0 < \omega)$.

It follows from our analysis of the curves in Figs. 3-5 that at small ultrasonic field amplitudes, when $P_m < P_{mk}$ (where P_{mk} is the threshold pressure corresponding to the onset of cavitation), the gas bubbles pulsate nonlinearly without collapsing. Whereas bubbles whose radii are smaller than resonance (see Fig. 3) or equal to it (see Fig. 4) pulsate with a period close to the ultrasonic period T, bubbles whose radii are larger than resonance (see Fig. 5) pulsate with a larger period T_0 roughly equal to the natural (linear) period of the bubble.

We shall consider the analysis of the proper pulsations of a cavitation bubble in further detail later on, resting at this point

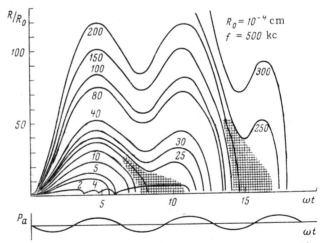

Fig. 3. Pulsations of a cavitation bubble at 500 kc for $R_0 = 10^{-4}$ cm.

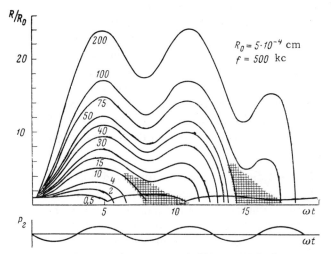

Fig. 4. Pulsations of a cavitation bubble at 500 kc for $R_0 = 5 \cdot 10^{-4}$ cm.

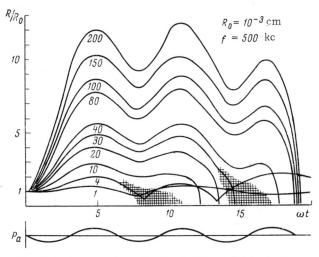

Fig. 5. Pulsations of a cavitation bubble at 500 kc for $R_0 = 10^{-3}$ cm.

with a discussion of the cavitation pulsations produced at ultrasonic field amplitudes greater than the threshold pressure P_{mk}.
In this case for small equilibrium gas bubbles, for which $\omega_0 \geq \omega$,
with increasing pressure P_m there is initiated and evolved cavitation which is characterized (see Figs. 3 and 4) by growth of the
bubble during the entire negative-pressure half-period of the ultrasonic field, as well as during a certain portion of the positive half-
period. The bubble grows to a certain maximum radius R_m, then
collapses. The time of growth to this radius is designated t_m,
and the collapse time is τ_m (the notation convention is illustrated
in Fig. 6).

Characteristically, as P_m is increased both R_m and the collapse time τ_m increase; in this case the collapse of the bubble is
initiated in the negative-pressure phase. Finally, beginning with
a certain value of P_m (see the dependence of R/R_0 on ωt for $P_m =$
25 atm in Fig. 3 and for $P_m = 15$ atm in Fig. 4), the bubble, while
growing to a certain maximum value and beginning to collapse,
does not collapse completely, but starts to grow again under the
influence of the pressure in the expansion phase to a certain maximum value R_{m2}, and only then does it finally collapse completely.

With a still greater increase in P_m, the formation of a third
and any i-th extremum with maximum value R_{mi} becomes possible, after which the gas bubble experiences cavitation collapse.
All of these situations are illustrated schematically in Fig. 6, along
with the notation to be used later, viz., the growth time t_m of the
cavitation bubble to the last extremal value R_{mi} and the collapse
time τ_m from that radius.

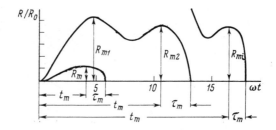

Fig. 6. Illustration of the notation convention.

For the pulsations of cavitation bubbles of larger initial radii (see Fig. 5), even for small amplitudes P_m corresponding to the onset of cavitation, several extrema are formed right away, their minimum number being determined by the stipulation that the lifetime of the bubble prior to its collapse must be at least equal to the natural period of the resonance pulsations. This is what differentiates the pulsations of larger-than-resonance cavitation bubbles from the pulsations of small subresonance and resonance bubbles. Therefore, one of the basic conditions for the similarity of the numerical solutions at different frequencies ω of the ultrasonic field for a given radius of the nucleus is the condition $\omega_0/\omega > 1$, as already mentioned.

It is apparent in Figs. 3-5 that, before the formation of each new extremum, the form of the functions $R/R_0(\omega t)$ exhibits the same qualitative character and only a slight quantitative variation under small variations of the pressure amplitude P_m, which is the parameter of the numerical solutions. The appearance of each new extremum occurs with a relatively slight change in a certain characteristic value P_m^* of that parameter. Here there occurs a substantial qualitative change in the form of the dependences of R/R_0 on ωt, which implies that for certain values of the coefficient $P_m = P_m^*$ the Noltingk–Neppiras equation is structurally unstable [19]. The domains of Figs. 3-5, characterizing structural instability of Eq. (15) in the plane of the solutions $(R/R_0, \omega t)$, are crosshatched. It is in these domains that the qualitative structure of the solutions is significantly altered under a slight variation of certain values of P_m.

We now investigate the domain of values of P_m for which the Noltingk–Neppiras equation is structurally unstable. The dependence of t_m/T (ratio of the growth time of a cavitation void to the ultrasonic wave period) on P_m for a bubble of radius $R_0 = 10^{-4}$ cm is shown in Fig. 7a as obtained on the basis of the numerical solutions illustrated in Fig. 3. Below that, in Fig. 7b, the solid curve indicates the dependence of the relative collapse time τ_m/T on P_m (the dependence of τ_m/T represented by the dashed curve will be discussed below). As the figure demonstrates, with an increase in P_m the value of t_m/T slowly increases, tending to the limiting value 0.75. This evinces the fact that in cavitation the expansion time of the cavity approaches $0.75T$ and changes little

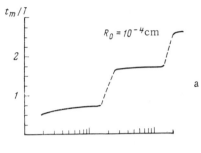

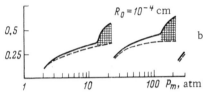

Fig. 7. Growth time and collapse time of a
cavitation void.

as the amplitude P_m is increased. As soon as t_m / T is equal to
0.75, for a certain value $P_m = P_m^*$ (in Fig. 7 the value of $P_m^* \simeq$
15-20 atm), a further slight increase in P_m causes the function
t_m / T to jump sharply, including the formation of a second extremum in the dependence $R/R_0(\omega t)$. A continued increase in P_m
brings about at a certain value a subsequent jump in t_m / T, etc.
Consequently, Eq. (15) is structurally unstable for values of P_m
such that the following relation is true for the growth time of the
cavitation void:

$$t_m = 0.75 \, T + (i - 1) \, T. \tag{59}$$

For values of P_m such that structural instability is manifested,
there is also a definite relation for the collapse time. It is evident from Fig. 7b that, in the domain of structurally unstable
solutions, the ratio t_m / T increases monotone with P_m; for the
same values of P_m at which Eq. (15) becomes structurally unstable a jump in t_m / T is observed, because the cavity is not able
to collapse. A new extremum is formed, and the collapse time is
greatly reduced, qualitatively mimicking the preceding dependence of τ_m/T on P_m. It is readily seen that, for the values of
P_m corresponding to structural instability, the following relation
is applicable for the collapse time

$$\tau_m \simeq 0.5 \, T. \tag{60}$$

The foregoing result is interesting in that Noltingk and Neppiras assumed, a priori, [11] that, with an increase in the collapse time of a cavitation void, the time could arrive when the parameter τ_m/T would be equal to 0.5, whereupon a reduction would have to be observed in the rate of collapse and intensity of the shock waves. Similar concepts were also invoked by Sirotyuk to interpret the extremum that he observed in the dependence of cavitation erosion on the electrical voltage on an ultrasonic concentrator [20].

The structural instability conditions (59) and (60) also apply to the Herring—Flynn and Kirkwood—Bethe equations, because, as shown in Fig. 2, the corresponding solutions R(t) of Eqs. (15), (25), and (36) are similar for $R \geq R_0$. A disparity shows up at the maximum rates of collapse, but does not affect the structure of the solutions.

Thus, for certain values of the pressure amplitude P_m of the ultrasonic field, the Noltingk—Neppiras, Herring—Flynn, and Kirkwood—Bethe equations are structurally unstable. This means that the cavitation bubbles described by the equations represent, according to the definition of Andronov [21], "delicate" systems. It is important to note (see Fig. 7) that "delicacy" is exhibited in a relatively small interval of P_m, separating considerably larger intervals in which the solutions are structurally unstable (i.e., in which the cavitation bubble comprises a "gross" system).

§3. Collapse of Cavitation Voids

The dashed curve in Fig. 7b represents the dependence of t_m/T on P_m on the basis of our proposed equation for the collapse time

$$\tau_m = 0.915 \, R_{mi} \left(\frac{\rho}{P_0 + P_m} \right)^{1/2}. \tag{61}$$

This equation differs from the classical Rayleigh equation (11) for the collapse of a vapor cavitation void of maximum radius R_m in a pressure field equal to P_0 at infinity in that the pressure P_0 in the denominator of the expression under the square root sign is replaced by $P_0 + P_m$. As shown in Fig. 7b, the value of τ_m, according to this equation, is reasonably consistent in the domain of structurally stable values of P_m, with the exact values obtained in the numerical solutions of the Noltingk—Neppiras equation.

This result is legitimate if we analyze the pulsations of cavitation bubbles as shown in Figs. 3-5. It is readily observed that for structurally stable values of P_m the cavitation bubble expands to a maximum value R_m, then begins to collapse; this happens at the instant the pressure in the liquid becomes equal to $P_0 + P_m$. The latter is in fact the pressure that determines the potential energy of the expanding bubble, and, according to Rayleigh [10], this energy not only governs the time, but also the rate of collapse. It is reasonable to assume, therefore, that the rate of collapse U of the cavitation void in an ultrasonic field is determined in the domain of structurally stable solutions by the asymptotic expression

$$U^2 = \frac{2}{3} \frac{P_0 + P_m}{\rho_0} \left(\frac{R_m^3}{R^3} - 1 \right), \tag{62}$$

which differs from the corresponding expression of Rayleigh (10) for the collapse of a vapor cavity under the influence of a pressure P_0 at infinity in the analogous replacement of P_0 by $P_0 + P_m$, so as to account for the action of the ultrasonic wave of pressure amplitude P_m.

The propriety of the expression (62) proposed above for estimating the rate of collapse of a cavitation void of maximum radius R_m in an ultrasonic field is confirmed in Fig. 8 by some specific sample calculations. Shown here is the dependence of the absolute value of M (i.e., the ratio $|U/c_0|$) for the collapse of a

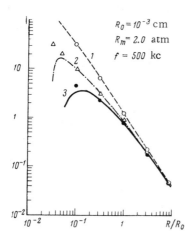

$R_0 = 10^{-3}$ cm
$R_m = 2.0$ atm
$f = 500$ kc

Fig. 8. Rate of collapse of a cavitation void. The ratio $|U/c_0|$ is plotted on the vertical axis.

cavitation bubble, i.e., with decreasing R/R_0. The dependence was obtained on the basis of the numerical solutions of Eqs. (15), (25), and (36) at frequency f = 10 kc for an equilibrium bubble of radius $R_0 = 10^{-3}$ cm and a pressure amplitude P_m = 2.0 atm. Curve 1 corresponds to the dependence of $|U/c_0|$ on R/R_0, according to the numerical solution of the Noltingk—Neppiras equation (15); curve 2 corresponds to the same for the Herring—Flynn equation (25); and curve 3 corresponds to the same for the Kirkwood—Bethe equation (36). Also shown in Fig. 8 are the calculated rates of collapse according to the various equations. The light circles indicate the calculations according to Eq. (62). It is seen at once that the results of the calculations exhibit good agreement with the exact velocities U determined by numerical solutions of the Noltingk—Neppiras equation.

In order to determine the rate of collapse of a void with regard for the compressibility of the liquid in the first approximation it is necessary to use the equation

$$U^2 = \frac{2}{3}\frac{P_0 + P_m}{\rho_0}\left(\frac{R_m^3}{R^3} - 1\right)\left(1 - \frac{4}{3}\frac{U}{c_0}\right)^{-1}, \tag{63}$$

where the additional factor is equal to the reciprocal of the factor associated with the second inertial term of the Herring—Flynn equation (25). The possibility of accounting in this fashion for the compressibility was noted by Flynn [12], who deduced an expression for the rate of collapse of a cavitation void for a pressure P_0 at infinity. The replacement of P_0 by $P_0 + P_m$ makes it possible to account for the influence of the sound field. The triangles in Fig. 8 indicate the results of calculations according to Eq. (63). Their concurrence with the numerically derived curve shows that Eq. (63) determines with sufficient accuracy the rate of collapse of a cavitation void with regard for the compressibility of the liquid in the first approximation.

With the compressibility of the liquid fully accounted for, the rate of collapse of the void can be determined from the equation

$$U^2 = \frac{2}{3}\frac{P_0 + P_m}{\rho_0}\left[\frac{R_m^3}{R^3}\left(1 - \frac{U}{3c_0}\right)^{-4} - 1\right], \tag{64}$$

which differs from the equation derived by Gilmore [17] by the appropriate substitution of $P_0 + P_m$ for P_0. The calculations based on Eq. (64) are represented by dots in Fig. 8. It is seen at once that the results of these calculations exhibit sufficient agreement with the numerical solutions of the Kirkwood − Bethe equation.

Unfortunately, Eqs. (62)-(64), insofar as they correspond to the rate of collapse of an empty cavity, do not permit one to determine the minimum radius R_{min} for which collapse is halted by virtue of compression of the gas in the cavity, and the velocity U becomes equal to zero. Close to R_{min} the velocities determined according to these relations can differ appreciably from the values predicted by the solutions of the corresponding equations.

As evident from Figs. 3-5, for amplitudes P_m close to the threshold values, the collapse of the bubble is initiated at a time when the pressure of the ultrasonic field has not yet reached its maximum value equal to the amplitude P_m. Therefore, in general, the effective pressure governing the collapse of the cavitation bubble will be the quantity $P_0 + \alpha P_m$ (the value of α falls between the limits $0 \le \alpha \le 1$).

§ 4. Analysis of the Solutions of the

Equations in the Phase Plane

As we know, in the qualitative analysis of nonlinear differential equations a very productive tool is the phase portrait method, which was developed in application to the problem of nonlinear oscillations by Andronov [21, 22]. As the Noltingk − Neppiras, Herring − Flynn, and Kirkwood − Bethe equations are not autonomous, it does not appear feasible to obtain their solutions with the help of the phase plane. However, considerable revelation is afforded by the representation of the numerical solutions in the phase plane, wherein it is possible to formulate certain inferences with regard to the structure of the numerical solutions and the kinematics of the cavitation bubbles described by those solutions.

Phase-plane representations of the numerical solutions of the Noltingk − Neppiras equations (15) for various pressure amplitudes are shown in Fig. 9 (the numerals attached to each curve represent the pressure amplitude in atm) at a frequency $f = 500$ kc for $R_0 = 5 \cdot 10^{-4}$ cm. The relative values of the bubble radius R/R_0 are plotted on the horizontal axis, and the velocity $d(R/R_0)/d\tau$, where

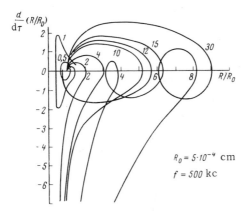

Fig. 9. Solutions of the Noltingk–Neppiras
equation, represented in the phase plane.

$\tau = \omega t$, is plotted on the vertical. The phase portraits are drawn for certain solutions shown in the coordinate system $(R/R_0, \omega t)$ in Fig. 4. As apparent from the phase portrait, for small values of the amplitude P_m the image point, on emerging from the coordinates corresponding to the initial conditions, continues to move along a closed nearly circular curve. Consequently, in this case the solution represents nearly sinusoidal oscillations [21].

With an increase in the pressure amplitude the form of the phase trajectory changes, and for values of P_m corresponding to onset of cavitation (at $P_m = 2$ atm in Fig. 9) the image point, on emerging from the initial point and moving clockwise, tends towards a point with coordinates $(0, -\infty)$, indicating the cavitation collapse of the gas bubble in accordance with the Noltingk–Neppiras equation. If the pressure is increased to 10 atm, the general form of the solutions remains the same in the phase portrait as for $P_m = 2$ atm. However, at a certain value of P_m (at $P_m = 12$ atm in the given instance), the image point does not tend toward a point at $(0, -\infty)$, but moves along a certain helical path around the initial point, indicating the absence of cavitation collapse for this value. The phase portrait of the solution in this case varies considerably with a slight change in P_m, indicating structural instability of the equation at the given pressure amplitude P_m.

With a further increase in P_m the equation is structurally stable; in the phase portraits the image point (see the curves for

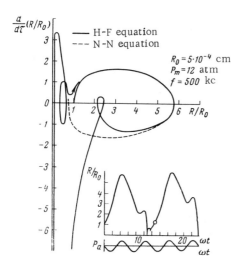

Fig. 10. Solution of the Herring—Flynn equation, represented in the phase plane.

P_m equal to 15 and 30 atm) tends ultimately toward $(0, -\infty)$, indicating cavitation collapse, but in the phase trajectories for the given values of P_m, the occurrence of a loop is readily perceived, reflecting the fact that for some smaller value of P_m structural instability is observed. It is seen at once that, with a further increase in P_m, the next domain of structural instability will leave a trail in the phase portraits of the stable solutions in the form of a second loop, etc.

The phase portraits reveal that for values of P_m for which structural instability is observed the gas bubble does not collapse, but pulsates nonlinearly with large amplitudes but small velocities.

It is instructive to investigate the phase portraits of the numerical solutions of the Herring—Flynn equation. The numerical solutions of Eq. (25) for $R_0 = 5 \cdot 10^{-4}$ cm at a frequency $f = 500$ kc and an ultrasonic pressure amplitude of 12 atm are presented in Fig. 10 as solid curves in the phase plane. As in the preceding figure, the coordinates of the phase plane are $d(R/R_0)/d\tau$ and R/R_0. The lower right-hand corner of the figure illustrates the corresponding numerical solutions in the coordinate system $(R/R_0, \omega t)$. The dot indicates the coordinates of the curve where the plot of the phase portraits in the phase plane end. The dashed curve in the phase plane indicates the corresponding solution of the Noltingk—Neppiras equation. The main conclusion drawn from the analysis of the phase portrait of Fig. 10 is that the image point in the representation of the solutions of the Herring–Flynn equation describes a very complex, but in the final analysis, closed trajectory, which determines in the phase plane a certain cycle characteristic of the given value P_m of the pressure amplitude of the ultrasonic field. In the investigation of the solutions in the succeeding time intervals new closed cycles are noticeable, differing very little from the first cycle, which constitutes a kind of "limit cycle" [21, 23]. As evident from Fig. 10, the phase trajectory can have complex and capricious configurations, but after a certain characteristic time period it rather faithfully mimics the "limit cycle." This attests to the fact that the pulsations of cavitation bubbles, according to the solutions of the Herring–Flynn equation, are almost periodic functions of the time. An analogous conclusion may be drawn with regard to the solutions of the Kirkwood–Bethe equation. The phase portraits of those solutions are qualitatively similar to the portraits of the corresponding Herring–Flynn solutions, and they also display "limit cycles," which are closely followed by the trajectories of the image point as it traces out a closed trajectory during a certain characteristic period that is a multiple of the ultrasonic field period. As shown in Chap. 3, this affords a means for explaining certain characteristic cavitation noise effects in ultrasonic cavitation.

§ 5. Experimental Investigation of the Pulsations of Cavitation Voids

The nonlinear differential equations describing the pulsations of cavitation bubbles in an ultrasonic field rest on several simplifying assumptions with regard to the motion of the bubble, the properties of the liquid, as well as the properties of the vapor and gas inside the bubble. It is entirely possible that some of these assumptions are too hypothetical, resulting in a substantial discrepancy between the pulsations of real cavitation bubbles and the solutions of the nonlinear differential equations describing those pulsations. It is instructive, therefore, to compare the numerical solutions of the above equations with the experimentally observed pulsations of actual cavitation bubbles.

The pulsations of cavitation bubbles have been investigated experimentally by means of a cylindrical focusing concentrator,

in whose focal zone cavitation was induced at a frequency of 15 kc.
The cavitation process was recorded with an SSKS-3 high-speed
photorecording apparatus, which is capable of producing frame-by-
frame pictures of fast processes at a speed of up to 300,000 frames
per second. Like other similar cameras, the SSKS-3 operates in
the driven-sweep mode, i.e., the image is formed on the nonmoving film, which is arranged in a circle with a rotating system
of mirrors at the center. Unlike the other cameras, however, this
apparatus enables one to obtain more detailed information regarding the evolution of the process with time, as the pictures are
recorded on 800 consecutive frames of the film.

The observed cavitation bubbles were illuminated with two
IFK-120 pulsed lamps hooked up in parallel. In order to enhance
the brightness, a condenser was used to concentrate the light from
the lamps in the focal zone of the cylindrical concentrator. The
cavitation bubbles were stabilized in the plane of maximum acuity
of the image by the insertion of a copper wire about 1.5 mm in
diameter in the focal zone of the concentrator; at the end of the
wire, the cavitation bubbles were less subject to the influence of
cavitation microstreaming than the free bubbles far from any types
of surface.

Photographs of developed (stationary) cavitation in standing
tap water at a frequency of 15 kc are shown in Fig. 11. The speed
of the film was 200,000 frames per second, so that the period of
the ultrasonic field spanned about twelve consecutive frames. The
tip of the wire near which the cavitation bubbles developed is
clearly visible in the photographs (which have a 20-fold magnification). A noticeable fact is the simultaneous occurrence in the ultrasonic field of a set of cavitation bubbles spaced at distances on
the order of their maximum radius. All the bubbles grow and collapse in phase, their dimensions at each instant remaining about
the same. According to the results obtained in the analysis of the
differential equations describing the pulsations of cavitation bubbles, this requires the fulfillment of two conditions. First, the
dimensions of the cavitation bubbles can be identical if the pressures of the ultrasonic field are identical. This was true in the
case investigated as the ultrasonic wavelength was approximately
100 times as large as the dimensions of the cavitation zone; hence,
the spatial pressure gradient was negligibly small. This condition

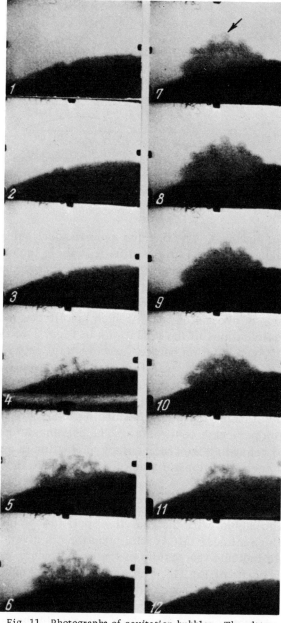

Fig. 11. Photographs of cavitation bubbles. The ultra-sonic frequency is 15 kc, the film speed 200,000 frames per second, and the ultrasonic pressure about 2.0 atm.

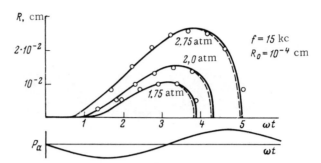

Fig. 12. Comparison between the pulsations of experimentally observed cavitation bubbles and the calculated pulsations.

is not enough, however. The bubbles grow to identical maximum radii independently of the initial radius R_0 only in the event the pressure amplitude of the ultrasonic field is much larger than the threshold pressure corresponding to that radius. The cavitation pattern illustrated in Fig. 11 corresponds to an ultrasonic pressure amplitude of about 2.0 atm. At such amplitudes (see the corresponding calculations of the pulsations in Figs. 3-5) the maximum radii of the cavitation bubbles are identical only if their equilibrium initial radii are at least 10^{-4} cm. Consequently, the illustrated cavitation develops in bubbles having an equilibrium radius greater than 10^{-4} cm. The number of cavitation nuclei of this radius in standing tap water, according to Strasberg [24], can be at most ten per cubic centimeter. The hypothesis developed by Sirotyuk [25] regarding the fragmentary nucleation of cavitation in a stationary cavitation domain tends to substantiate the result obtained.

A comparison of the pulsations of experimentally observed real cavitation bubbles with the analytical pulsations obtained by solution of the nonlinear differential equations is given in Fig. 12. Here the dots indicate the results of the experiments portrayed in the film records of Fig. 11, which were obtained under analogous conditions for various pressures of the ultrasonic field. The solid curves indicate the numerical solutions of the Herring –Flynn and Kirkwood –Bethe equations, which coincide in the cases investigated; the dashed curves indicate the solutions of the Noltingk –Neppiras equation. The calculations were made for bubbles having an initial equilibrium radius $R_0 = 10^{-4}$ cm at ultrasonic pressure amplitudes of 1.75, 2.0, and 2.75 atm, consistent with the experimental conditions. The results evince the fact that the Noltingk–Neppiras equation is adequate, not only qualitatively, but also quantitatively, for describing the behavior of actual cavitation bubbles. The time variations of the cavitation bubble radius represented in Fig. 12 for $P_m = 2.0$ atm correspond to the bubble marked with the arrow in Fig. 11. This bubble is situated on the boundary of the cavitation zone, but the distance from it to the other bubbles is clearly too small to allow it to be regarded as an individual cavitation bubble. The fact that the behavior of such a bubble is sufficiently consistent with the solutions of the nonlinear differential equations without regard for interaction between the bubbles encourages the notion that these equations adequately describe the pulsations of real cavitation bubbles in an ultrasonic wave field.

Chapter 3

Relationship of the Pulsations of Cavitation Voids to the Emission of Shock Waves and Cavitation Noise

§ 1.   Shock Waves in Cavitation

During the collapse of a cavitation void, the pressure on its surface can attain extremely large values. The void radiates spherical waves of finite amplitude, which in the course of propagation through the liquid are converted into shock waves, and the latter are responsible for the action of cavitation on various substances and structures [26, 27].

As mentioned before, the investigation of finite-amplitude spherical waves radiated by a cavitation void is tantamount to the assessment of the function $G(R, t_R)$ on the surface of a void of radius R and the computation of the arrival time of the wave at the investigated point with coordinate r according to Eq. (33). From (30) and (32) it is inferred that

$$G(R, t_R) = R\left(H + \frac{U^2}{2}\right),$$ (65)

where H and U are determined from the solution of the Kirkwood−Bethe equation (36) for pulsations of a cavitation void. For the calculation of the time t we represent G and r in the form [16]

$$G = rc_0 u(1 + \beta u),$$ (66)

$$\widetilde{c} = c_0(1 + 2\beta u),$$ (67)

where $\beta = (n + 1)/4c_0$. Relations (66) and (67) are valid near the shock wave front, where the Rankine–Hugoniot conditions prevail [28], whereas far from the front the same relations admit error. However, it is obvious that the vicinity of the shock wave front is the region of greatest concern. Expressing dr from (66), substituting it together with (67) into (33), and recognizing that the function G remains constant during propagation, we obtain [29]

$$t = t_R + \frac{2G}{c_0^3} \left[\frac{1 + 2\beta u}{\beta u (1 + \beta u)} - \frac{1 + 2\beta U}{\beta U (1 + \beta U)} - 2 \ln \frac{(1 + \beta u)\beta U}{\beta u (1 + \beta U)} \right], \qquad (68)$$

where

$$\beta u = \frac{1}{2} \left[\left(1 + \frac{n + 1}{rc_0^2} G \right)^{1/2} - 1 \right], \qquad (69)$$

$$\beta U = \frac{1}{2} \left[\left(1 + \frac{n + 1}{Rc_0^2} G \right)^{1/2} - 1 \right]. \qquad (70)$$

Consequently, the delay $(t - t_R)$ in the propagation of the function G is determined by the values of G, R, and r.

The function G proves useful in the sense that its value at a point r is determined by the magnitude of the pressure p in the liquid at the point. It follows from (27) and (67) that

$$c = c_0 \left(1 + \frac{n - 1}{2 c_0} u \right). \qquad (71)$$

Expressing the value of $(\rho/\rho_0)^{n-1}$ from (40) and (71) and substituting it into (38) with regard for (69), we obtain [29]

$$p = A \left[\frac{2}{n + 1} + \frac{n - 1}{n + 1} \left(1 + \frac{n + 1}{rc_0^2} G \right)^{1/2} \right]^{\frac{2n}{n-1}} - B. \qquad (72)$$

Consequently, the pressure p at point r is determined as a function of the value of $G(R, t_R)$ on the surface of the cavitation void. The propagation time of a spherical wave in this case is calculated according to Eq. (68).

The resulting relations afford a means for analyzing spherical waves of finite amplitude radiated during the collapse of a cavitation void. We consider some of the ensuing characteristics in an example of a concrete special case. Suppose that a gas bubble having

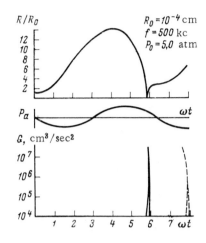

Fig. 13. Calculation of the function G
for a cavitation void.

an equilibrium initial radius $R_0 = 10^{-4}$ cm pulsates in a 500-kc
ultrasonic field with a pressure $P_m = 5.0$ atm.

The time variation of the radius of the bubble according to the
numerical solution of the Kirkwood–Bethe equation (36) is illustrated in Fig. 13a. Due to the negative pressure of the ultrasonic field, the bubble grows to a maximum radius $R_m = 14R_0$ and
collapse in the compression phase to a radius $R_{min} = 1.3 \cdot 10^{-2}R_0$.
The solid curve in Fig. 13b represents the value of $G(R, t_R)$ on the
surface of a cavitation void of radius R as calculated according to
Eqs. (39) and (65). As Fig. 13 shows, in the collapse of the cavitation void the value of $G(R, t_R)$ experiences a catastrophic increase.
This is because in the final stage of collapse, the pressure of the
gas in the void attains extremely large values, while the function G,
according to (39) and (65), bears a nearly linear relation to the
pressure.

According to the Kirkwood–Bethe theory, G is propagated in
the liquid at a velocity $\tilde{c}$, remaining constant the whole time. Here,
according to (33), the form of the function $G(r, t)$ is distorted
during propagation; larger values of G lead smaller values.

In Fig. 13b the dashed curve indicates the time dependence
of G in the liquid at a distance $r = 10^3R_0$, i.e., at a distance of 0.1
cm from the collapsing cavitation bubble. It is quickly seen that
propagation in the liquid causes an appreciable distortion of the
form of $G(r, t)$.

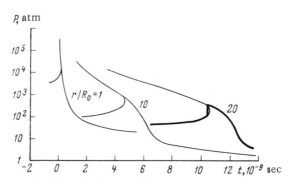

Fig. 14. Calculation of the pressure radiated by a
cavitation void.

It is a simple matter on the basis of the values of G, according to Eqs. (68) and (72), to calculate the time variation of the pressure p of a spherical wave propagating in a liquid at various distances r. The pressure p is shown in Fig. 14 as a function of the time at points with coordinates r equal to R_0, $10R_0$, and $20R_0$ for the case of cavitation void collapse depicted in Fig. 13 (the initial time t = 0 in Fig. 14 corresponds to the time of total collapse of the cavitation void in Fig. 13). As indicated by Fig. 14, with increasing r distortion of the pressure waveform takes place. If the function p(t) represents a spiked pulse near the cavitation void, similar in form to the function $G(R, t_R)$ on the surface of a sphere, then by the time r = $10R_0$ the pressure p becomes a multivalued function of t. This is physically inadmissible and, as we are aware [28], denotes the formation of a discontinuity of the function p(t). When the spherical wave is given in the coordinate system u(r) (where u is the hydrodynamic velocity), the position of the discontinuity in space and the wave amplitude can be determined from the condition of equal areas set off by the line of discontinuity in accordance with the conservation of the flux of mass, momentum, and energy across the discontinuity [28]. At a sufficient distance from the radiating sphere, such that $r \gg R_{min}$, the determination of the position of the discontinuity from the condition of equal areas is also possible in the case when the wave is represented in the form p(t) [29].

In Fig. 14 the heavy line indicates the variation of the pressure p(t) for r = $20R_0$ with allowance for the discontinuity, whose position t_s and maximum value p_s are determined from the conditions formulated above. Physically the discontinuity denotes the front of a shock wave. Thus, the propagation of finite-amplitude spherical waves radiated by a cavitation void leads to the generation of shock waves.

As implied by Fig. 14, the shock wave amplitude p_s is considerably lower than the pressure amplitude p_m, determined without regard for the existence of a discontinuity. Figure 15 shows how large this discrepancy can become. This figure illustrates the variation of the pressure amplitude in the liquid as a function of the coordinate r for the above case of radiation of a wave from a cavitation void. Curve 1 corresponds to the pressure amplitude p_m, which decreases with increasing r due to spherical divergence, and curve 2 corresponds to the amplitude p_s, which decreases with increasing r due to, besides spherical divergence, the irreversible thermodynamic losses at the shock wave front, which are indirectly accounted for by the construction of the discontinuities [28]. Even at a distance r of about (10 to 10^2)R_0 (in the given case), the pressure amplitude p_s of the shock wave is almost two orders of magnitude smaller than the amplitude p_m, obtained from calculations of the pressure according to spherical divergence (on the basis of the maximum pressure p_{max} in a totally collapsed gas cavity) without regard for the irreversible losses at the shock wave front. It is readily seen that the shock wave amplitude is large only near the cavitating void, so that logically the various processes corresponding to cavitation (erosion and cavitation

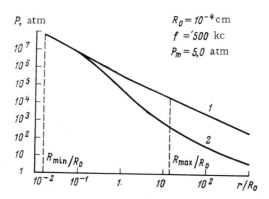

Fig. 15. Variation of the maximum pressure in a liquid during the propagation of a shock wave.

noise) are most effectively manifested in the immediate vicinity of the collapsing cavitation voids, a result that has indeed been confirmed experimentally [30, 31].

Of paramount practical importance is the investigation of the effect of the initial equilibrium radii of the cavitation voids on their collapse. The results obtained on the basis of numerical solutions of the Kirkwood—Bethe equation for the case of cavitation bubbles pulsating at identical pressure amplitudes of the ultrasonic field (P_m = 10 atm, frequency 500 kc) are summarized in Table 1 for three initial equilibrium bubble radii R_0: 10^{-3}, $5 \cdot 10^{-4}$, and 10^{-4} cm.

As the table indicates, all the bubbles, despite the considerable variability of their radii, grow to roughly the same maximum radius R_m, as the latter is stipulated by the ultrasonic field amplitude and depends very little on the initial radius [12]. However, as the minimum radii R_{min}, to which the cavitation bubble collapses, becomes smaller, the smaller becomes the initial radius R_0. This is attributable to the influence of the gas in the bubble. The gas content δ of the three cavitation bubbles is shown in Table 1. The gas content is interpreted according to [8, 11] as the ratio of the gas pressure in the bubble at maximum dilation to the pressure P_0, i.e.,

$$\delta = \frac{P_0 + 2\,\sigma/R_0}{P_0} \left(\frac{R_0}{R_m}\right)^3. \tag{73}$$

The smaller the equilibrium radius R_0, the smaller is the gas content for one and the same R_m (determined by the ultrasonic pressure). It follows from the table that with decreasing δ the maximum pressure p_{max} on the surface of the cavitation bubble in

TABLE 1. Collapse of Cavitation Voids Having Various Initial Equilibrium Radii

R_0, cm	R_m, cm	R_{min}, cm	δ	p_{max}, atm	$(U/c)_{max}$	$c\,c_0$
10^{-3}	$2.9 \cdot 10^{-3}$	$5.8 \cdot 10^{-5}$	$4.8 \cdot 10^{-2}$	$7.4 \cdot 10^{4}$	0.43	4.5
$5 \cdot 10^{-4}$	$2.6 \cdot 10^{-3}$	$2.5 \cdot 10^{-5}$	10^{-2}	$1.9 \cdot 10^{5}$	0.52	6.2
10^{-4}	$2.3 \cdot 10^{-3}$	$0.9 \cdot 10^{-6}$	$2 \cdot 10^{-4}$	$2.8 \cdot 10^{8}$	2.7	13.3

the final stage of collapse increases considerably, as does the
maximum rate of collapse of the bubble. With a decrease of R_0
from 10^{-3} to 10^{-4} cm the gas content suffers about a 10^2-fold decrease, which gives rise (for the same ultrasonic pressure) to a
more than 10^4-fold increase in the pressure p_{max} and an almost
20-fold increase in the rate of collapse U. A qualitatively-similar
pattern of the influence of the gas content on the variation of the
cavitation efficiency as estimated in terms of erosion and luminescence has been observed experimentally by Sirotyuk [32].

§ 2. Discrete Spectral Components
of Cavitation Noise

As mentioned earlier, in the present part of the book we are
interested in the pulsations of a stationary cavitation bubble containing a definite quantity of gas, which is constant on the average
over a characteristic period of time and stipulates a certain equilibrium radius R_0. Numerous experiments [3-6, 33-36] have
shown that such a bubble pulsates for several periods T of the
ultrasonic field. However, the Noltingk–Neppiras equation does
not allow us to determine the nature of the pulsations of the cavitation bubble after its collapse, and the assumption of incompressibility of the liquid introduces an indeterminacy in the numerical
solutions at the time corresponding to the final stage of collapse.
This indeterminacy has the effect that the rate of collapse determined in the solutions tends to a very large value. All of these
difficulties can be circumvented by using the Kirkwood–Bethe
equation to describe the pulsations of the cavitation bubbles. The
numerical solutions obtained on the basis of that equation yield
a number of conclusions that are consistent with the experimental

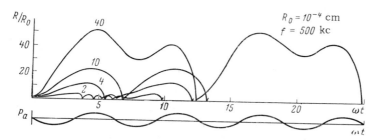

Fig. 16. Periodicity of the pulsations of cavitation voids.

facts. The dependences of R/R_0 on ωt obtained on the basis of the numerical solutions of the Kirkwood–Bethe equation for a bubble having an initial radius $R_0 = 10^{-4}$ cm and an ultrasonic field of frequency f = 500 kc are shown in Fig. 16. The curves represent various pressure amplitudes of the ultrasonic field, which are designated by the proper numerals (in atm). It is apparent from the curves that at small ultrasonic pressure amplitudes the cavitation bubble, after collapsing in the compression phase, executes several pulsations with a period smaller than the period T of the ultrasonic field (this will be discussed in further detail below) and, as soon as the expansion phase of the ultrasonic field begins, starts to grow again. The interesting fact here is that, after the period T, the growth and subsequent pulsations of the cavitation bubbles repeat (see the pulsations in Fig. 16 at amplitudes P_m equal to 2, 4, and 10 atm). The resulting functions are not strictly periodic, but they at least meet all the conditions governing almost-periodic functions [37]. It has been possible on the basis of an analysis of many numerical solutions obtained for various R_0 and frequencies ω to verify that the functions $R(t)$ are almost-periodic functions for which at a constant period T the form of the function changes so that in general $R(t) \neq R(t + T)$. But the difference between the values of the function after one period is very slight, and the following relation holds

$$|R(t) - R(t + T)| \ll R_{\max}, \tag{74}$$

where $R_{\max}$ is the maximum value of the function $R(t)$ on the entire interval of observation. It may be assumed on the basis of condition (74) that the dependence $R(t)$ is in the first approximation a periodic function with period T. Consequently, the sound pressure radiated by the cavitation void to produce cavitation noise is also a periodic function in the first approximation. The coefficients of the expansion of this function in a Fourier series determine the amplitude of the discrete components of the cavitation noise (the so-called cavitation noise harmonics). The slight deviations of the function from strict periodicity may be regarded as time fluctuations of the amplitude. According to Macfarlane [38] and Bohn [39], such fluctuations add to the spectrum of the cavitation signal, besides adding the discrete harmonic components, and a continuous white noise segment. However, inasmuch as in our case, according to condition (74), the amplitude fluctuations are relatively very small, the contribution of this mechanism to the continuous cavitation noise spectrum will be very small. As shown below, the main portion of the continuous noise in cavitation is attributable to other factors. Consequently, the periodicity of the functions in the pulsations of cavitation bubbles makes it possible to explain the existence of the discrete harmonic components of cavitation noise. It is essential to note that in the measurement of cavitation noise under real-life conditions the hydrophone almost always detects the sound pressure from many cavitation bubbles of varying radii. At a given steady-state pressure amplitude P_m of the ultrasonic field, bubbles of different radii emit signals of different form, but, inasmuch as the signals from each bubble are periodic functions with a period T equal to that of the ultrasonic field, the integral field perceived by the hydrophone must also in this case exhibit discrete harmonic cavitation noise components at frequencies nf, where n = 1, 2, 3,... .

It is evident from Fig. 16 that, with increasing pressure amplitude P_m of the ultrasonic field, the time will come when the cavitation bubble has increased to such large dimensions that it cannot collapse in the compression phase of the first period T, executing instead a secondary pulsation until finally, at the end of the second period it collapses. These pulsations are rather clearly reproduced subsequently, but the period of variation of the function $R(t)$ in this case becomes twice as large as the ultrasonic period, i.e., equal to 2T. An analogous twofold increase in the period of the pulsations is observed when the initial bubble radius R_0 increases at a constant, even fairly small, amplitude P_m of a few atmospheres. The pressure radiated in the pulsations of such a cavitation bubble has a period exactly twice the period T of the ultrasonic field; hence, in the Fourier series spectral representation of the signal, there will be spectral components at frequency $(m/2)f$, where m = 1, 2, 3,... . For even m we obtain discrete components whose frequencies are equal to those of the discrete harmonic components already present in the cavitation noise spectrum. For odd m, on the other hand, we obtain completely new discrete components at frequencies $nf - \frac{1}{2}f$. These are the so-called subharmonic components of the cavitation noise spectrum, which have been observed by several researchers [40, 41], but whose origin has been the target of many conflicting hypotheses [12].

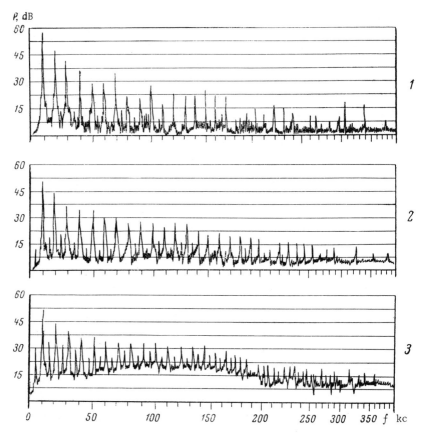

Fig. 17. Subharmonic spectral components of cavitation noise.

In Fig. 17 we trace the onset of the subharmonics of the cavitation spectrum with increasing ultrasonic pressure amplitude P_m. A block diagram of the experimental setup is shown in Fig. 18. Cavitation was excited in fresh tap water at a frequency $f = 10$ kc in the focal zone of a hollow cylindrical barium titanate ceramic radiator with a height of about 15 cm and mean diameter of about 22.5 cm. A miniature hydrophone was placed in the cavitation zone produced in the focal zone of the focusing radiator, where it received the cavitation noise with a practically uniform sensitivity up to 500 kc. The signal was transmitted to a wideband amplifier, in which the necessary correction of the hydrophone frequency characteristic was made; then it was analyzed by means

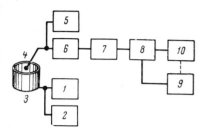

Fig. 18. Block diagram of the spectral analysis of cavitation noise. 1) Ultrasonic generator; 2) voltmeter; 3) cylindrical focusing radiator; 4) wideband hydrophone; 5) voltmeter; 6) rejection filter; 7) wideband amplifier; 8) heterodyne analyzer; 9) detector; 10) recording unit.

of a heterodyne spectral analyzer, whose output filter had a bandwidth of about 0.4 kc. The spectrograms from the output of the analyzer were recorded on a Bruel and Kjaer recorder, whose tape drive mechanism was mechanically linked to the heterodyne analyzer. The spectrograms in Fig. 17 are arranged in order of increasing ultrasonic pressure amplitude; the frequency is plotted on the horizontal axis, and the pressure of the corresponding spectral components are plotted on the vertical on log scale. Spectrogram 1 corresponds to well-developed cavitation in fresh tap water at an ultrasonic pressure amplitude $P_m = 0.4$ atm. In the given case the cavitation noise coefficient K [42], which is equal to the ratio of the effective noise pressure P_m to the pressure P_0 of the fundamental tone, was equal to 10%. Numerous discrete harmonic components at frequencies nf are evident in the cavitation noise spectrum. Also in evidence is very small-amplitude white noise as a predecessor to the continuous part of the cavitation noise spectrum.

Spectrogram 2 corresponds to very highly developed cavitation (K $\simeq$ 20%) at an ultrasonic pressure amplitude $P_m = 0.6$ atm, and spectrogram 3 characterizes vigorous cavitation (K $\simeq$ 35%) at an amplitude $P_m = 0.8$ atm. Already in spectrogram 2 we see, in addition to the discrete harmonic components and a fairly distinct white noise segment, the occurrence of subharmonic components at frequencies of 15, 25, 35 kc, etc. In spectrogram 3 the subharmonics have grown almost to the size of the harmonic components. The principal subharmonic at 5 kc, preceding the spectral components at the fundamental frequency of 10 kc on the horizontal axis, is readily noted. It is important to point out that the 5-kc subharmonic component in the spectrograms is lower in amplitude than its actual value by about 20 dB, because the spectral analyzer that we used had a lower frequency limit of 8 kc and attenuated the 5-kc spectral component by 1/10 the output relative to the higher frequencies. The principal subharmonic is almost always the largest of all the subharmonic components of the cavitation noise spectrum; at certain ultrasonic pressure amplitudes, it can be of the same order as the harmonic components.

With a further increase in the ultrasonic pressure amplitude or initial radius R_0, the cavitation bubble cannot collapse even at the end of the second period of the ultrasonic field, in which case the characteristic period of the pulsations of the bubble becomes equal to $3T$ (as shown in Chap. 2, § 2). In the spectrum of the noise pressure radiated by such a bubble, therefore, there appear subharmonic discrete components at frequencies $(l/3)f$, where $l = 1, 2, 3,....$ An analogous mechanism stipulates the occurrence of subharmonic components at frequencies $(q/4)f$, where $q = 1, 2, 3,....$. The indicated spectral components have been observed experimentally by many authors. Using focused ultrasound of sufficiently high intensity, Esche [40] was apparently the first to observe the subharmonic component at frequency $\frac{1}{2}f$ in the cavitation noise spectrum. Somewhat later Kikuchi [41] succeeded in producing and observing subharmonics at $f/3$ and $f/4$.

§ 3. The Continuous Cavitation

Noise Spectrum

As already mentioned, the amplitude fluctuations of the pressure radiated by a pulsating cavitation bubble are too small to account for the origin of the continuous part of the cavitation noise spectrum. It is apparent from Fig. 17 that considerable energy can be concentrated in that part of the spectrum; hence, there is definite importance attached to the investigation of the mechanism of its generation. For this purpose we turn to an analysis, not only of the pulsations of the cavitation bubble, but also of the sound pressure radiated by the bubble during pulsation. The time variations of the bubble radius R/R_0 are shown in Fig. 19 (curve a), along with the pressure P_a of the ultrasonic field (curve b) and the pressure p radiated by the oscillating bubble in a liquid at some distance r (curve c). All of the curves were obtained by numerical solutions of the Kirkwood−Bethe equation for the case of adiabatic pulsations of a bubble having an initial radius $R_0 = 2 \cdot 10^{-3}$ cm in an ultrasonic field of amplitude $P_m = 1.0$ atm and frequency $f =$

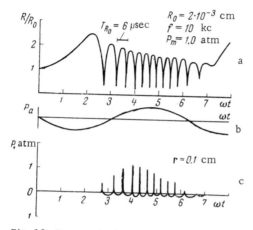

Fig. 19. Pressure in the liquid during pulsations
of a cavitation void.

10 kc. The radiated pressure was determined from Eq. (72) by
the analytical method described above.

We notice particularly in Fig. 19a that the cavitation bubble
begins after its first collapse to pulsate unsteadily with a period
T_R smaller than the period T of the ultrasonic field. It turned out
that this period has a specific physical sense and can be determined fairly precisely. At the end of the compression phase, at
the instant $\omega t = \pi$, when the pressure in the liquid is equal to the
hydrostatic pressure P_0, the pulsation period of the cavitation
bubble was equal to T_{R_0}, i.e., the natural resonance period of a
bubble of equilibrium radius R_0 as determined on the basis of the
Minnaert formula [18]

$$T_{R_0} = \frac{2\pi R_0}{(3\gamma)^{1/2}} \left(\frac{\rho}{P_0 + 2\sigma/R_0} \right)^{1/2}. \tag{75}$$

Above the curve in Fig. 19a is indicated the value of T_{R_0} computed
according to Eq. (75) for the case investigated in the appropriate
scale. We readily note the very good agreement of this quantity
with the period of the numerically obtained pulsations. It is also
apparent from the same equation that at other times the pulsation
period $T_R < T_{R_0}$ when the pressure in the liquid $P_0 + P_a > P_0$, or
$T_R > T_{R_0}$ when $P_0 + P_a < P_0$ due to the negative pressure of the
ultrasonic field. This is all consistent with the notions afforded by

the linear theory of Minnaert, provided we recognize the fact that the resonance properties of the bubble are determined by the pressure in the liquid. For the hypothetically nonlinear effect investigated here of the pulsations of a cavitation bubble, Eq. (75) can only be used for coarse estimates.

It may be inferred on the basis of the results presented in Fig. 19 that for the pulsations of a cavitation bubble, in addition to the general periodicity of the functions R(t) with period T, one can discern a finer-structured periodicity of the natural pulsations of the cavitation bubble with period T_R. As evident in Fig. 19c, this causes the sound pressure radiated by the bubble after the first collapse to have a rather complex form. In the first approximation it may be deduced on the basis of an analysis of the pulsations that the cavitation bubble is an oscillatory system subjected periodically to shock excitation. Compared with this simplified model, however, the cavitation bubble is a more complex nonlinear oscillatory system, whose natural pulsation period T_R varies with time. Such pulsations are very clearly seen in Fig. 20, which shows photographs of an integral cavitation signal and cavitation noise without the fundamental signal. The photographs correspond to cavitation in fresh tap water in the focal zone of a cylindrical concentrator.

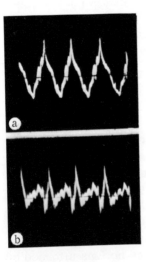

Fig. 20. Oscillograms of a cavitation signal. Cavitation in tap water at an ultrasonic frequency of 10 kc and pressure of about 1.0 atm: a) signal in the frequency band below 0.5 Mc; b) signal after a rejection filter attenuating the signal at 10 kc by 40 dB.

It is perfectly clear that the spectrum of the signal received from such a bubble will display, in addition to the harmonic (and subharmonic) discrete spectral components, other components due to nonstationary pulsations at the natural frequencies. Inasmuch as the amplitude and phase (frequency) of the natural modes vary considerably with time, the amplitude of the continuous part of the cavitation noise spectrum will increase in a certain frequency band located in an interval determined by the resonance frequency of the pulsating bubble.

We have already mentioned the fact that every hydrophone perceives the cavitation noise radiated by several cavitation bubbles of various dimensions. Each cavitation bubble radiates, besides harmonic discrete components of the frequency, a sound pressure in the form of a continuous portion of the spectrum in a certain frequency band. But inasmuch as cavitation bubbles of different sizes are present in the cavitation zone, ranging from very large (on the order of the resonance radius for the given ultrasonic frequency) to very small (set by the bubbles of "threshold radii" that can still cavitate at the given pressure amplitude of the ultrasonic field), the continuous part must take up a very broad frequency interval. Consequently, the continuous part of the spectrum will carry information on the size distribution function of the cavitation bubbles in the cavitation zone. In no case is it possible to link to the distribution of the cavitation bubbles (which have been produced from nuclei and have grown during a certain buildup time of the cavitation zone to definite radii as a result of diffusion) a distribution function with respect to the sizes of the cavitation nuclei to characterize the cavitation properties of any liquid prior to the onset of cavitation. It would be possible to determine the cavitation nucleus size distribution from the cavitation noise spectrum if one could experimentally measure this spectrum at the exact instant the ultrasonic field is applied and cavitation is initiated at the nuclei, but before the stationary cavitation zone has had time to build up. Clearly, this is impossible in principle, because the instrument determination of the spectrum of a cavitation signal requires a definite analysis time, which covers at least several periods of the ultrasonic field. However, as experiments have shown [36], several periods are fully adequate for the formation of a stationary cavitation zone, i.e., during this period of time,

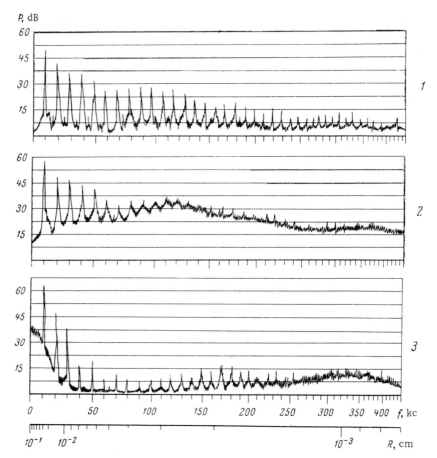

Fig. 21. Spectrograms of cavitation noise for water containing various amounts of gas.

due to rectified diffusion, the cavitation nuclei have already formed into cavitation bubbles of significantly large equilibrium radii.

Next we set out to determine the distribution function of the equilibrium radii of cavitation bubbles in a cavitation zone. Here we interpret the equilibrium radius R_0 of a gas bubble as the radius that the same quantity of gas, as contained in the pulsating cavitation bubble in the stabilized stationary cavitation zone, would have in the absence of the ultrasonic field.

Cavitation noise spectrograms obtained experimentally by the arrangement described earlier and illustrated in Fig. 18 are shown in Fig. 21. As in the preceding case, cavitation was excited at an ultrasonic frequency f = 10 kc, but was observed under different conditions. Spectrogram 1 corresponds to fairly well-developed cavitation (K $\simeq$ 25%) in fresh tap water with a relatively high gas constant G_g = 58% (at a temperature of 18°C). Spectrogram 2 also corresponds to well-developed cavitation (K $\sim$ 25%) in the same tap water after degassing, when the gas content was lowered to G_g = 30% (at the same temperature). The degassing was carried out by evacuation of the water; the cylindrical focusing radiator had its ends attached to flanges and rarefaction was created in this inner cavity over the surface of the water for two hours by means of a forepump. With a reduction in the gas content the cavitation zone consists of cavitation bubbles of smaller equilibrium radii than the cavitation zone in gas-saturated water. This is confirmed by direct visual observations of cavitation in water with various gas contents and is a familiar fact to every experimenter. It turns out that the effect has a counterpart in the behavior of the envelope of the continuous spectrum of cavitation noise. It is apparent in spectrogram 1 that, in water having a relatively high gas content, the form of the envelope of the continuous spectrum is almost independent of the frequency in the frequency interval from 10 to 150 kc, diminishing appreciably only at higher frequencies. If we assume that the damping decrement associated with natural pulsations of the cavitation bubbles has the same order of magnitude for bubbles of various sizes, then, in the case represented in the spectrogram, the cavitation zone consists of sufficiently large bubbles, whose radii vary from $3 \cdot 10^{-2}$ to $2 \cdot 10^{-3}$ cm (in the lower part of the figure, parallel to the frequency axis, is shown the axis of the corresponding radii of resonance bubbles according to the Minnaert equation on the assumption of adiabatic pulsations at a hydrostatic pressure P_0 = 1 atm). As apparent from spectrogram 2, a reduction in the gas content of the water, decreasing the equilibrium radii of the cavitation bubbles, is clearly reflected in a change in the form of the envelope of the continuous cavitation noise spectrum. There is a distinct extremum in the envelope in the vicinity of 120-130 kc, indicating a sizable increase in the number of bubbles of radius $2 \cdot 10^{-3}$ to $3 \cdot 10^{-3}$ relative to larger bubbles of radius 10^{-2} cm, which have tended to shrink in the cavitation zone.

The reduction of the cavitation envelope at frequencies of 300 to 400 kc attests to the fact that cavitation bubbles having radii 10^{-3} to 10^{-4} cm are present in much smaller quantities in the cavitation zone.

The above-postulated relationship between the form of the envelope of the cavitation noise spectrum and the distribution of the equilibrium radii of the cavitation bubbles is very graphically illustrated by spectrogram 3. Here we see the cavitation noise spectrum obtained after evacuation, when cavitation was excited in water on whose surface the pressure was equal to several millimeters on the mercury column. The hydrostatic pressure in the focal zone of the concentrator was determined solely by the weight of the 7.5-cm column of water and was therefore about 0.1 atm. With the excitation of the ultrasonic field, cavitation was observed with fairly large cavitation bubbles of radius 10^{-1} to 10^{-2} cm. The observed process should more properly be called, rather than cavitation, nonlinear pulsations of large bubbles in an ultrasonic field. The effect in question is fairly clearly seen in spectrogram 3, from which it is evident that the dominant part of the continuous spectrum of the nonlinear pulsations of the bubbles is concentrated in the interval of relatively low frequencies up to 30 kc (in which case allowance must be made for the fact that the particular analyzer used attenuated the spectral components considerably at frequencies below 8 kc). This indicates a significant relationship between the form of the envelope of the continuous cavitation noise spectrum and the size distribution function of the cavitation bubbles.

The aforementioned increase in the frequency of the extremum in the cavitation noise envelope with decreasing gas content of the water has been observed experimentally by Bohn [39]. Another experimental fact was brought out in the latter paper; even with a constant gas content in the same sample of water, as the pressure of the ultrasonic field is increased, the frequency of the extremum in the cavitation noise envelope increases considerably. From the point of view of the notions set forth above, this fact is entirely legitimate. In fact, the cavitation bubbles grow from the cavitation nuclei in the cavitation zone to equilibrium size due to rectified diffusion into the bubble of air dissolved in the water. If the intensity of the ultrasonic field is increased, cavitation occurs in the cavitation zone at a large number of nuclei, whose separation is much smaller than the previous case, so that the effective domain of the liquid that supplies the cavitation bubble with gas dissolved in it becomes smaller. Consequently, the amount of gas diffused into the bubble is smaller, and the most probable equilibrium radius of the bubble decreases.

Conclusion

The results obtained in the present study bring us to the conclusion that the equations for the pulsations of cavitation voids (despite the restrictions and assumptions admitted in their derivation) adequately describe the behavior of actual voids. Particularly significant is the fact that these equations can be used for the analysis of the pulsations of cavities in a cavitation zone; as noted, in this case the action of the neighboring cavities on the pulsations is so slight as to be negligible. The application of the theory of Kirkwood and Bethe in the form outlined above makes it possible to calculate not only the generation, but also the dissipation of shock waves induced by the collapse of cavitation voids. Other prior applications of the Kirkwood – Bethe theory (as, e.g., in [43]) have proved less effective.

The foregoing analysis of cavitation noise has shown that the spectral components of cavitation noise provide a means for acquiring valuable information on the evolution of cavitation process. Very extensive theoretical and experimental studies are needed in this direction, however.

References

1.   D. Y. Hsieh and M. S. Plesset, Theory of rectified diffusion of mass into gas bubbles, J. Acoust. Soc. Am., 33(2):206 (1961).
2.   O. A. Kapustina and Yu. G. Statnikov, Influence of acoustic microstreaming on the mass transfer in a gas bubble – liquid system, Akust. Zh., 13(3):383 (1967).
3.   W. Güth, Kinematographische Aufnahmen von Wasserdampfblasen [Motion pictures of water vapor bubbles], Acustica, 4(5):445 (1954).
4.   A. T. Ellis, Techniques for Pressure Pulse Measurements and High-Speed Photography in Ultrasonic Cavitation, Cavitation in Hydrodynamics, London (1956).
5.   E. Mandry and W. Güth, Kinematographische Untersuchungen der Schwingungskavitation [Motion picture studies of vibration-induced cavitation], Acustica, 7(4):241 (1957).
6.   I. Schmid, Kinematographische Untersuchung der Einzelblasen-Kavitation [Motion picture investigation of the individual cavitation bubble], Acustica, 9(4):321 (1959).

7.   M. Kornfel'd, Elasticity and Strength of Liquids, Moscow—Leningrad (1951).

8.   W. Güth, Zur Entstehung der Stosswellen bei der Kavitation [Formation of shock waves in cavitation], Acustica, 6(6):526 (1956).

9.   H. Lamb, Hydrodynamics, New York (1945).

10.   Rayleigh, On pressure developed in a liquid during the collapse of a spherical cavity, Phil. Mag., 34:94 (1917).

11.   B. E. Noltingk and E. A. Neppiras, Cavitation produced by ultrasonics, Proc. Phys. Soc., 63B:675 (1950), 64B:1032 (1951).

12.   H. G. Flynn, Physics of Acoustic Cavitation in Liquids, Physical Acoustics (W. P. Mason, ed), Vol. 1B, Academic Press, New York (1964).

13.   M. I. Vorotnikova and R. I. Soloukhin, A calculation of the pulsations of gas bubbles in an incompressible liquid subject to a periodically varying pressure, Akust. Zh., 10(1):34 (1964).

14.   L. Trilling, The collapse and rebound of a gas bubble, J. Appl. Phys., 23(1):14 (1952).

15.   C. Herring, Theory of the Pulsation of the Gas Bubbles Produced by an Underwater Explosion, OSRD Rept. No. 236 (1941).

16.   R. H. Cole, Underwater Explosions, Princeton Univ. Press (1948).

17.   F. R. Gilmore, The Growth or Collapse of a Spherical Bubble in a Viscous Compressible Fluid, California Inst. Technology Rept. No. 26-4 (1952).

18.   M. Minnaert, On musical air-bubbles and the sounds of running water, Phil. Mag., 16(17):235 (1933).

19.   W. Y. Cunningham, Introduction to Nonlinear Analysis, New York (1958).

20.   M. G. Sirotyuk, On the behavior of cavitation bubbles at large ultrasonic intensities, Akust. Zh., 7(4):499 (1961).

21.   A. A. Andronov, A. V. Vitt, and S. Z. Khaikin, Theory of Oscillations, Moscow (1953).

22.   A. A. Andronov, Mathematical problems in the theory of self-sustained oscillations, Collected Works, Izd. AN SSSR (1956).

23.   J. J. Stoker, Nonlinear Vibration in Mechanical and Electrical Systems, New York (1950).

24.   M. Strasberg, Onset of ultrasonic cavitation in tap water, J. Acoust. Soc. Am., 31(2):163 (1959).

25.   M. G. Sirotyuk, Energetics and dynamics of the cavitation zone, Akust. Zh., 13(2):265 (1967).

26.   L. A. Glikman, V. P. Tékt, and Yu. E. Zabachev, On the physical nature of cavitation destruction, Zh. Tekh. Fiz., 25(2):280 (1955).

27.   E. Meyer, High-Intensity Sound in Liquids, Underwater Acoustics (V. M. Albers, ed.), Plenum Press, New York (1961), pp. 139-158.

28.   L. D. Landau and E. M. Lifshits, Mechanics of Continuous Media, Moscow (1953).

29.   V. A. Akulichev, Yu. Ya. Boguslavskii, A. I. Ioffe, and K. A. Naugol'nykh, Radiation of finite-amplitude spherical waves, Akust. Zh., 13(3):321 (1967).

30.   R. H. Mellen, An experimental study of the collapse of a spherical cavity in water, J. Acoust. Soc. Am., 28(3):447 (1956).

31. R. T. Knapp, Recent investigation of the mechanics of cavitation and cavitation damage, Trans. ASME (Annual Meeting), p. 106 (1954).

32. M. G. Sirotyuk, Effect of the temperature and gas content of the liquid on cavitation processes, Akust. Zh., 12(1):87 (1966).

33. T. Lange, Methoden zur Untersuchung der Schwingungskavitation in Flüssigkeiten mit Ultraschall [Methods for the investigation of vibration-induced cavitation in liquids with ultrasound], Akust. Beih., 2:75 (1952).

34. G. W. Willard, Ultrasonically induced cavitation in water, J. Acoust. Soc. Am., 25(4):667 (1953).

35. L. O. Makarov and L. D. Rozenberg, On the mechanism of ultrasonic cleaning, Akust. Zh., 3(4):37 (1957).

36. M. G. Sirotyuk, Experimental investigation of the growth of ultrasonic cavitation at 500 kc, Akust. Zh., 8(2):216 (1962).

37. G. G. Malkin, Selected Problems in the Theory of Nonlinear Oscillations, Moscow (1956).

38. G. T. Macfarlane, On the energy spectrum of an almost-periodic succession of pulses, Proc. IRE, 37:1139 (1949).

39. L. Bohn, Schalldruckverlauf und Spektrum bei der Schwingungskavitation [Sound pressure variation and spectrum of vibration-induced cavitation], Akust. Beih., 2:201 (1952).

40. R. Esche, Untersuchung der Schwingungskavitation in Flüssigkeiten [Investigation of vibration-induced cavitation in liquids], Akust. Beih., 4:208 (1952).

41. Y. Kikuchi (ed.), Engineering Aspects of Ultrasonic Cavitation, Research Group of Ultrasonic Cavitation, Japan (1961).

42. V. A. Akulichev and V. I. Il'ichev, Spectral indication of the origin of ultrasonic cavitation in water, Akust. Zh., 9(2):158 (1963).

43. R. Hickling and M. S. Plesset, Collapse and rebound of a spherical bubble in water, Phys. Fluids, 7(1):7 (1964).

PART V

EXPERIMENTAL INVESTIGATIONS OF ULTRASONIC CAVITATION

M. G. Sirotyuk

Introduction

If cavitation is to be generated specifically for the purpose of doing useful work (cleaning of parts, dispersion of liquids and solids, etc.), one must know how to control the attendant cavitation processes. With this in mind it is little wonder that so many researchers have turned their attention to these effects in recent times. This is evidenced by the constantly growing number of publications both in the periodical literature and in separate book form. The cumulative wealth of theoretical and experimental material has prompted the publication of several survey papers in the last three years (e.g. [1-4]) on the topic of acoustic cavitation.

In the present section of the book we examine some problems associated with the onset of cavitation, the feasibility of "controlling" its intensity, and the balance of energy in cavitation; we also consider the dynamics of the cavitation zone. The study is devoted primarily to experimental research conducted for the most part with high-intensity focusing concentrators [27] in the focal region of which strongly developed cavitation can be produced. The resulting variation of the wave impedance in the cavitation zone, owing to its localization, does not affect the parameters of the radiator itself as has been observed when cavitation occurs right next to the radiator [52]. A series of high-intensity focusing concentrators that we developed [27, 28, 60] has afforded highly useful tools for the investigation of both the cavitation zone itself and the processes engendered in that zone.

Inasmuch as many of the problems discussed here do not as yet have a rigorous theoretical solution, several approximations

are often made, sometimes rather crude, in order to interpret the
results. Nevertheless, the satisfactory agreement in most cases
between crude theory and experiment justifies the assumption
that the main physical factors governing one effect or another are
properly accounted for in the approximations. In particular, looking ahead, it is important to state that the widespread notion of the
collapsing cavitation bubbles being acted upon, not only by the hydrostatic pressure, but also by a positive acoustic pressure, is
clearly valid only for the individually isolated bubble. Experiments
have shown that this is not so in a "loosely structured" cavitation
zone comprising a set of cavitation bubbles. A number of the most
widely variegated experiments supports the fact that the pressure
acting on the cavitation bubbles in the positive half-period is roughly equal to the hydrostatic pressure.

Chapter 1

The Strength of Liquids

§ 1. Theoretical and Actual Strength

Under the influence of pressure or temperature the average distances between the molecules in a liquid change. For each liquid there exists a specific negative pressure or temperature at which these distances attain certain critical values; when the latter are exceeded, the liquid breaks down. According to the kinetic theory of liquids, the strength of absolutely pure liquids is determined by the molecular bonds. This theory states that liquids must have tensile strengths up to several tens of thousands of atmospheres (depending on the type of liquid).

The surface energy, i.e., the energy formed by the surface tension σ, can be determined as the excess potential energy per unit surface due to the deficit of neighbors about the particles of the surface layer. This energy can be compared with the vaporization energy of the liquid per unit volume to obtain the distance R between adjacent particles such that the surface tension forces are still present. Thus, for water this distance is equal, of course, to $R \simeq 10^{-8}$ cm.

In order for the liquid to break down, its particles must be separated by a distance on the order of twice the spacing of the adjacent molecules. Consequently, the minimum work per unit area 2σ can be set equal to the breaking strength. The liquid can therefore sustain a maximum tensile stress

$$P_t \simeq \frac{2\sigma}{R}.$$

Setting $R \cong 2 \cdot 10^{-8}$ cm, for water (σ = 75 dyn/cm) we obtain $P_t \cong$ 10,000 atm.

It is true, of course, that Zel'dovich [6] has indicated the possibility of the strength of liquids being lowered due to the spontaneous occurrence of a vapor bubble in it, which can be formed as a result of thermal fluctuations. An analysis of the probability of formation of the new phase and the work required to do so yields the expression

$$P_t \simeq P_v - 44\sqrt{\frac{\sigma^2}{T}}, \tag{1}$$

where P_v is the vapor pressure in the cavity and T is the absolute temperature.

Calculations performed for pure water have shown that the value of P_t, for which breakdown of the liquid takes place, must be about 10^3 atm, which is about one order of magnitude smaller than predicted on the basis of the molecular strength.

However, the value of P_t calculated according to expression (1) is still significantly higher than the experimental strength. A composite graph is presented in Fig. 1, showing the cavitation strength* $P_K = P_t$ for distilled water (dark symbols), fresh tap water (light circles), and air-impregnated water (stars) as measured by various authors at different frequencies. As apparent from Fig. 1, the cavitation strength of real water is no more than a few hundred atmospheres, which is far below the theoretical prediction, and the experimental data are widely divergent from one another; the latter consideration is discussed in §3.

The discrepancy between the experimental and theoretical strengths is attributable to the presence in real water of various insoluble impurities, which comprise "weak spots" and sharply diminish the strength.

---

*We define the cavitation strength of a liquid as the minimum negative acoustic pressure at which there is formed in the liquid a cavity that collapses or implodes rapidly in the positive half-period and gives rise to what are known as cavitation effects.

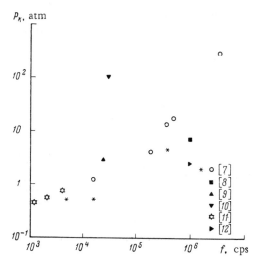

Fig. 1. Experimental threshold for the onset of
cavitation in water at various acoustic frequencies
according to the data of several authors.

§ 2 .   Cavitation Nuclei

The production of absolutely pure water is impossible. One
of the strongest existing solvents, it dissolves the walls of the
container and, on coming into contact with any gas, dissolves that
gas.

Inevitably present in liquids, besides dissolved impurities,
are assorted suspended particles, which fall into the liquid from
the atmosphere or from the walls of the containers.

It is known from the kinetic theory of liquids that dissolved
or thoroughly wetted impurities cannot lower the strength of the
liquid appreciably, because their contact angle is close to zero.
Even surfaces that are initially poorly wetted on falling into water,
with contact angles of 50-60°, after just a few days have contact
angles approaching zero [13]. This is caused by the gradual dissolution of the layer of air adsorbed on the surface. However, a
vapor-gas nucleus often remains trapped in crevices and recesses
[13]. If the crevice is poorly wetted, the liquid in contact with it
forms a convex meniscus relative to the gas, and if, similarly, it
has a sharp bottom, then under no amount of pressure can the

liquid completely fill the crevice. Consequently, the nucleation of a bubble under the action of tension in the liquid is always initiated from the vapor-gas phase contained in such a crevice or recess.

Finally, as a result of the enormous permeability of high-energy particles generated in outer space, all liquids that we are likely to encounter are subjected to irradiation by these particles, which initiate microbubbles. It is assumed at the present time that the initiation of vapor bubbles by a charged particle in a liquid is induced either by the action of ions formed along the particle trajectory, which produce bubbles, or, more probably, by the local evolution of heat from the transformation of energy dissipated by δ-electrons created by the high-energy particle [17]. The vapor bubbles thus initiated along the particle trajectory have radii on the order of 10^{-7} to 10^{-6} cm. The radii of the bubbles vary with time until they attain the critical radius corresponding to the thermodynamic equilibrium of the liquid-vapor bubble system

Lieberman [20] and Sette [21], who subsequently continued their work with a staff of co-workers [8, 22, 23], discovered experimentally that a decrease in the number of high-energy particles penetrating the liquid container (by lead shielding of the container) enhances its cavitation strength. Removal of the shield or artificial irradiation of the liquid in the shielded container with high-energy particles (for example, a flux of neutrons), on the other hand, lowers the cavitation strength. The authors of the cited papers showed that the microbubbles produced by high-energy particles are indeed cavitation nuclei.

Consequently, of all the diverse impurities contained in a liquid, only vapor-gas bubbles lodged in fine crevices of suspended insoluble particles or independently existing bubbles can exert a significant effect on the strength of the liquid. The prolonged existence of independent bubbles elicits doubt at first glance; large bubbles are rapidly occluded to the surface, and the elevated pressure in the small bubbles,

$$P_g = P - P_v + \frac{2\sigma}{R_0},\qquad(2)$$

where P is the hydrostatic pressure and P_v is the saturation vapor pressure of the liquid (and increases with diminishing R_0), should result in the dissolution of its contents. Experiments have shown,

however, [9, 10, 14, 15], that degassing of the liquid or the preliminary application of an elevated hydrostatic pressure increases
its strength. This indicates that stable bubbles must exist in the
liquid, their number and radii diminishing by comparison with normal conditions.

The stable existence of vapor gas bubbles in a liquid is most
credibly accounted for by ion theory [16, 17]. It is postulated that
on the surface of the bubble are uniformly distributed like charges
due to ions present in the liquid. The repulsion of these charges
prohibits implosion of the bubble. The presence of electrical
charges on the bubble surface was experimentally verified some
time ago [18], but the mechanism by which the ions impinge on the
bubble surfaces in water has only recently been studied by Akulichev [19].

Ions injected into water form local inhomogeneities. It is well
known that water molecules densely surround such ions as Mg^{++},
Li^+, and Na^+, forming a hydrate group with a local increase in the
density of the water. As a result, the ions behave as hydrophilic
inclusions in water (positive hydration). However, there exist
other ions, such as K^+, Cs^+, F^-, Cl^-, Br^-, and I^-, which decrease
the local density of the water. These ions, which tend to move the
water molecules farther apart, are like hydrophobic inclusions
(negative hydration). Logically, the probability of the existence of
a hydrophilic inclusion on the bubble surface is smaller than the
probability of a "well-adhering" hydrophobic inclusion. Consequently, an electrical charge can occur on the bubble surface only
if ions inducing negative hydration are present in the water. In
ordinary liquids there are almost always various ions, including
those which produce electrical charges on the surface of a bubble
and stabilize its radius. If a solution of substances that enhance
negative hydration is artificially injected into the liquid, it is to be
expected that the radius of the bubble will increase and the cavitation strength of the liquid will decrease under the influence of the
growing Coulomb forces.

As a matter of fact, the experiments conducted by Akulichev
[19] showed that the addition of solutions of substances promoting
negative hydration in water lowers its cavitation strength. To be
sure, this effect can be clearly observed with a reduced gas content in the water (< 20%), when its cavitation strength (even without the addition of ions) is fairly high.

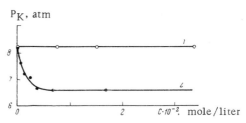

Fig. 2. Influence of the concentration of
solutions on the cavitation strength. 1)
Solution of LiOH promoting positive hydration; 2) solution of KBr promoting negative
hydration.

The dependences of the cavitation strength P_K of water on the
concentration C of LiOH ions (curve 1) and KBr ions (curve 2) dissolved in it, where the former produces positive and the latter
produces negative hydration, are shown in Fig. 2 [19]. Both dependences were obtained for water with an air content of 13%. As
anticipated, the addition of ions promoting positive hydration is
not related to the cavitation strength of the water (curve 1), whereas the addition of ions promoting negative hydration lowers it
(curve 2). It is also evident from Fig. 2 that the influence of the
ions asserts itself only up to a definite saturation concentration
of the solution (0.005 mole/liter); this is clearly attributable to
the fact that ions of like charge are capable of approaching the
bubble at distances such that the force of the translational dis-
placements of the ion toward the bubble do not become equal to
the electrostatic repulsion force.

If electrically charged bubbles are present in the liquid, the
existence of an electric field in the liquid must set these bubbles
in motion in the direction of the field. With the prolonged application of a constant electric field the bubbles, or cavitation nuclei,
tend to abandon the local domain between the electrodes creating
the field, becoming concentrated near the electrodes themselves,
whereupon this domain of the liquid gains an increased cavitation
strength.

An experiment of the type indicated would clarify the existence
of electrical charges on bubbles.

A schematic diagram of our experiment is shown in Fig. 3;
the focal region O of the focusing concentrator [27] was placed in

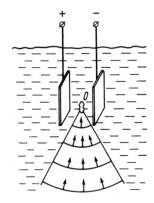

Fig. 3. Diagram of an experiment to exhibit the presence of electrical charges on bubbles.

the electric field produced by two plane insulated electrodes (wire mesh in chlorovinyl insulation). The liquid used was distilled water or water containing dissolved KBr (the replacement of water by KBr solution with concentrations up to 0.06 g/mole scarcely altered the results of the measurements). The distance between the electrodes was 20 mm, and the electrical potential difference was 3 kV.

The variation of the strength P_K (the measurement procedure is described in §4 of the present chapter) of ordinary distilled water as a function of the application or absence of an electric

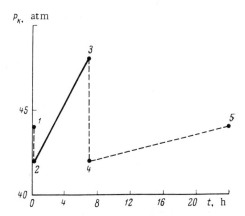

Fig. 4. Variation of the cavitation strength of water as a function of its incumbency time in the electric field.

field during a period of time t is shown in Fig. 4 (solid and dashed curves, respectively). Point 1 corresponds to the strength of water standing quiescently for 17 h after the preceding experiment. Inasmuch as the previously cavitating liquid contains weaker nuclei comprising the "products" of cavitation (growth of the bubbles due to the diffusion of gas during their pulsations, and the formation of new nuclei; see Chap. 1, §4, and Chap. 2, §6), the first measurements were followed up with repeated measurements, yielding point 2, which corresponds to the strength just preceding the application of the field. Then an electrical voltage was applied to the electrodes for seven hours, after which the strength of the liquid was again measured (with the field turned off), yielding point 3. Point 4 represents the repeated measurement of the strength. As expected, this strength is equal to that at point 2. Point 5 corresponds to the liquid after standing for 17 h; in this period of time the gas which had diffused into the bubble nuclei was partially dissolved, and the strength of the liquid returned to its initial value (at point 1). Each point represents the average of the measurements with a maximum scatter of ±15%.

As apparent from Fig. 4, the presence of the electric field increases the strength of the liquid. Under the experimental conditions described, the strength is increased by 14%, which supports the ion theory accounting for the stabilization of vapor-gas bubbles in a liquid.

Consequently, the condition of equilibrium of a vapor-gas bubble in a liquid must be described in the form

$$P_g = P - P_v - P_e + \frac{2\sigma}{R_0},\tag{3}$$

where $P_e = \frac{e^2 n^2}{8\,\pi\varepsilon R_0^4}\left(1 - \frac{\varepsilon}{n}\right)$ is the pressure elicited by the Coulomb repulsion forces [17] (e is the elementary charge, n is the number of charges on the bubble surface, and ε is the dielectric constant of the liquid).

It may be stated in summary that bubbles (initiated in a liquid by high-energy particles or formed by air entrapped by the liquid during its motion), on whose surface there exist at least a few like charges induced by ions present in the liquid, are no longer in contact, because the surface tension forces are offset by the Coulomb repulsion forces. With the passage of time gas diffuse into the bubble from the liquid, and a stable vapor-gas bubble is formed. On penetrating such a bubble, a particle produces a sharp rise in the temperature, which leads to vaporization of the liquid from the bubble walls and growth of the bubble [22].

Bearing in mind the foregoing model of the formation of a nucleus, it is reasonable to expect that the number of very small bubbles (10^{-7} to 10^{-6} cm) left in the wakes of entering high-energy particles will considerably exceed the number of large nuclei (10^{-4} to 10^{-3} cm) formed by the repeated penetration of particles into bubbles of small radius. As revealed below (see Fig. 7), the experiments confirm this hypothesis.

§3. The Cavitation Strength

It was shown in the preceding section that the only possible cavitation nuclei are vapor-gas bubbles existing as stable entities in the liquid or in crevices and recesses of solid particles. We now consider the strength of a liquid containing such actual nuclei. With a reduction in pressure in the liquid, the gas-containing crevices can act as a source of bubbles in a manner similar to boiling water. This occurs in situations such that the bubble escapes the edge of the recess or crevice, retaining an almost-spherical configuration with a radius of curvature R_0 equal to that of the crevice opening. For a wide interval of contact angles and recess geometries, the radius R_0 is a critical quantity governing the strength of the liquid [13]. We therefore orient our ensuing discussion around a vapor-gas bubble of radius R_0 in a liquid, where R_0 is the equivalent of the radius of the crevice opening.

If a tensile (or compressive) stress $P_a = P_m \sin \omega t$ is applied to the liquid, its amplitude varying with the frequency ω, the bubble is acted upon by $P \pm P_a$, which alters the radius to a value R. Neglecting the diffusion of the gas, as we are allowed to do for one or more pulsation cycles of the bubbles (see Part IV), and bearing in mind that the pressure of the gas contained in the bubble varies inversely as R^3, we write

$$P - P_a = P_g \left(\frac{R_0}{R}\right)^{3\gamma} + P_v - \frac{2\sigma}{R} - \frac{e^2 n^2}{8\pi \varepsilon R^4}\left(1 - \frac{\varepsilon}{n}\right),$$

where γ is the polytropy index and characterizes the equation of state of the vapor-gas in the bubble.

Substituting expression (3) in the above equation, we obtain

$$P - P_a = \left[P - P_v + \frac{2\sigma}{R_0} - \frac{e^2 n^2}{8\pi\varepsilon R_0^4} \left(1 - \frac{\varepsilon}{n} \right) \right] \left(\frac{R_0}{R} \right)^{3\gamma} +$$
$$+ P_v - \frac{2\sigma}{R} + \frac{e^2 n^2}{8\pi\varepsilon R^4} \left(1 - \frac{\varepsilon}{n} \right). \tag{4}$$

In the final stage of collapse of the cavitation bubble, the state of the vapor-gas mixture in it varies adiabatically, so that $\gamma = 4/3$ [3]. It is readily seen that in this case the terms in Eq. (4) accounting for the electrical charge cancel out. In the expansion stage of the bubble, however, when the state of the vapor-gas mixture inside varies isothermally, $\gamma = 1$. Now one of the electrical charge terms in the same equation is equal to $\frac{e^2 n^2}{8\pi\varepsilon R_0 R^3} \left(1 - \frac{\varepsilon}{n} \right)$, while the other is equal to $\frac{e^2 n^2}{8\pi\varepsilon R^4} \left(1 - \frac{\varepsilon}{n} \right)$.

At the initial instant of formation of cavitation, R cannot yet increase too appreciably and differs from R_0 at most by one order of magnitude; it may be approximately assumed, therefore, that $R_0 R^3 \simeq R^4$. Then, Eq. (4) may be written

$$P - P_a \simeq P_v + \left(P - P_v + \frac{2\sigma}{R_0} \right) \left(\frac{R_0}{R} \right)^3 - \frac{2\sigma}{R}. \tag{5}$$

The latter equation makes it possible to explicate the dependence of R on the tensile stress $R_t = P_a$ for various values of R_0. The analysis shows [5] that, until P_a attains a certain critical value P_K, the bubble is stable. With a further increase in P_a the bubble loses its stability and begins to grow abruptly, until it breaks. Differentiating (5) with respect to R and setting the derivative equal to zero, we find the value of the critical radius

$$R_K = \sqrt{3} R_0 \sqrt{ \frac{R_0}{2\sigma} \left(P - P_v + \frac{2\sigma}{R_0} \right) }.$$

The substitution of R_K into (5) gives us an expression for the strength of the liquid at a bubble of radius R_0

$$P_K \simeq P - P_v + \frac{2}{3\sqrt{3}} \sqrt{ \frac{\left(\frac{2\sigma}{R_0} \right)^3}{P - P_v + \frac{2\sigma}{R_0}} }. \tag{6}$$

Inasmuch as the condition $P_v \ll P$ normally holds, the quantity P_v may be neglected in expression (6) and the latter rewritten in the form

$$R_0^3 + \frac{2\sigma}{P} R_0^2 - \frac{32\sigma^3}{27P(P - P_K)^2} \simeq 0. \tag{7}$$

Equation (7) has also been deduced in [24] from a differential equation describing the behavior of a gas bubble in a sound field [25] when the inertial terms are disregarded.

Curve 1 of Fig. 5 represents the quantity P_K corresponding to the cavitation threshold in water at a nucleus of radius R_0 according to Eq. (6). The solution of Eq. (7) gives a curve that is essentially (correct to plotting error) congruent with the dependence obtained from expression (6). The solid part of curve 1 corresponds to cavities having radii up to 10^{-7} cm. Such small bubble radii would appear at first glance to complicate their theoretical analysis. The use of macroscopic thermodynamic parameters for the description of a system of several thousand molecules merits little justification. However, the auspicious development of the theory of initiation for bubble chambers demonstrates that such a description is workable even for smaller systems [17]. It is not surprising, therefore, that an analysis carried out for cavities of radius smaller than 10^{-7} cm, even down to the intermolecular scale (dashed part of curve 1), should lead approximately to the limiting strength derived from the kinetic theory of liquids (see §1). Curve 1 does not allow for the effect of the acoustic frequency on the cavitation threshold, although this effect does occur. Above all, the differential equations describing the time behavior of a cavitation bubble, such as, for example, those in [25] or others given in Part IV, Chap. 1, imply that the variation of the radius of the cavitation bubble is affected by the kinetic energy of the apparent mass of the liquid. In the indicated differential equations, this energy is included in the inertial terms. Curve 2 depicts the dependence of P_K on R_0 with allowance for the apparent mass of the water and its effect on the pulsation of the bubble. This curve is drawn through points corresponding to an average acoustic pressure of frequency 500 kc and inducing the onset of cavitation at bubbles of various radii. Some of the points (up to $R_0 < 10^{-4}$ cm) were obtained on the basis of numerical solutions given in Part IV, Chap. 1, for the differential equation of [25] characterizing the time variation of the bubble radius, while other points ($R_0 > 10^{-4}$ cm) were obtained analogously at a lower frequency, below the resonance frequency of the bubble. As a matter of fact, if the acoustic frequency exceeds the resonance frequency of the bubble, the oscillations acquire a complex character, and collapse occurs only at an acoustic pressure far in excess of P_K. This is obvious from the curves characterizing the pulsations at various acoustic pressures of bubbles whose resonance frequencies are lower than the acoustic frequency (see Part IV, Chap. 2).

It follows from curves 1 and 2 that the slight difference ($\sim 10\%$) in the values obtained for the cavitation strength with and without regard for the inertia of the water surrounding the bubble permits that inertia to be neglected if the water does not contain bubbles having radii $>10^{-3}$ cm (as, for example, in the case of standing water). But if the liquid contains nuclei of larger radius ($>10^{-3}$ cm), it is required to include the inertia of the liquid surrounding the bubble, because in this case the calculated strength of the water turns out to be considerably lower than when the inertia is omitted and corresponds more nearly to reality in view of our knowledge that sea water or fresh tap water in which bubbles of the indicated large radius are present has a cavitation strength lower than 1 atm (see Fig. 1).

We now consider in more detail the relationship between the acoustic frequency and the resonance frequency of the bubble. The minimum value of the tensile stress that induces cavitation at a nucleus of radius R_0 (with the above-mentioned errors) is determined by Eqs. (6) or (7). If the acoustic frequency is increased until it exceeds the resonance frequency of the bubble, the bubble oscillations become complex, and collapse does not occur at this acoustic pressure. Consequently, for an assigned alternating pressure and acoustic frequency there exists a minimum and a maximum bubble radius at which cavitation can take place. The minimum radius is determined by Eqs. (6) and (7), and the maximum is determined by the resonance radius R_r, which can be found from the familiar expression derived by Minnaert [26]:

$$f = \frac{1}{2\pi R_r} \sqrt{\frac{3\gamma}{\rho}\left(P + \frac{2\sigma}{R_r}\right)}, \tag{8}$$

where f is the resonance frequency of a bubble of radius R_r, $\gamma = c_p/c_v$ is the specific heat ratio for the gas in the bubble, and ρ is

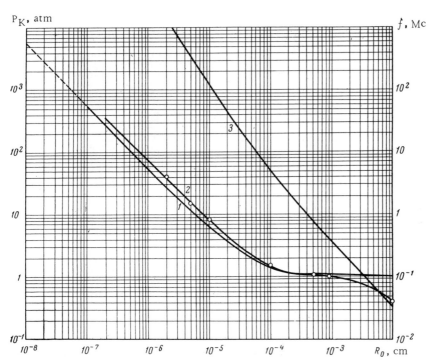

Fig. 5. Theoretical strength of water as a function of the radius of the bubble nucleus. 1) Without regard for the acoustic frequency; 2) with regard for the acoustic frequency; 3) resonance bubble radius.

the density of the liquid. Equation (8) was deduced, of course, for harmonic oscillations of the bubble at small amplitudes. In the event of cavitation, when the oscillations of the bubble are no longer harmonic, this expression is not strictly applicable. However, it can be used to estimate the order of magnitude of the radius R_r characterizing the frequency limit of cavitation [25]. Curve 3 in Fig. 5 indicates the acoustic frequency above which a bubble of radius $R_0 = R_r$ cannot collapse. The illustrated dependence was calculated according to Eq. (8).

As the frequency is increased, the only bubbles that can cavitate are those whose resonance frequencies are higher, i.e., bubbles of ever-diminishing radii. The cavitation threshold in this case increases. Applying expressions (6) and (8) simultaneously, we can roughly estimate the cavitation threshold for a nucleus in bubble form with variation of the acoustic frequency.

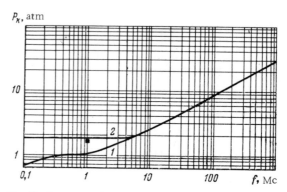

Fig. 6. Frequency dependence of the cavitation strength
of water. 1) Nuclei from 10^{-5} to 10^{-2} cm in radius present in the water; 2) only nuclei of radius smaller than
10^{-5} cm present in the water.

This can also be accomplished graphically with the aid of Fig. 5.
The same dependence, plotted graphically by means of curves 2
and 3 of Fig. 5, is shown in Fig. 6 (curve 1).

The frequency dependence illustrated by curve 1 of Fig. 6 for
the cavitation strength of water occurs if the water contains nuclei
of all possible radii, up to 10^{-3} cm, as in the case of sea water
or fresh tap water. For liquids in which there are no nuclei of
such large dimensions (for example, standing water) the form of
the frequency dependence is somewhat different, because the cavitation threshold is higher. Thus, about 0.5 liter of standing distilled water, as will become apparent presently (Fig. 7), contains
just one nucleus having a radius of about 10^{-5} cm, so that the cavitation threshold for this water increases to 1.8 atm (see Fig. 5).
Accordingly, the frequency dependence of the cavitation strength
of standing water acquires the form represented by curve 2 of
Fig. 6.

The cavitation threshold and its frequency dependence change
if the volume of the liquid also changes. In fact, if the acoustic
pressure is concentrated in a small domain, as in the case of
focusing concentrators, the cavitation threshold increases sharply
relatively to the threshold for a larger volume of liquid, because
the probability of a nucleus of large radius being present in the
focal spot is small. The latter consideration is an implication of
the bubble distribution shown in Fig. 7.

Therefore, the value of the cavitation strength of a liquid is determined by the largest of all the nuclei present in the investigated volume of liquid (if $R_0 \le R_r$).

Unfortunately, the many experiments aimed at determining the cavitation strength of one and the same water medium, for example [7], have been conducted under dissimilar experimental conditions, specifically with different volumes of irradiated water and, hence, with dissimilar concentrations of nuclei in the sample, a fact that was not taken into account. This oversight imposes a very broad scatter in the experimental values obtained by different authors for the cavitation threshold, from a few to hundreds of atmospheres (see Fig. 1).

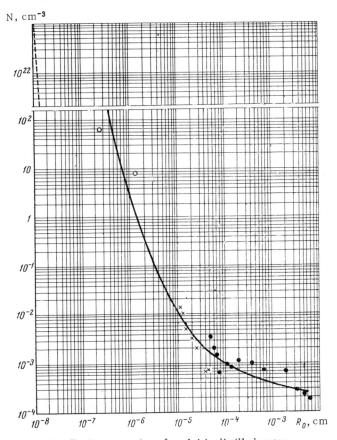

Fig. 7. Concentration of nuclei in distilled water.

The one orthodox experiment conducted recently to determine the cavitation strength of distilled water [8] demonstrated that it is equal to 1.7 atm at a frequency of 1 Mc [this value is represented in Fig. 6 by the small square and is very consistent with the theoretical value for the indicated frequency (curve 2)].

§ 4. Size Distribution of the Nuclei

If we let N denote the number of cavitation nuclei whose radii are between R_0 and $R_0 + \Delta R$, then the probability that these nuclei will occur in a volume V comprising a small part of the volume V_1 is equal to $\omega_N = 1 - (1 - V/V_1)_N$. But the probability of cavitation setting in at ω_K for an average acoustic pressure P_K in the volume V is equal to the probability of finding one nucleus in that volume:

$$P_{\kappa} = \frac{\Sigma P_{\kappa i} (\omega_{\kappa})_i}{\Sigma (\omega_{\kappa})_i},$$

where i is the order number of the pressure variation subinterval.

The probability of the onset of cavitation can be determined by metering batches of liquid from a volume V_1 into a container of volume V with an acoustic pressure P_{Ki} acting on the liquid, so that $\omega_N = \omega_K$, i.e.,

$$(\omega_{\kappa})_i = 1 - (1 - V/V_1)^{N(P_{\kappa i})}, \tag{9}$$

where $N(P_{Ki})$ is the number of nuclei in V capable of producing cavitation in a sound field at a pressure between P_{Ki-1} and P_K.

The number of nuclei as a function of the acoustic pressure

$$N_i = N (P_{\kappa i}) \tag{10}$$

is easily obtained from expression (9).

The foregoing recommendation for the determination of $N(P_{Ki})$ was suggested by Messino, Sette, and Wanderlingh [8]. They performed a great many experiments, which enabled them to obtain the number N of nuclei per unit volume of distilled water that would be capable of eliciting cavitation under variation of the acoustic pressure between P_{Ki} and P_{Ki-1} (the pressure was varied stepwise in steps of 0.1 atm).

Using the relation indicated in Fig. 5 between R_0 and P_K, we can go from the function (10) to the new function

$$N = N (R_{0i}). \tag{11}$$

The quantitative distribution of the nuclei with respect to their radii can also be deduced in the event of a sharply delineated region in which a known acoustic pressure has been set up equal to the cavitation threshold. Then on the basis of the same relation, as that shown in Fig. 5, between the cavitation threshold and the radius of the nuclei we can find the radii of the nuclei in the volume V (and their number N per unit volume N = 1/V). It is assumed, accordingly, that the presence of one nucleus in the focal zone is sufficient for the initiation of cavitation and that there is no displacement of the liquid in that zone. The sharply delineated region may be interpreted as the focal spot of a focusing system with a large aperture angle.

Our experiments to determine the number of nuclei per unit volume of standing distilled water (with a gas content of 0.025 cm^3/ml) as a function of the radii of the nuclei were carried out at frequencies of 0.5 and 1 Mc with focusing concentrators characterized by focal spots of volume $1.2 \cdot 10^{-1}$ cm^3 [27] and $1.5 \cdot 10^{-2}$ cm^3 [28].

The method of determining the cavitation threshold was analogous to that described in [8]. The average pressure in the focal spot was elevated in steps up until the instant of cavitation initiation. Also to be considered here is another factor that affects the strength of the liquid, namely, the irradiation time. Thus, due to the nonuniformity of the sound field, a certain time is required in order for the nuclei to enter a region of higher pressure. With prolonged irradiation of the liquid, the bubbles in it are caused by the diffusion of gas into them to grow and weaken accordingly. Our investigations confirmed this hypothesis; a long irradiation time (more than 20 sec) brought about a reduction in the strength of the water; for a short irradiation time (3-8 sec) the threshold is raised somewhat, becoming stable in this time interval with a maximum scatter of 10%. Consequently, the time in which the cavitation threshold was determined amounted to about 5 sec.

Movement of the liquid in the focal spot was excluded by the placement in front of the focal spot (at a distance of 25 mm) of a thin acoustically transmissive film representing the bottom of the reactor vessels of the concentrators [27, 28], and blocking any steady currents in the liquid.

TABLE 1. Experimental and Theoretical Results

Cavitation threshold	Pressure P_K = 0.02U atm (U = voltage on radiator) from [8] (see Fig. 6)	$5.4 \cdot 10^{-1}$	$6.4 \cdot 10^{-1}$	$7.4 \cdot 10^{-1}$	$8.4 \cdot 10^{-1}$	$9.4 \cdot 10^{-1}$	1.04	1.14	1.24	1.3
	Acoustic intensity (in W/cm²) from [8] (see Fig. 12)									
	Pressure $P_K \simeq (3I)^{1/2}$, atm									
	Pressure P_K (in atm) meas. with concentrators [27, 28]									
Radius of nuclei R_0 (in atm) (from Fig. 5)		$6.4 \cdot 10^{-3}$	$4.9 \cdot 10^{-3}$	$3.5 \cdot 10^{-3}$	$2 \cdot 10^{-3}$	$7 \cdot 10^{-4}$	$4 \cdot 10^{-4}$	$2 \cdot 10^{-4}$	$1.5 \cdot 10^{-4}$	$1.25 \cdot 10^{-4}$
Volume (in cm³) containing one nucleus	From [8] (see Fig. 12)									
	Volume of concentrator focal spots [27, 28]									
Number of nuclei per unit volume N, cm⁻³	From [8]	$1.8 \cdot 10^{-4}$	$2.2 \cdot 10^{-4}$	$2.8 \cdot 10^{-4}$	$7 \cdot 10^{-4}$	$7.5 \cdot 10^{-4}$	$1 \cdot 10^{-3}$	$1.1 \cdot 10^{-3}$	$8.5 \cdot 10^{-4}$	$9.5 \cdot 10^{-4}$
	1/V — From [8]									
	1/V — Our experiments									

The results of our experiments are summarized in Table 1, along with data borrowed from the above-mentioned paper [8] for P_K, V, and N; Fig. 5 provides a means for transforming from the function (10) to (11), i.e., to obtain the distribution of the nuclei with respect to their radii.

Figure 7 shows the quantitative distribution of nuclei in standing distilled water;* the dots, crosses, and circles denote the experimental data corresponding to the first, second, and third

* The quantitative distribution of the nuclei with respect to their radii in Fig. 7 differs somewhat from the distribution we presented in [29]. This is because, first, allowance was not made in [29] for the influence of the kinetic energy of the liquid on the oscillation of the bubble nucleus and, second, an unfortunate arithmetic error occurs in the last row of Table 2 in [29] (the last five figures are too large by a factor of 10^2).

Connecting the Radius and Density of Cavitation Nuclei

1,56	1,64	1.8	1.94								
				2	4	6	9	15	18		
				2.45	3.46	4.25	5.2	6.7	7.33		
										42,4	238
$8\cdot10^{-5}$	$7\cdot10^{-5}$	$6\cdot10^{-5}$	$5.2\cdot10^{-5}$	$4.5\cdot10^{-5}$	$2.5\cdot10^{-5}$	$2\cdot10^{-5}$	$1.55\cdot10^{-5}$	$1.2\cdot10^{-5}$	$1.1\cdot10^{-5}$	$1.4\cdot10^{-6}$	$2.5\cdot10^{-7}$
				$1.2\cdot10^{3}$	$5\cdot10^{3}$	$3\cdot10^{2}$	2.10^{2}	$1\cdot10^{2}$	80		
										$1.2\cdot10^{-1}$	$1.5\cdot10^{-2}$
$6.5\cdot10^{-4}$	$1.6\cdot10^{-3}$	$2\cdot10^{-3}$	$3.45\cdot10^{-3}$								
				$5.9\cdot10^{-4}$	2.10^{-3}	$3.3\cdot10^{-3}$	$5\cdot10^{-3}$	1.10^{-2}	$1.25\cdot10^{-2}$		
										8.35	67

rows of the entry N in Table 1. The average curve drawn through those experimental data tends to the concentrations of nuclei whose radii correspond to the intermolecular separations (dashed part of the curve). Whereas the solid curve characterizes the concentration of nuclei in the form of bubbles with radii up to about 10^{-7} cm, the dashed part pertains to nuclei to be interpreted as inhomogeneities having a lifetime on the order of 10^{-14} sec, which are created by δ-electrons permeating the liquid [17], as well as inhomogeneities generated by thermal fluctuation processes [6].

It is apparent from Fig. 7 that the distribution of nuclei in the form of vapor-gas bubbles in standing distilled water is exceedingly nonuniform; the number of nuclei of small radius ($< 10^{-6}$ cm) is approximately 10^{5} times the number of nuclei of large radius

($> 10^{-3}$ cm). This has an important consequence. The nonuniform distribution of the nuclei implies a dependence of the cavitation strength of water on the volume of the liquid irradiated. The cavitation strength increases as the latter volume is decreased, because the smaller the irradiated volume, the smaller the probability of a large nucleus existing therein. The strength of the liquid increases, however, as the size of the nucleus diminishes.

Chapter 2

Characteristics of Cavitation Bubbles and Properties of the Cavitation Zone

§ 1. Oscillations of Cavitation Bubbles in a Sound Field

The equations describing the behavior of a cavitation bubble in a sound field that varies according to the law $P_a = P_m \sin \omega t$ (where P_m is the acoustic pressure amplitude and ω is the acoustic frequency) relate the kinetic energy of the additional mass of the liquid and the sum of the work performed by the surface tension, pressure of the gas in the bubble, and pressure in the liquid [3, 25] (see Part IV, p. 207). The numerical solutions of the differential equations show that a bubble acquires a certain velocity in the expansion half-period and, after expanding inertially to a maximum radius R_{max}, collapses at an ever-increasing rate under the influence of the positive pressure in the liquid. This behavior on the part of the cavitation bubble is qualitatively confirmed by high-speed motion pictures showing the variation of the bubble radius at separate instants of time [30, 31].

The actual correspondence between the theoretical and experimental variation of the diameter 2R of a cavitation bubble in the time T (T is the acoustic period) is shown in Fig. 8. Here curve 1 represents the computer solution of Eq. (26) of Part IV describing the behavior of a cavitation bubble in a real liquid (with regard for its compressibility and viscosity). The initial

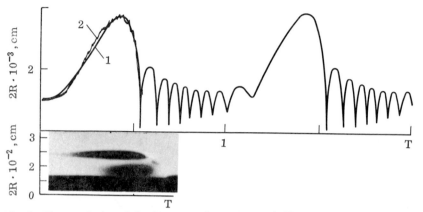

Fig. 8. Time variation of the diameter of a cavitation bubble. 1) Theoretical; 2) experimental. Below the graph is a photograph through a microscope, obtained with a high-speed motion picture camera, showing the track of the cavitation bubble.

conditions were as follows: equilibrium radius* $R_0' = 10^{-3}$ cm; $P_m = 1.2$ atm; $f = 28$ kc. After the first collapse the cavitation bubble pulsates at a frequency close to the natural frequency determined by Eq. (8) (see Part IV for the details). Superimposed on the theoretical curve (1) is the experimental variation of the diameter of the cavitation bubble (curve 2, which represents the result of the photometric analysis, using an MF-4 microphotometer, of a photograph showing the behavior with time of a single cavitation bubble). The photograph itself, which we obtained by means of an SFR ultrahigh-speed motion picture camera operating in the photorecorder mode with continuous scanning of the process on stationary film, is shown in the lower part of Fig. 8. The light source (an IFK-120 pulse lamp) was located behind the cavitation bubble so that the wake of the latter (shadow) would be distinctly visible against the light background. A little further below we see the track of another bubble outside the focus of the objective. Below the bubbles is seen the edge of the tip of the concentrator (tip dimensions 1 × 4 mm), which was driven at a frequency

---

*The equilibrium radius R_0' differs from the initial radius R_0 due to the elevated content of gas that diffuses into the bubble during its pulsation, i.e., so-called rectified diffusion; R_0' is the radius that would be measured immediately after the ultrasonic field is turned off.

of 28 kc by a ferrite transducer. As indicated from Fig. 8, the
theoretical and experimental curves exhibit good agreement,
evincing the fact that Eq. (26) in Part IV adequately describes the
true behavior of a cavitation bubble in a sound field. The fine-
structured periodicity determined by the natural pulsation frequency of the bubble was not observed in the experiment due to the
inadequate magnification of the optical system (these pulsations
are partially visible in Fig. 9).

The phase relations between the variation of the acoustic
prssure and the diameter of the cavitation bubble are clearly revealed in Fig. 9. The middle section of the figure shows the time
behavior of a cavitation bubble produced directly on the tip of a
working vibrator; the upper section shows the result of photometric analysis of the track left by the bubble. For better recording of the amplitude and phase of the oscillations of the tip
of the concentrator, the latter was additionally illuminated from
the front with a pulse lamp. The time oscillations of the concentrator are seen in the photograph as light sinusoidal lines. The
scale is shown on the left; one large division is equal to 10^{-2} cm.

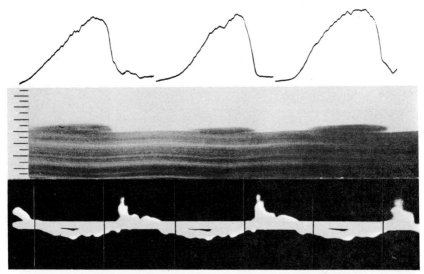

Fig. 9. Photograph of a cavitation bubble at the tip of a working vibrator (middle
section). The upper section gives the result of photometry of the track, and the lower
section shows the acoustic pressure near the cavitation bubble.

The variation of the pressure near the cavitation bubbles was recorded with a miniature hydrophone (with a spherical piezoceramic about $3 \cdot 10^{-2}$ cm in diameter) with a flat frequency characteristic up to 10 Mc. The broad frequency curve of the subsystem made it possible to ascertain the pressure buildup time in the shock wave with little distortion. The hydrophone was placed at a distance of $\sim 2R_{max}$ from the tip of the concentrator over the site of initiation and collapse of the cavitation bubbles. The pressure pulses, thus generated, are shown in the oscillogram in the lower part of Fig. 9.

For greater clarity vertical lines are drawn through the points in Fig. 9 corresponding to the passage of the vibrator tip through the neutral plane (zero acoustic pressure). It is evident that the bubble expands in the negative half-period and attains maximum radius R_{max} at the very end of that period, then collapses until the beginning of the positive half-period, when $P_a \simeq 0$ (Fig. 9 was obtained for a small acoustic pressure $P_m < 1$, when only one cavitation bubble was formed in the field of view of the microscope).

With an increase in the acoustic pressure, the number of cavitation bubbles increases. The upper part of Fig. 10, which was

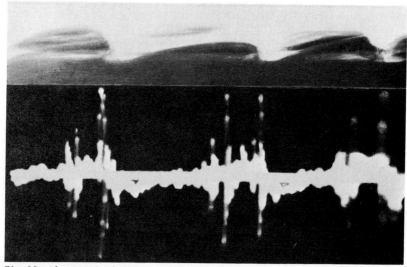

Fig. 10. Photograph of several cavitation bubbles. The lower section shows an oscillogram of the pressures induced by the bubbles.

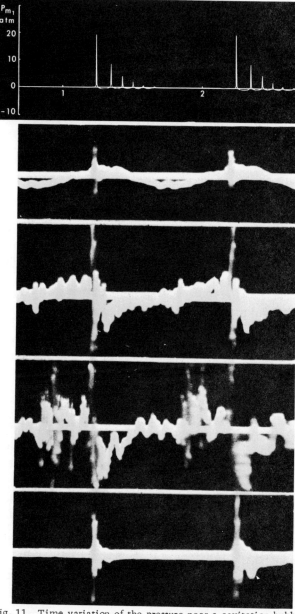

Fig. 11. Time variation of the pressure near a cavitation bubble. The upper section shows the theoretical variation; the lower photographs give the experimental data for various amplitudes of the vibrator tip oscillations.

obtained as Fig. 9, but at a larger oscillation amplitude, shows images of several cavitation bubbles (some of them are outside the focus of the microscope); the lower part shows an oscillogram of the pressures corresponding to the oscillation amplitude. It is clear that the bubbles do not collapse simultaneously and that the time shift between them can attain 0.5T (see also Fig. 11).

It follows from Figs. 9 and 10 that the collapse of the cavitation bubbles can begin either at the end of the negative or at the beginning of the positive half-period.

§2. Acoustic Pressure in the Cavitation Zone

The upper part of Fig. 11 shows the variation in time T of the pressure p_m near a cavitation bubble according to a numerical solution of the Kirkwood–Bethe equation [see Eq. (36) of Part IV]. The calculation applies to the pressures near the bubble for the case when $r/R_0 = 100$ (r is the distance from the center of the bubble). The pulsations of the bubble are represented by the heavy curve in Fig. 8. In the first collapse of the bubble, the pressure p_m attains a very large value; the amplitude of the succeeding pulses generated by the resonance pulsations of the bubble is monotone-decreasing.

Also shown in Fig. 11 are photographs of oscillograms indicating the variation of the acoustic pressure near cavitation bubbles, as recorded with a miniature hydrophone by the procedure described above (the hydrophone was situated at a distance of $\sim 10^{-2}$ cm beneath the tip of the vibrator). The upper oscillogram was obtained for a vibrator tip displacement amplitude of $\sim 0.1\ \mu$. Each succeeding oscillogram corresponds to an oscillation amplitude $0.3\ \mu$ larger than the preceding one.

For small oscillations of the vibrator tip, such that a single cavitation bubble is formed on it (first oscillogram), good qualitative agreement is observed between the real and theoretical relations between p_m and T. An increase in the oscillation amplitude leads, first, to the formation of a large number of cavitation bubbles, which do not collapse in phase, and, second, to an abatement of the vibrator-induced acoustic pressure P_a in the liquid with the development of cavitation (the oscillograms are shown in the same scale).

From Fig. 11 we draw a conclusion of practical import, that in the event of weakly developed cavitation the collapse of the bubbles terminates at the beginning of the positive half-period, when the amplitude of the acoustic pressure is close to zero (see Fig. 9); for well-developed cavitation the actual value of P_a drops appreciably.

Thus, the hydrophone measurement of the acoustic pressure showed that the average acoustic pressure P_C in the cavitation zone decreases with increasing radiated acoustic power [33]. The dependence of the average acoustic pressure P_c in the cavitation zone on the electrical voltage on the piezoelectric radiator is shown in Fig. 12, where the voltage U on the quartz mosaic transducer of a high-intensity focusing concentrator operating at 500 kc [27] is plotted on the horizontal, and the average pressure P_c is plotted on the vertical.

The acoustic pressure was measured by means of a waveguide receiver protected (in the absence of the miniature hydrophone [32]) against the destructive action of cavitation [33]; the emf of the hydrophone, located at the center of the focal spot of the concentrator, was measured with a vacuum-tube voltmeter, which averaged its readings. Curve 1 was obtained with a receiver having a flat frequency characteristic up to 3 Mc, and curve 2 was obtained with the same receiver, but with the insertion of a filter passing only the radiated sound frequency of 500 kc (pass band of the filter, 10 kc). In the absence of cavitation the dependence of P_c on U would be linear (dashed line). With the onset of cavitation in the focal spot (P_c =

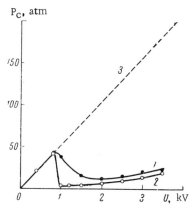

Fig. 12. Average acoustic pressure in the cavitation zone.

$P_K \simeq 42$ atm [29]), the acoustic pressure P_c, despite the growth of U, at first drops abruptly, increasing slightly only with a further increase in U but remaining at all times smaller than P_K. As expected, the wideband receiver (curve 1), which permits recording of the higher harmonics of the spectrum, measured an acoustic pressure higher than that measured by the narrowband receiver (curve 2).

It is impossible, however, to make a precise estimate of P_c with the hydrophone placed in the cavitation zone, even if its frequency characteristic allowed it to perceive the entire spectrum of the shock wave. The hydrophone does not afford information on the energy of shock waves or the true acoustic pressure; hence, it perceives only the total spectrum from a very small number of cavitation bubbles collapsing on the immediate surface. The highest-frequency components of the spectrum, in which the main portion of the shock wave energy is concentrated, are absorbed along the path to the hydrophone. The energy of the remaining spectral components is reflected diffusely from the container walls and surface of the liquid to form a roughly uniform diffuse field throughout the entire volume. This can be verified by placing the hydrophone in the cavitation zone or outside it. The situation

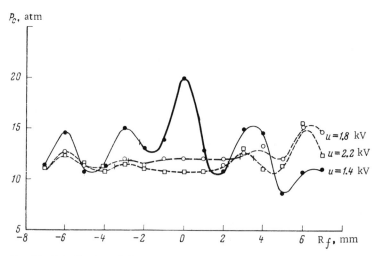

Fig. 13. Distribution of the average acoustic pressure along an axis intersecting the focal spot of the concentrator at various stages of development of the cavitation zone.

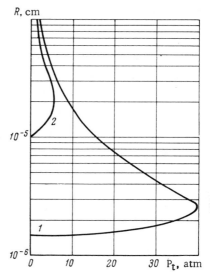

Fig. 14. Cavitation bubble radius versus the tensile stress in the liquid. 1) $R_0 = 1.5 \cdot 10^{-6}$ cm; 2) $R_0 = 10^{-5}$ cm.

is confirmed in Fig. 13, which shows the average acoustic pressure P_c perceived by a waveguide receiver moving along the axis R_f, which lies in the focal plane of the concentrator and intersects the cavitation zone (point O corresponds to the center of the focal spot). The diameter $2R_K$ of the visible cavitation zone for concentrator voltages U = 1.4, 1.8, and 2.2 kV (see Fig. 21) are indicated on the P_c curves by vertical thick marks, and the pressure at the limits of this zone are marked by heavy lines. As apparent from Fig. 13, the displacement of the receiver from the cavitation zone to the zone in which cavitation is absent does not give a pressure jump. This was pointed out by Rozenberg [34]. Whereas in the case of weakly developed cavitation, when $2R_K < 4$ mm [27], the structure of the focal spot is still apparent (solid curve in Fig. 13); for well-developed cavitation, when $2R_K > 4$ mm, the structure of the focal spot breaks down, and the acoustic pressure becomes approximately uniform; only far from the cavitation zone are the vestiges of secondary maxima noticeable (dashed curves).

The reduction in the acoustic pressure with the development of cavitation may be interpreted as follows. The tensile stress around a nucleus in the liquid can increase until the nucleus is destroyed. Equation (5) makes it possible to find the dependence of the radius R of the bubble nucleus on the tensile stress (pressure) P_t. The relations calculated according to that equation be-

tween R and P_t are shown in Fig. 14 for two nuclei: $R_0 = 1.5 \cdot 10^{-6}$
(curve 1) and $R_0 = 1 \cdot 10^{-5}$ cm (curve 2) ($P_v = 0.02$ atm; $P = 1$ atm;
$\delta = 75$ dyn/cm). It is apparent from Fig. 14 that the cavitation
bubble is stable until the tensile stress P_t reaches a maximum
(cavitation threshold for $R_0 = 1.5 \cdot 10^{-6}$ cm, $P_t = P_K \simeq 40$ atm;
for $R_0 = 1 \cdot 10^{-5}$ cm, $P_t = P_K \simeq 5.5$ atm). Just a slight further
increase in the tension causes an unlimited growth of the bubble
and breakdown of the liquid, whereupon the liquid essentially no
longer offers any resistance to expansion (apart from energy
losses in the work of vaporization of the liquid in the cavity). As
the bubble distends, the pressure around it drops, as indicated by
the branches running to the left of the point of discontinuity in Fig.
14. The presence of a large number of cavitation bubbles induces
a drop in the tensile pressure over the entire cavitation zone, as
disclosed by a hydrophone situated in that zone. For well-developed cavitation the cavitation bubble concentration (see §6 of the
present chapter) increases, and the formation of cavitation bubbles
occurs approximately throughout the entire negative half-period
(see Fig. 10), so that the cavitation-inducing acoustic pressure
near the bubbles diminishes with the development of the cavitation zone (Figs. 11 and 12). This is particularly graphic in
curve 2 of Fig. 12, which corresponds to an average acoustic
pressure $P_c \simeq P_t$ in the cavitation zone (a certain increase in P_c
is brought about by the reduction in concentration of cavitation
bubbles for large values of U on the concentrator; see Fig. 28).

§3.   Maximum Cavitation Bubble Radius

On the basis of the considerations set forth in the preceding
section, we can draw a qualitative picture of the pressure variation near a cavitation bubble. The upper part of Fig. 15 illustrates
this variation as a function of time.

When $P_a = P_K$ the liquid breaks down, and the resulting cavity expands during the time $T/4$, and if $P_a > P_K$ it does so in the
period $T/4 + t_1$. After breakdown the pressure near the bubble, as
indicated in Fig. 14, drops and, as the bubble expands, continues
to diminish. The magnitude $P_t(R)$ of the tensile stress acting at
the given instant can be found from Eq. (5). The pressure drop
near the individual bubble is clearly visible in the oscillogram in
the lower part of Fig. 15; the negative half-period is not symmetric, as its left side has a slight dip. This oscillogram, which

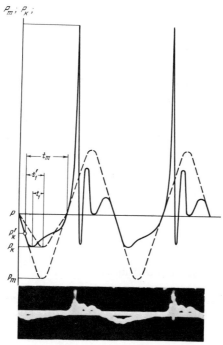

Fig. 15. Schematic graph and oscillogram of
the pressure variation near a cavitation bubble.

bears out the qualitative picture just portrayed, was obtained
with a miniature hydrophone according to the procedure outlined
in §1 of the present chapter.

During the next half-period of the sound field, the bubble rapidly collapses under the effect of the positive pressure, and the pressure inside it attains a value $p_m \gg P_m$. Subsequently, the bubbles
execute damped natural oscillations and, accordingly, acquire pressure inside.

It is evident from Fig. 15 that the time in which the bubble is
acted upon by the tensile stress P_t depends on P_m and P_K; an increase in the pressure P_m induces an increase in t_1; a decrease
in the cavitation threshold P_K to a value P'_K (this is possible in
stationary cavitation by virtue of growth of the nuclei) also increases this time to a value t'_1. It is readily seen that the time t_m
in which the bubble is acted upon by the tensile stress (without

regard for the inertial forces) is equal to

$$t_m = \frac{T}{4} + \frac{T}{4} \cos^{-1} \frac{P_K}{P_m} . \tag{12}$$

The function $\cos^{-1}(P_K/P_m)$ is approximately linear, as long as $P_K/P_m \le 0.5$. On this linear interval

$$t_m \simeq \frac{T}{4} + \frac{T}{4} \left(1 - \frac{P_K}{P_m} \right) .$$

We know [1] that, if a tensile stress P_t acts on the bubble for a time t_m, the radius of the bubble increases by the law

$$R \simeq \sqrt{\frac{2}{3} \frac{P_t}{\rho}} \, t_m,$$

where ρ is the density of the liquid.

This expression allows us to estimate the maximum cavitation bubble radius R_{max}. Bearing Eq. (12) in mind and setting $P_t = P_t(R)$, we obtain

$$R_{max} \simeq \sqrt{\frac{2}{3} \frac{P_t(R)}{\rho_K}} \frac{T}{4} \left(1 + \cos^{-1} \frac{P_K}{P_m} \right) , \tag{13}$$

or for the case when $P_K \ll P_m$,

$$R_{max} \simeq \sqrt{\frac{2}{3} \frac{P_t(R)}{\rho_K}} \frac{T}{4} \left(2 - \frac{P_K}{P_m} \right) ,$$

where ρ_K is the average density of the liquid in the cavitation zone [42]; $\rho_K = \rho(1 - K) + \rho_n K$; here ρ_n is the density of the vapor-gas mixture in the cavity, and K is the cavitation index [64]. Inasmuch as $K \simeq 0.1$, even for strongly developed cavitation, it is reasonable to assume that $\rho_K \simeq \rho$.

Curve 1 in Fig. 16 represents the dependence of R_{max} on P_m according to Eq. (13) in the cavitation zone of a focusing concentrator [27] for which $T = 10^{-6}$ sec and $P_K \simeq 42\sqrt{2}$ [29]. Figure 12 permits us to estimate $P_t(R) \simeq P_c/2$ (curve 2) for various values of $P_m = \sqrt{2} P_c'$ (line 3) corresponding to a definite radius R_{max} (the electrical voltage transmitted from the hydrophone to an averaging

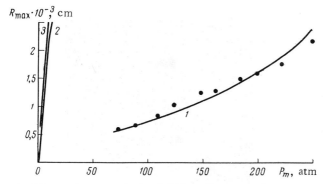

Fig. 16. Maximum cavitation bubble radius versus the acoustic
pressure amplitude.

double half-period diode-capacitor meter and MVL voltmeter is
larger in the positive than in the negative half-period, so that the
average value of the half-period is equal to $P_c / 2$).

The points in Fig. 16 correspond to the experimental data obtained in the cavitation zone of a focusing concentrator. The value
of R_{max} was determined by taking photographs of the cavitation
bubbles through a microscope with the light pulse from a spark
having an average flash time ($\sim 3 \cdot 10^{-6}$ sec) somewhat longer than
the acoustic period, so that the bubble could always attain the
maximum radius [35].

As an example, Fig. 17 shows one of the photographs taken of
bubbles in the focal zone of the concentrator. During processing
of the film, the arithmetic average of about ten of the most distinctly visible bubbles was taken. The maximum scatter or their
radii was ±7%. The photographing of R_{max} was carried out for
several voltages U on the concentrator.

The dependence of R_{max} on U is shown in Fig. 18; also plotted
on the horizontal axis is the scale of the acoustic pressure $P_m =
\sqrt{2}P_c$ that should exist by hypothesis if cavitation were not present.

It is evident from Fig. 16, therefore, that the foregoing estimate of R_{max} from the proposed equations (13) is in sufficiently
good agreement with the absolute value of R_{max} determined experimentally. The attendant discrepancy is attributable to the replacement of $P_t(R)$ by $P_c/2$ in Eq. (13), as well as to the supposi-

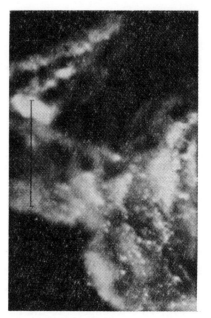

Fig. 17. Photograph of cavitation bubbles.

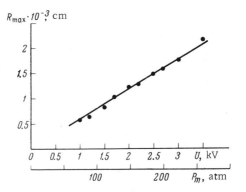

Fig. 18. Experimental dependence of the maximum cavitation bubble radius on the electrical voltage on the piezoelectric radiator (a focusing concentrator).

tion that P_K = const, whereas in point of fact P_K diminishes with the development of cavitation. The latter situation is abetted by the formation of new, weaker nuclei (see §6).

Curve 2 in Fig. 16 illustrates the variation of R_{max} as a function of P_m according to an analysis of the numerical solutions of the differential equation [25] describing the pulsations of cavitation bubbles in a sound field (see Part VI, Fig. 3). The solution was obtained for $R_0 = 10^{-4}$ cm and f = 500 kc. It is clear that the curve departs markedly from the actual dependence of R_{max} on P_m. This discrepancy is attributable to the fact that the equation describing the pulsations of cavitation bubbles [25] does not allow for the variations of the tensile stress after breakdown of the liquid; if prior to breakdown and the formation of cavitation bubbles the pressure was $P_m \sin \omega t$, then after breakdown the pressure around the bubbles would diminish to $P_t(R)$.

If we put $P_t(R) = P_m$ in Eq. (13) ($P_K \simeq 1.5$ atm for $R_0 = 10^{-4}$ cm), the resulting dependence of R_{max} on P_m (curve 3) satisfactorily agrees with curve 2 in Fig. 16.

§4. Gas Pressure in a Bubble at the Time of Collapse; the Gas Content Parameter

If we assume that the vapor-gas mixture contained in a real cavitation bubble under adiabatic compression behaves as an ideal gas, then during the collapse of the bubble from its maximum radius R_{max}, the pressure of the vapor-gas mixture in it increases inversely as the radius to the power 3γ (where γ is the adiabatic index of the vapor-air mixture and may be assumed equal to $\gamma = 4/3$ [3]). Then the pressure p_m, during collapse of the cavitation bubble to the minimum radius R_m, is equal to

$$p_m = P_{vg} \left(\frac{R_{max}}{R_{min}} \right)^4, \tag{14}$$

where P_{vg} is the pressure inside the cavity at maximum radius R_{max} and is equal to the sum of the partial pressures P_v of the water vapor and P_g of the gas (air). The pressure P_{vg} depends on the quantity of vapor-gas mixture and on R_{max}, which is determined in turn by the acoustic pressure P_a. The relationship of P_{vg} to these variables can be found by comparison of the work A_l done by the liquid in compressing the bubble from radius R_{max} to R_{min} with the work A_g expended in compressing the vapor-gas mixture in that bubble.

The work of the hydrostatic pressure P and the acoustic pressure P_a acting during compression of the bubble is equal in general to

$$A_l' = {}^4/_3\pi (P_a + P)(R_{max}^3 - R_{min}^3). \tag{15}$$

However, as implied by §§1 and 2 of the present chapter, the bubble collapses in the interim between the end of the negative half-period, when P_a is still negative, and the beginning of the positive half-period, when P_a is small. Therefore, for the case of periodic acoustic cavitation, expression (15) can be written with a certain approximation in the form

$$A_l \simeq {}^4/_3\pi P (R_{max}^3 - R_{min}^3). \tag{16}$$

The work A_g done in compression of the vapor-gas mixture in the bubble during its collapse from the maximum radius R_{max} to R is equal to

$$A_g = \int_{R_{max}}^{R} P_{vg}\left(\frac{R_{max}}{R}\right)^{3\gamma} 4\pi R^2 dR. \tag{17}$$

The computation of the integral (17) for the case $R = R_{min}$ yields

$$A_g = - 4\pi P_{vg} R_{max}^3 \left(\frac{R_{max}}{R_{min}} - 1\right) \tag{18}$$

(the minus sign, which accounts for the direction of the force, will be omitted henceforth).

Equations (15) and (17) do not allow for the action of the surface tension forces, which are felt only at exceedingly small values of the radius.

Inasmuch as $A_l = A_g$ and $R_{max} \gg R_{min}$, we obtain the relation

$$\frac{R_{min}}{R_{max}} \simeq 3\frac{P_{vg}}{P}. \tag{19}$$

By analogy with [36] we refer to the ratio P_{vg}/P as the gas content parameter δ, which characterizes the amount of vapor-gas mixture contained in the bubble; then from (19) we have

$$\frac{R_{min}}{R_{max}} \simeq 3\delta. \tag{20}$$

The ratio R_{min}/R_{max} may be treated as the dimensionless minimum radius.

In [36] the dimensionless minimum radius is determined according to the semi-empirical formula

$$\frac{R_{min}}{R_{max}} \simeq \frac{3\delta}{1 + 3\delta - \delta^{1.6}}. \tag{21}$$

Ordinarily $\delta \ll 1$ or, equivalently, $R_{max} \gg R_{min}$; then (21) goes over to our expression (20).

Using (20), we can write Eq. (14) in the form

$$p_m \simeq \frac{1}{81\delta^3}\, p\,, \tag{22}$$

from which it is clear that the pressure in the bubble at the time of collapse is completely determined for given P by the vapor-gas content parameter $\delta = P_{vg}/P$. For the estimation of the real

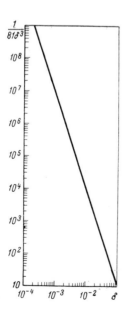

Fig. 19. Dimensionless factor $1/81\,\delta^3$ versus the gas content parameter.

values of p_m, Fig. 19 shows the dimensionless factor $(1/81)\,\delta^3$ in Eq. (22) as a function of the parameter δ.

§ 5. The Collapse Time

Investigating the problem of the collapse of an empty bubble in an incompressible liquid, Rayleigh [37] equated the kinetic energy of a liquid of density ρ admitted into a bubble of radius R

$$A = \frac{\rho}{2} \int_R^\infty v^2 4\pi R^2 dR = 2\pi\rho v^2 R^3 \tag{23}$$

to the work A_l done by the liquid when the bubble is shrunk from a radius $R_{\max}$ to R under the influence of the hydrostatic pressure P [Eq. (15), in which $P_a = 0$] and obtained the average rate of collapse of the empty bubble:

$$v^2 = \frac{2}{3}\frac{P}{\rho}\left(\frac{R_{\max}^3}{R^2} - 1\right).$$

From this Rayleigh invoked some straightforward transformations to find the total collapse time of the bubble:

$$\tau = 0.915 R_{\max}\sqrt{\frac{\rho}{P}}. \tag{24}$$

Expression (24) fails to take account of, first, the energy A_g of compression of the vapor-gas mixture that is always present in a real bubble and, second, the action of the acoustic pressure, which can take place in acoustic cavitation. The latter consideration is all the more justified by the fact that the collapse time determined from the empirical formula

$$\tau = F R_{\max}\sqrt{\frac{\rho}{P_m + P}},$$

which was proposed by Akulichev [38], closely coincides with the collapse time computed directly from the theoretical radius-time curves (see Part IV, Figs. 3-5) obtained by numerical solution of the equation describing the pulsations of a cavitation bubble [25] [even better agreement is obtained if the pressure amplitude P_m in Akulichev's formula is replaced by the average pressure over the half-period, $(2/\pi)P_m$].

However, as already demonstrated, in the collapse of a bubble under real conditions the acoustic pressure P_a is small (see §§1 and 2 of the present chapter), so that the work done by the liquid on the bubble will be described by Eq. (16), and the balance of energy in the collapse of a cavitation bubble assumes the form

$$A = A_l + A_g.$$

Bearing (23), (16), (18), and (20) in mind, we obtain the average rate of collapse of a bubble in acoustic cavitation

$$v = \frac{2}{3} \frac{P}{\rho} \left[\left(\frac{R_{\max}^3}{R^3} - 1 \right) - 3\delta \frac{R_{\max}^3}{R^3} \left(\frac{R_{\max}}{R} - 1 \right) \right]^{1/2}. \tag{25}$$

Inasmuch as $v = dR/dt$, after the substitution of $R/R_{\max} = z$ and $R_{\min}/R_{\max} = z_{\min}$ and the transformation to new limits of integration, which are equal to one and $z_{\min}$ for R varying from $R_{\max}$ to $R_{\min}$, from Eq. (25) we arrive at the total collapse time

$$\tau = R_{\max} \sqrt{\frac{3}{2} \frac{\rho}{P}} \int_{z_{\min}}^{1} \frac{z^2 dz}{[(-z^4 + z) - 3\delta (1 - z)]^{1/2}}. \tag{26}$$

This equation is identical to the expression deduced for τ by Khoroshev [36], who computed the integral on the right-hand side. The dependence of this integral, denoted by F, on the gas content parameter $\delta = P_{vg}/P$ is shown in Fig. 20.

With regard for the dependence so obtained, the collapse time of the cavitation bubble is equal to

$$\tau = R_{\max} \sqrt{\rho/P} \, F. \tag{27}$$

For $\delta = 0$ this expression goes over to the Rayleigh formula. For $\delta > 0$ the length of the cavitation bubble collapse period increases. In real cavitation bubbles, the gas content parameter is usually small ($\delta \simeq 0.02$ or 0.03 [36]); hence, the coefficient F is essentially always close to unity.

As implied by expression (14), the pressure p_m inside the cavitation bubble during its collapse is higher, the larger the value of $R_{\max}$. However, this increase is not unlimited.

Back in 1950 Noltingk and Neppiras indicated [25] that for a high acoustic pressure, such that the bubble grows to a very large radius, its collapse time τ can prove to be so large as to prevent

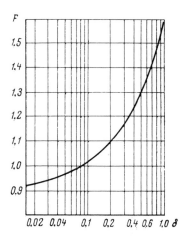

Fig. 20. Coefficient F versus the gas content parameter.

the collapse of the bubble from going to completion and to cause the acoustic pressure to change sign and become negative. This effect takes place when

$$t_m + \tau > \omega^{-1}(2\pi + \sin^{-1}\frac{P}{P_m}),$$

where t_m is the expansion time of the bubble to R_{max}.

An analysis [38] of the numerical solutions of the differential equation describing the behavior of a cavitation bubble in a sound field [25] has shown that if the acoustic pressure exceeds a certain limit, the bubble will pulsate for a time before eventually collapsing (see Part IV, Fig. 3). Here, of course, shock waves will not be formed in every period of oscillation, and their total effect will be less (for example, cavitation erosion). This logical conclusion has been confirmed by our experiments in [35], where we showed that when the bubble collapse time is equal to or greater than half the acoustic period ($\tau \geq 0.5T$), the bubble cannot collapse before the tensile stresses become effective.

In the radiators used for various engineering applications, the sound intensity is not usually sufficient to initiate pulsation of the bubble without collapse. However, in systems that concentrate ultrasonic energy the effect can indeed take place. It is readily observed experimentally [35]. We have estimated the shock wave intensity from the collapse of cavitation bubbles on the basis of the cavitation damage to an aluminum cylinder (of dimensions 2×2 mm^2) set in the focal spot of a concentrator [27] operating at 513 kc. The small cylinder was subjected to the action of cavitation at various concentrator voltages under a constant irradiation time (5 min). The weight loss ΔG of the cylinder, determined as the difference in weight before and after the application of ultrasound, served as a measure of the intensity of the cavitation bubble shock waves. The dependence of ΔG on U at the concentrator is shown in Fig. 21 (U is given on the lower horizontal scale). Each point represents the average of several measurements. An increase in U and, hence, in the acoustic pressure in the primary sound wave, as indicated by Fig. 18, produces an increase in R_{max} The value of p_m in the shock wave increases accordingly [see Eq. (14)]. It will be shown in §6 that the total number of cavitation bubbles grows concurrently. Consequently, the maximum of the curve in Fig. 21 may be attributed to these two factors. For now we consider one of them, namely, the influence of R_{max}.

It is reasonable to expect that as R_{max} increases, the pressure in the shock wave will increase at first, reaching a maximum at $\tau = 0.5T$ and then decreasing to the minimum value. Also shown in Fig. 21 is the dependence of the cavitation damage ΔG on the ratio $\tau / 0.5T$ (upper horizontal scale). The bubble collapse time τ was calculated from Eq. (27). The quantity R_{max} involved in that equation was found from Fig. 18 for the voltages V corresponding to the damage ΔG.

Fig. 21. Cavitation damage versus the ratio of the cavitation bubble collapse time to the acoustic half-period.

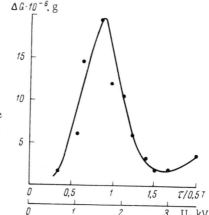

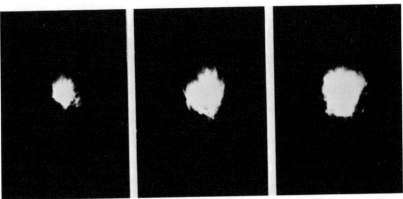

Fig. 22. Photographs of a cavitation zone in a focusing

Figure 21 reveals that as $\tau/0.5T$ increases, the cavitation damage actually increases, reaching a maximum when

$$\frac{\tau}{0.5T} \simeq 1. \tag{28}$$

With a further increase in $\tau/0.5T$ the damage, like the integral effect of the shock waves, diminishes, reaching a minimum at $\tau/0.5T \simeq 2$, when the cavitation bubbles produced in the negative half-period are unable to collapse in the second, positive part of the period.

§ 6. The Cavitation Zone and Number

of Cavitation Bubbles

The shape of the cavitation zone depends on the nature of the sound field. Thus, in a container whose dimensions are comparable with the acoustic wavelength, cavitation occurs both at the interfaces between the liquid and solid phases, where gas cavitation nuclei are always present, and in the liquid proper as strands and filaments consisting of a large cluster of cavitation bubbles. In devices that focus acoustic energy, cavitation occurs in the focal spot, where very large sound intensities are concentrated. The local cavitation zone thus created with a large density of cavitation bubbles is highly active and well-suited to research purposes. Photographs of the cavitation zone in a focusing concentrator operating at ~500 kc [27] with various voltages on

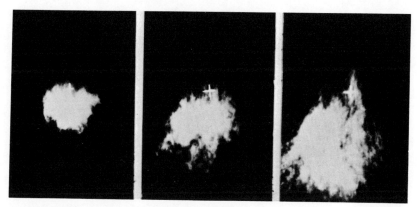

concentrator for various initial sound intensities.

the mosaic transducer (1.4, 1.8, 2.2, 2.6, 3.0, and 3.5 kV) are
shown in Fig. 22. The photographs had an exposure of 0.5 msec,
so that the photographs show the time-average (250 periods) configuration of the cavitation zone. It is seen that the cavitation
zone in a focusing system has an approximately spherical shape.
It is also apparent that as the voltage is increased the center of the
focal spot moves away from the focus of the system (marked by a
cross in the photographs) toward the radiating mosaic. We will
discuss this effect later. The variation of the radius R_K of the
cavitation zone as a function of the voltage squared u^2, which
characterizes the acoustic power radiated to the focal spot of
the concentrator, is shown in Fig. 23. The points correspond to
the values of R_K determined from the photographs of Fig. 22. As

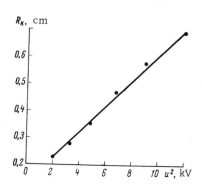

Fig. 23. Average radius of the cavitation
zone versus the square of the voltage on
the mosaic transducer of a focusing concentrator.

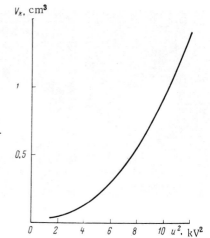

Fig. 24. Variation of the volume of the cavitation zone in a focusing concentrator.

shown in Fig. 23, the radius R_K of the cavitation zone increases approximately linearly as the acoustic power delivered to it. The dependence of the volume $V_K = \frac{4}{3}\pi R_K^3$ on the voltage squared is shown in Fig. 24. We recall that the acoustic pressure in the zone roughly fits a uniform distribution (see Fig. 13).

The efficiency of the cavitation processes induced by cavitation bubbles is determined both by the number of cavitation bubbles N participating in the process and by the pressure p_m in the shock wave. The contribution of the latter, which is a function of R_{max}, has been discussed in the preceding section.

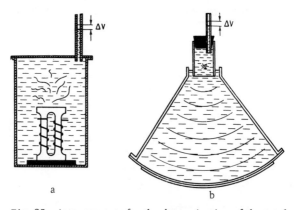

Fig. 25. Arrangements for the determination of the total volume ΔV of all the cavitation bubbles.

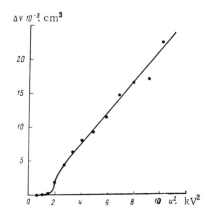

Fig. 26. Volume ΔV in a focusing concentrator versus the square of the voltage on the mosaic transducer.

The number of cavitation bubbles can be found once the radius R_{max} of one cavitation bubble and the total voluume ΔV of all the cavitation bubbles existing in the liquid in the stage of maximum expansion are known

$$\Delta V = {}^4/_3\pi R^3_{max}N. \tag{29}$$

The volume ΔV can be measured directly, for example, by the procedure described in [39]. For this it suffices to enclose the radiator together with the liquid in which cavitation is to take place (or only the cavitation zone) in a hermetic volume equipped with a graduated capillary (Fig. 25a). Then the liquid "broken up" by cavitation bubbles spans a volume equal to ΔV in the capillary.

For the determination of the volume ΔV of all the cavitation bubbles in a specific focusing concentrator [27], a small glass cyl-

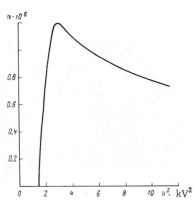

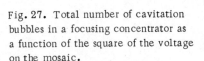

Fig. 27. Total number of cavitation bubbles in a focusing concentrator as a function of the square of the voltage on the mosaic.

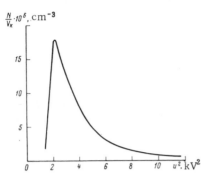

Fig. 28. Variation of the concentration
of cavitation bubbles in a cavitation
zone.

inder was placed in the reaction vessel of the concentrator. The
bottom of the cylinder had a thin acoustically transmissive poly-
ethylene film stretched over it, through which all the acoustic
energy delivered to the focal spot was transmitted (see Fig. 25b).
The variation of the water volume was read by means of a calibrated capillary sealed into the hermetic cover of the cylinder.
With the formation of cavitation bubbles the level of the liquid in
the capillary was raised by an amount that could be easily read
from the capillary scale. The dependence of the total volume on
the voltage squared is shown in Fig. 26. Using Eq. (29) and the
data of Figs. 26 and 18, one can obtain N as a function of u^2 in the
focal spot of the concentrator. The relation thus computed between the total number of cavitation bubbles in the concentrator
and u^2 is illustrated in Fig. 27. Notice the very large absolute
number of cavitation bubbles. The concentration N/V_K of cavitation bubbles in the cavitation zone for various values of u^2 is
shown in Fig. 28 (the curve was plotted on the basis of the data of

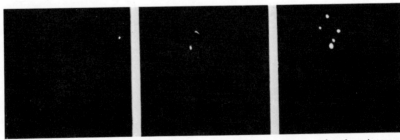

Fig. 29. Dynamics of the development of a cavitation zone. The time interval

Figs. 27 and 24). The most interesting fact is that the number of cavitation bubbles per cubic centimeter is on the order of 10^6 to 10^7 [40]. We obtained values of the same order in a direct computation of the number of bubbles from photomicrographs of the cavitation zone, one of which is shown in Fig. 17. According to the data of Fig. 7, the number of potential cavitation nuclei per cubic centimeter does not exceed a few tens. Consequently, the number of cavitation bubbles is greater than the number of nuclei by at least a factor of 10^5. This is only possible in the event the cavitation bubble generation process constitutes a chain reaction.

The foregoing hypothesis is also corroborated by the result of our earlier [41] high-speed film study of the process of the nucleation and growth of the cavitation zone. We found that cavitation generated at a single nucleus in the course of a few tens of acoustic periods develops into a stable zone comprising a set of cavitation bubbles. A typical cavitation nucleation and development process in water, as photographed with the SFR high-speed camera, is shown in Fig. 29. The exposure time per frame was $2 \cdot 10^{-6}$ sec, i.e., one frame of the picture corresponds to one ultrasonic period. The time interval between frames is three periods. The linear scale is shown on the left side; one division corresponds to 1 mm. As apparent from Fig. 29, the exposure of the film began before the arrival of sound in the focal region (first frame), then after the arrival of sound in the focal spot a cavitation bubble appears, which after three periods initiates the formation of several closely spaced bubbles. This cluster of bubbles is seen in the second frame as a light dot. The size and number of clusters comprising a set of cavitation bubbles grows steadily

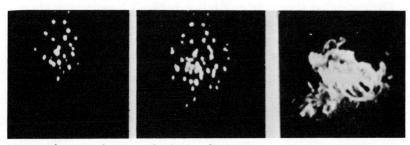

between frames is three periods; the last frame depicts stationary cavitation.

from one frame to the next, and in a time of $(20 \text{ to } 60) \cdot 10^{-6}$ sec they form into a stable cavitation zone* (sixth frame).

A similar experiment was conducted recently by Akulichev [42], who also investigated the onset and development of a cavitation zone, but at a lower frequency (15 kc). The development process was photographed with an SSKS-3 motion picture camera at a film speed of 200,000 frames per second, making it possible to study in detail the behavior of a single cavitation bubble in from one to many periods of the sound field.

The result of the processing of films of the onset and development of a cavitation zone for $P_m \simeq 2$ atm is shown in Fig. 30. The figure illustrates the dependence of the number of cavitation bubbles (N) formed at the tip of a thin (1.5 mm) wire (placed in the sound field for stabilization of the position of the cavitation zone) on the order number n of the acoustic period ($T = 6.65 \cdot 10^{-5}$ sec). It is apparent from the figure that the number of cavitation bubbles grows from one period to the next, reaching saturation after about ten periods. The average number of cavitation bubbles in a fixed domain depends on P_m. Thus, an increase of P_m from 2 to 2.5 atm yields an increase in the average number of cavitation bubbles N from 16 to 35 [42].

Considerations of the possibility of the "multiplication" of cavitation bubbles were advanced earlier by Willard [43]. According to his hypothesis, the shock wave from a collapsing bubble, as it propagates, strongly compresses closely adjacent nuclei, which expand explosively after passage of the wave. However, even if all the nuclei in the cavitation zone take part in the process, their number, as indicated, will still be much smaller than the observed value.

The cascaded growth of cavitation bubbles forming the stationary cavitation zone at a given acoustic pressure may be portrayed as follows. We are aware that in the collapse of a cavitation void stability can be lost, and the void can break up into parts

_____

*The finite rise time of steady-state oscillations in the concentrator due to the Q of the latter is $\sim 3 \cdot 10^{-4}$ sec; therefore, the rate of growth of the cavitation zone ($\sim 3 \cdot 10^{-5}$ sec) exceeds the stabilization rate of the oscillations in the concentrator by approximately tenfold.

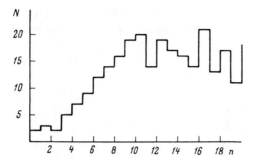

Fig. 30. Number of cavitation bubbles versus the order number of the acoustic period ($P_m \simeq$ 2 atm).

[3, 5, 44]. Inasmuch as the destruction process occurs in stages when the pressure and temperature in the void are maximal, it becomes obvious that the pressure and temperature of the vapor-air mixture in the resulting "fragments" will be elevated. In the expansion half-period they expand easily and become new cavitation nuclei, which are weaker than those which are present in stable fashion in the liquid. The cavitation voids occurring at these nuclei, in turn, generate new members, doing so until the strength of the nuclei exceeds the existing acoustic pressure. An increase in the latter causes a further increase in the number of cavitation voids, as well as their radii and, hence, the size of the cavitation zone. This continues as long as condition (28) is met,

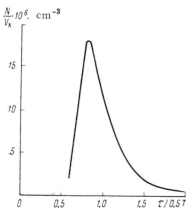

Fig. 31. Concentration of cavitation bubbles versus the ratio of their collapse time to the acoustic half-period.

i.e., $\tau/0.5T = 1$. With a further increase in R_{max}, such that the condition $\tau/0.5T > 1$ prevails, the bubbles do not collapse in every acoustic period. The increased time between collapse events can induce partial coagulation of the "fragments" (for example, as a result of the action of the Bjerknes forces [45]). The number of "fragments" and, accordingly, of cavitation bubbles diminishes. These arguments are borne out by Fig. 31, which shows the dependence of the concentration N/V_K of cavitation bubbles on the ratio $\tau/0.5T$ [Eq. (27) and Fig. 18 make it possible to obtain τ for various values of U]. It is clear that the cavitation bubble concentration does in fact exhibit a maximum in the vicinity of $\tau/0.5T \simeq 1$.

The formation of a large number of bubbles in the cavitation zone which are weaker than the pre-existing nuclei means that, as the acoustic energy decreases, cavitation must cease at lower values of the acoustic pressure relative to the cavitation threshold. This peculiar "hysteresis" effect takes place and has been observed experimentally [10, 46].

Chapter 3

Energy Balance of
the Sound Field in Cavitation

§ 1. Cavitation Formation Energy;
Streaming

A certain fraction of the energy of the sound field is consumed in the formation of cavitation. In the first stage of the
cavitation process, the field energy goes for the formation and
growth of cavitation bubbles. Then, as the latter collapse, the
energy stored in them is restored to the medium, mainly as shock
wave energy. However, the main portion of the shock wave energy,
as shown in Chapt. 2, §2, is very rapidly attenuated. Therefore,
a hydrophone situated in the cavitation field does not yield information on the shock wave energy, which would characterize the
energy spent in the formation of cavitation.

Due to the absorption of acoustic energy in the medium, an
energy density gradient is formed, generating a unidirectional
motion of the medium, i.e., streaming of the liquid [47]. Logically, the energy spent in the formation of cavitation must produce
an equivalent liquid flow [48]. This flow acts on an obstacle in the
sound field with a force F_s.

Applying the principle of conservation of momentum in a
closed domain, Borgnis [49] deduced an approximative theorem
stating that, during the propagation of a sound wave, the sum of
the energy density of the sound field and the flow kinetic energy is

constant. This sum is easily measured with a radiometer. The average force F acting on the radiometer consists of two components [49]

$$F = F_r + F_s = \frac{IS}{c} = \frac{W}{c},$$

where F_r is the force due to the energy of the sound field in the plane of the radiometer (acoustic radiation pressure), F_s is the force due to the streaming kinetic energy of the liquid in the plane of the radiometer, I is the sound intensity, W is the acoustic power received by the radiometer, S is the cross-sectional area of the sound beam in the plane of the radiometer, and c is the sound propagation velocity.

This theorem has been corroborated experimentally [50].

If the radiometer captures all of the energy radiated, the acoustic power in water (c = 1.47 · 10^5 cm/sec), expressed in watts, is equal to the following for a reflective radiometer surface:

$$W' = 7.35F; \tag{30}$$

and to the following for an absorptive radiometer surface:

$$W'' = 14.7F; \tag{31}$$

here F is the force (in grams).

By measuring the sound field energy before and after the cavitation zone, one can determine the energy spent in the formation of cavitation.

If the radiometer receives all of the energy radiated, the total energy of the sound wave plus streaming can be measured, and if an acoustically transmissive film is placed in front of the instrument for insulation against the steady flow of the liquid, its readings will correspond to the energy of the sound wave.

The proposed method enables one to use a radiometer and acoustically transmissive film for an exceedingly simple determination of the energy spent in the formation of cavitation.

The results of measurements in a focusing concentrator [27] are shown in Fig. 32. The square of the voltage of the mosaic transducer is plotted on the horizontal, and the radiometer-

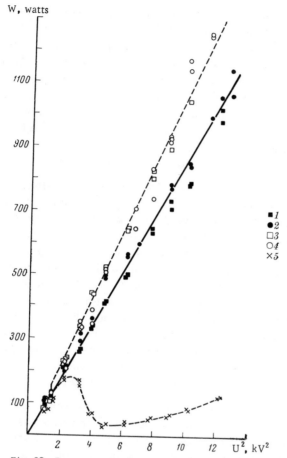

Fig. 32. Power measured with a radiometer versus the square of the voltage on the mosaic transducer of a focusing concentrator.

measured power is plotted on the vertical axis (with regard to the correction for the sphericity of the wave). The measurement procedure was similar to that described in [48].

The nomenclature of the figure is as follows: 1) total acoustic power radiated by the concentrator with the radiometer situated below the focus, where the sound intensity is not yet sufficient to produce cavitation; 2) readings of the radiometer in the same position, but with a thin film placed in front of it (at a distance of 5 mm) to insulate it against the streaming of the liquid; clearly the experimental results provide a good fit to a straight line (solid in the figure) with a small scatter for both cases, indicating the very slight streaming velocity of the liquid in front of the focus; 3, 4) readings of the radiometer when situated above the focal spot; 3) without the film; 4) with the film inserted below the focus, approximately where it was situated in the preceding experiment, i.e., 50 mm below the radiometer. Again, the experimental data fit a straight line (dashed), which at first coincides with the curve from the preceding experiment, but then at the onset of cavitation experiences an abrupt jog and rises more steeply.

The increased readings of the radiometer after the onset of cavitation may be attributed, clearly, to the fact that after passing the cavitation zone the wave front loses some of its sphericity, and the resulting acoustic streaming tends accordingly to lose its purely radial direction, so that the correction introduced for the sphericity of the front in the case of a purely spherical front must become smaller. Moreover, the velocity of sound diminishes in the cavitation zone, so that the radiometer, which measures the energy density $E = I/c$, can give readings that are somewhat too high. These experiments indicate that the absence or presence of a film between the radiating surface of the concentrator and the cavitation zone does not exert an appreciable influence on the streaming of liquid in front of the radiometer.

The nomenclature 5 corresponds to the readings of the radiometer when it is placed above the cavitation zone (as in the last experiment), but is protected against the action of streaming by a film inserted at a distance of 5 mm in front of it. As the results of this experiment indicate, the sound field energy behind the focus rises linearly only until the onset of cavitation, and with the development of the latter the intensity of the field falls off.

The difference between the total radiated power W and the power of the sound field behind the cavitation zone determines the power W_0 spent in the formation of streaming in the liquid, and the kinetic energy of streaming characterizes the acoustic energy spent in the formation of cavitation. Curve 1 in Fig. 33 shows the dependence of this energy $A_0 = W_0 T$ (where T is the acoustic period) on the voltage squared; for comparison the dashed curve 2 indicates the same dependence of the total acoustic energy $A = WT$. It is evident from Fig. 33 that, as the radiated acoustic energy

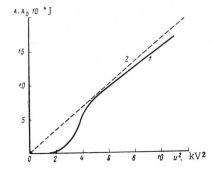

Fig. 33. Total acoustic energy (1) and energy spent in the formation of cavitation (2) versus the square of the voltage on the focusing concentrator.

(which is proportional to the voltage squared) is increased, beginning with the time of onset of cavitation, the energy A_0 spent in the formation of the growing cavitation zone also increases.

A similar result can be obtained with conventional technological radiators.

The dependence of the total radiated power (curve 1) and the power of the sound field (curve 2) on the square of the voltage on a ferrite radiator [51] with a resonance frequency of 26 kc is illustrated in Fig. 34. The power was measured with a radiometer in the form of a hollow disk, which was suspended on a torsion balance. For the elimination of streaming of the liquid during measurement of the acoustic power, a thin acoustically transmissive film was placed, as before, between the radiator and the radiometer.

Inasmuch as the tank in which the transducer operated had small dimensions and was not deadened, and as the radiated wave

Fig. 34. Variation of the total radiated power (1) and power of the sound field (2) versus the square of the voltage on a ferrite radiator.

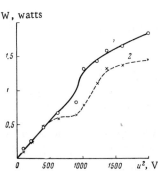

had a divergent front, these variations are merely of a qualitative character. It is apparent from the curves that, after the onset of cavitation (which took place at 20 V) and its subsequent development, the increase of the radiated power slows down (curve 1); this is explained by the familiar reduction of the acoustic impedance of the medium [52]. About 20% of the total radiated power is used for the formation of streaming; curve 2 corresponds to a measurement in which a film was placed in front of the radiometer to cut off the flow of liquid.

§ 2. Cavitation Energy

The work A_K done by N identical cavitation bubbles of radius R_{max} can be calculated with the aid of Eq. (15):

$$A_{\text{к}} = {}^4/_3\pi\,(P_a + P)\,(R_{max}^3 - R_{min}^3)\,N, \tag{32}$$

or its equivalent expression (18)

$$A_{\text{к}} = 4\pi P_{vg} R_{max}^3 \left(\frac{R_{max}}{R_{min}} - 1 \right) N. \tag{33}$$

Since it is reasonable to assume in the cavitation regime that P_a is nearly equal to zero (see Chap. 2, §2) and to allow for the fact that $R_{min} \ll R_{max}$, Eq. (32) may be rewritten in the form

$$A_{\text{к}} \simeq {}^4/_3\pi R_{max}^3 PN. \tag{34}$$

Taking account of (21), we write Eq. (33) as

$$A_{\text{к}} = {}^4/_3\pi R_{max}^3 PN\,(1 - \delta^{1.6}). \tag{35}$$

Under normal conditions $\delta \ll 1$; hence, expression (35) goes over to (34).

Using Eq. (29), we rewrite relation (34) in the form

$$A_{\text{к}} \simeq P\Delta V, \tag{36}$$

where ΔV is the total volume of all the cavitation bubbles in the stage of maximum expansion and is readily measured experimentally by the procedure outlined in Chap. 2, §6.

Making use of the data obtained in Fig. 26 for ΔV in a focusing concentrator, we find the energy developed by all the cavitation bubbles in the cavitation zone.

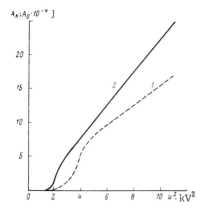

Fig. 35. Energy spent in the formation of cavitation (1) and the energy
developed by the cavitation bubbles
(2).

In Fig. 35, curve 2 corresponds to the relation calculated according to (36) between the energy A_K developed by all the cavitation bubbles and the voltage squared. Curve 1 indicates the energy
spent by the sound field in the formation of cavitation (curve A_0 in
Fig. 33). Inasmuch as the energy spent in the production of cavitation bubbles is equal to the energy developed in their collapse (the
energy of shock waves, luminescence, chemical reactions, noise,
etc.), both curves in Fig. 35 should exhibit little disparity, as is
indeed the observed case. Their disparity may be attributed to
the disregard for the gas content parameter δ, which lowers the
value of A_K, as well as to experimental error. Despite the fact
that these curves were obtained by independent techniques, the
difference between them is relatively slight.

§ 3. Cavitation Efficiency

It would seem that the efficiency of cavitation process could
be assessed in terms of the potential work A_K stored in the cavitation bubbles. Referring this work to the collapse time τ or, more
properly, to the short time interval τ' in the final stage of collapse when the main power is released, we obtain the average
power delivered by all the cavitation bubbles during their collapse.
However, the collapse time τ and certainly the time τ' are not
generally known.

On the other hand, the efficiency of cavitation processes
(erosion, luminescence, the formation of chemical radicals, etc.)
depends both on the pressure p_m produced at the instant of col-

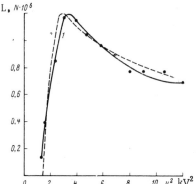

Fig. 36. Relation between the luminescence brightness (1) and total number of cavitation bubbles (2).

lapse of the cavitation bubbles and on the number of bubbles, per se, N. Inasmuch as N is a function of p_m (see Chap. 2, §6), the efficiency of cavitation processes may be characterized by the number of cavitation bubbles contributing to the process. For example, such variables as the total light flux of luminescence, the total quantity of suspension formed, the density of chemical radicals, etc., depend on the total number of cavitation bubbles, whereas factors such as erosion depend on the bubble concentration per unit volume.

Curve 1 in Fig. 36 represents the variation of the luminescence L (in relative units) of all cavitation bubbles in the cavitation zone of a focusing concentrator [27] as a function of the square of the voltage on the mosaic transducer. The integral luminescence brightness L was recorded with an FÉU-18A photomultiplier mounted directly on the reaction vessel of the concentrator. Curve 2 (from Fig. 27) indicates the total number of cavitation bubbles N as a function of the voltage squared.

Curve 1 of Fig. 37 (plotted from the data of Fig. 21) shows the cavitation erosion ΔG of an aluminum specimen smaller than the cavitation zone, and curve 2 (from Fig. 28) shows the concentration of cavitation bubbles in that zone as a function of the square of the voltage on the concentrator mosaic. The presence of the aluminum specimen in the focal spot of the concentrator facilitated the onset of cavitation, so that the graph of ΔG proved somewhat wider than the graph of N/V_K, which was obtained in the "pure" focal region with its higher cavitation threshold.

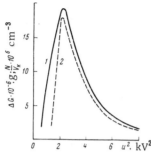

Fig. 37. Relation between the erosion (1) and cavitation bubble concentration (2).

The foregoing Figs. 36 and 37 reveal that the efficiency of cavitation processes as a function of the radiated acoustic power is in fact determined solely by the number of cavitation bubbles participating in the process.

The dependence of the cavitation energy density A_K/V_K on the voltage squared in the cavitation zone of a focusing concentrator is shown in Fig. 38, curve 1) according to the data of Figs. 35 and 24. It is apparent that the energy density A_K/V_K increases rapidly at first with the radiated acoustic power (as the square of the voltage), attaining a maximum at ~1.6 kV, and then dropping. It follows from Fig. 39, which illustrates the same density A_K/V_K, but as a function of $\tau/0.5T$ [the substitution can be made by means of Eq. (27) and Fig. 18], that this maximum corresponds

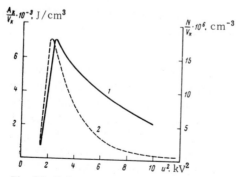

Fig. 38. Relation between the cavitation energy density (1) and cavitation bubble concentration (2).

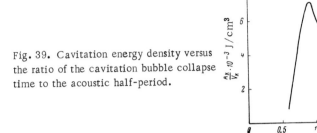

Fig. 39. Cavitation energy density versus the ratio of the cavitation bubble collapse time to the acoustic half-period.

to the condition $\tau \simeq 0.5T$, when the cavitation bubbles can succeed in collapsing during the period $0.5T$.

Since for $\tau/0.5T > 1$ the bubble can no longer collapse in every acoustic period, but pulsates for a certain period of time before collapsing (see Chap. 2, §5), some of the energy A_K is spent on these pulsations before the final collapse of the bubble. Consequently, the energy A_k stored by the cavitation bubbles in the stage of maximum expansion can be used to characterize the cavitation processes associated with the collapse of the bubbles only when $\tau/0.5T \leq 1$. This is borne out by a comparison of the curves for A_K/V_K (1) and N/V_K (2) in Fig. 38. The curve for N/V_K, as implied by Fig. 37, characterizes the shock wave intensity in the collapse of the cavitation bubbles.

Therefore, when $\tau/0.5T \leq 1$ the energy stored by the cavitation bubbles in the stage of maximum expansion is completely transformed during collapse of the bubbles into the specific energy of cavitation, viz., shock waves, luminescence, chemical processes, etc.

If $\tau/0.5T > 1$ the energy stored by the cavitation bubbles is only partially transformed into cavitation energy, while a certain fraction of it is lost in pulsations of the bubbles prior to their ultimate collapse. As a result of rectified diffusion [9], which takes place during the pulsations of the bubbles, their gas content is increased, and the pressure p_m at the time of collapse of the bubbles must be reduced. However, nothing is known at this time about the quantitative aspect of this effect.

Chapter 4

Influence of the Acoustic Pressure and Characteristics of the Liquid on the Shock Wave Intensity

It is evident from Eq. (22) and Fig. 19 that the pressure in the shock wave, formed at the time of collapse of a cavitation bubble at a given P, is fully* determined by the gas content parameter of the bubble

$$\delta = \frac{P_{vg}}{P} = \frac{P_v + P_g}{P},$$

where P_v is the saturation vapor pressure and P_g is the pressure of the gas in the bubble for $R = R_{max}$.

The vapor pressure P_v in the bubble does not depend on the bubble radius, but is a function of the liquid temperature t. The pressure P_g of the gas in the bubble, on the other hand, depends not only on the temperature of the liquid and its gas content

$$\alpha = \frac{V_g}{V_w}$$

(where V_w is the volume of water and V_g is the volume of dissolved gas, reduced to atmospheric pressure), but also on the

---

*The viscosity of the liquid, which diminishes the rate of collapse [1], is disregarded here; in ordinary low-viscosity liquids its effect is small, and high-viscosity liquids (oils) are not of practical significance.

radius of the bubble. As the sound intensity and, accordingly, the bubble grow, the pressure of the gas inside it for a fixed quantity of gas decreases as $1/R^3$.

We propose to discuss the influence of each of the aforementioned parameters.

§ 1. The Acoustic Pressure

If we assume that the diffusion of gas from the bubble into the surrounding liquid and in the opposite direction does not occur, we can proceed from the bubble equilibrium condition

$$P = P_v + P_g - \frac{2\sigma}{R_0},$$

and obtain

$$P_g = \left(P - P_v + \frac{2\sigma}{R_0}\right)\left(\frac{R_0}{R_{max}}\right)^3,$$

where R_0 is the initial radius and σ is the surface tension. Then the parameter δ may be written

$$\delta = \frac{P_v + \left(P - P_v + \frac{2\sigma}{R_0}\right)\left(\frac{R_0}{R_{max}}\right)^3}{P}. \tag{37}$$

It is clear that, for sufficiently large R_{max} ($R_{max} \gg R_0$), the first term proves much smaller than P_v, i.e., $\delta_{(R_{max} \gg R_0)} \simeq P_v / P$.

However, as already demonstrated, the maximum cavitation bubble radius determined by the expression (13)

$$R_{max} \simeq \sqrt{\frac{2}{3} \frac{P_t(R)}{\rho}} \frac{T}{4}\left(1 - \frac{P_к}{P_m}\right),$$

is interpreted as increasing until the bubble collapse time τ becomes equal to half the acoustic period (28):

$$\tau = \frac{T}{2}.$$

Considering (27) and (28) as well, we obtain the limit $P_{m\,max}$, above which a reduction in the cavitation efficiency is observed

$$P_{m\,max} = \frac{P_к}{1 - \dfrac{1}{F\sqrt{\dfrac{2}{3}\dfrac{P_t(R)}{P}}}}. \tag{38}$$

Hence, it follows that the bubble can expand as long as

$$0.75 \sqrt{\frac{P_t(R)'}{P}} > 1.$$

When $P_g \simeq 0$ the parameter δ is numerically equal to the vapor pressure P_v of the liquid; for water at room temperature and $P = 1$ atm, we have from standard tabulated data $P_v \simeq 2.5 \cdot 10^{-2}$ and $\delta \simeq 2.5 \cdot 10^{-2}$. The pressure in the bubble at the time of compression, as indicated from Fig. 19, attains a value $p_m \simeq 10^3$ atm. It would appear that p_m no longer increases with R_{max}, as now δ is determined only by the saturation vapor pressure of the liquid, which remains invariant at a given temperature. However, this happens only for relatively slow expansion of the bubble. If, on the other hand, the rate of expansion v_e of the bubble is large, the rate of vaporization v_v of water from the surface of the bubble can begin to depart from the velocity of the walls, and the vapor pressure in the bubble becomes smaller than the saturation value. It is well known that the rate of vaporization from the surface of a liquid is equal to [53]

$$v_v = 4 \cdot 10^{-2} \sqrt{\frac{RT}{2\pi\mu}},$$

where R is the universal gas constant, T is the absolute temperature, and μ is the molecular weight.

Hence, the critical velocity of the bubble wall in water under standard conditions ($P = 1$ atm, $t = 20\,^\circ$C) is equal to

$$v_\kappa = v_v \simeq 7 \cdot 10^2 \text{ cm/sec.} \tag{39}$$

On the other hand, the time in which the bubble acquires its maximum radius is about $0.75T$ [38], so that the average rate of expansion of the cavitation bubble is $R_{max}/0.75T$. Consequently, if the following condition holds

$$\frac{R_{max}}{0.75T} = v_c > v_\kappa, \tag{40}$$

the vapor pressure P_v in the bubble becomes lower than the saturation vapor pressure, effecting a reduction in δ and, accordingly, an increase in the pressure p_m inside the bubble.

Let us now estimate the possibility of condition (40). At a frequency of about 20 kc ($T = 5 \cdot 10^{-5}$ sec), the maximum bubble radius commonly used in practice at acoustic pressures inducing well-developed cavitation in a liquid ($P_m \simeq 2$ atm) is on the order of (1 to 10) $\cdot$ 10^{-3} cm (see, e.g., Figs. 9 and 10), corresponding to a bubble expansion rate no greater than $v_e \simeq 2.7 \cdot 10^2$ cm/sec < v_K. Consequently, under these standard conditions the water vapor pressure in the expanding bubble is always equal to the saturation pressure. Only an appreciable increase in the frequency and acoustic pressure can produce an increase in the expansion rate.

Condition (40) is easily met in focusing systems. Thus, in a focusing concentrator operating at a frequency of about 500 kc [27] the maximum cavitation bubble radius has a value $R_{max} > 1 \cdot 10^{-3}$ cm (see Fig. 18). The expansion rate of the bubble in this case begins to exceed the critical velocity $v_e > v_K$, and the parameter δ decreases.

This implies that at high frequencies it is possible to obtain a lower value of δ for cavitation bubbles even if R_{max} is smaller than at low frequencies.

§ 2.   Temperature and Gas Content

of the Liquid

A reduction in the temperature t and gas content of the liquid α brings about a decrease in the pressure P_{vg} of the vapor-gas mixture in the bubble and an increase in p_m. Unfortunately, the calculation of the absolute values of the pressure p_m in the bubble is made exceedingly difficult by the fact that the quantity $P_{vg} = P_v + P_g$ is not usually known. The water vapor pressure P_v can be readily determined from suitable tables, and the air or gas pressure P_g in the bubble at maximum radius remains essentially constant. Hence, the shock wave intensity under variation of t and α has been estimated experimentally from the cavitation damage done to a small aluminum cylinder placed in the cavitation zone. The weight loss of the cylinder ΔG, found as the difference in weight before and after the action of cavitation, was used as a measure of the relative intensity of the shock waves from cavitation bubbles.

The experiments were carried out with a constant acoustic pressure and at various temperatures in ordinary standing water and degassed distilled water. The reduction in gas content in degassed water should cause a decrease in the pressure inside the cavitation void due to the reduction in the quantity of gas entering the void.

A ferrite transducer was used in the experiments, operating at a frequency of 28.5 kc and a sound intensity of about 2 W/cm². Each test lasted 10 min. The aluminum cylinder was placed in the zone of well-developed cavitation, which was considerably larger than the small cylinder (2 × 2 mm). In this way any variations in the dimensions of the cavitation zone (for example, growth due to a temperature increase) did not affect the erosion.

The application of a ferrite transducer (ferrite 21) made it possible to heat the water simultaneously with the radiator placed in it, because this particular ferrite is capable of operating in the temperature range to 400°C without appreciable modification of its properties [54]. The water level in the vessel with the transducer situated on its bottom was chosen so as to produce a standing wave antinode at the surface of the transducer, i.e., to create suitable conditions for the maximum development of cavitation.

The variation of the cavitation erosion ΔG of the aluminum cylinder is shown in Fig. 40 as a function of the water temperature t. Curve 1 refers to ordinary standing water with an air

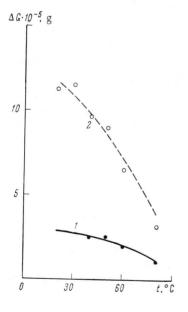

Fig. 40. Cavitation erosion as a function of the temperature of the liquid.

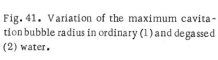

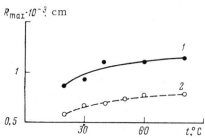

Fig. 41. Variation of the maximum cavitation bubble radius in ordinary (1) and degassed (2) water.

content $\alpha \simeq 22$ cm^3/liter, and curve 2 refers to the degassed water with an air content $\alpha \simeq 6.8$ cm^3/liter. The degassing of the water was realized by evacuation, and the air content was determined by Winkler's technique [55].

As anticipated, the cavitation erosion in degassed water was higher than in standing water; with an increase in the water temperature it diminished in both cases. In degassed water a smaller quantity of air is admitted to the bubble than in standing water, lessening the damping effect of the vapor-gas mixture in the bubble during its collapse, as well as an increase in the pressure in the shock wave. With an increase in temperature the pressure inside the bubble, being determined by the pressure of the vapor and gas, increases, and the shock wave is attenuated.

This is confirmed by Fig. 41, which shows the variation of the maximum radius R_{max} of the cavitation bubbles in standing (1) and degassed (2) water as a function of the temperature. The radius R_{max} was determined by filming through a microscope with a light pulse whose duration slightly exceeded half the ultrasonic period. Consequently, during the flash time the bubble always managed to reach its maximum radius. Each experimental point in Fig. 41 corresponds to the average obtained from the processing of three to five photographs of a great many bubbles near the cylinder.

Inasmuch as the sound intensity in every case was constant, the smaller value of the maximum radius of cavitation bubbles in degassed water relative to the radius in standing water is due to the smaller content of air in the degassed water, while the growth of R_{max} with increasing temperature is caused by the increased pressure P_{vg} of the vapor-gas mixture inside the bubble. Therefore, the relative decrease of the gas content parameter of the bubble in degassed water relative to the same parameter in standing water is equal to

$$\frac{\delta_d}{\delta_s} = \frac{R_{max\,d}}{R_{max\,s}}. \tag{41}$$

Making use of Eq. (22), we can find the increase in pressure in the cavitation bubble during its collapse in degassed water relative to the pressure in standing water

$$\frac{p_{m\,d}}{p_{m\,s}} = \frac{\delta_s}{\delta_d^3}. \tag{42}$$

Substituting the value of δ_s / δ_d from (41), we find

$$\frac{p_{m\,d}}{p_{m\,s}} = \left(\frac{R_{max\,s}}{R_{max\,d}}\right)^9.$$

The temperature dependence of p_{md}/p_{ms}, calculated according to the averaged data of Fig. 41, is presented in Fig. 42. The points indicate the corresponding ratio of the degree of cavitation damage of the cylinder in degassed and standing water, $\Delta G_d /\Delta G_s$, as a function of the temperature according to the experimental curves in Fig. 40.

This is particularly apparent in Fig. 43, which illustrates the dependence of $\Delta G_d/\Delta G_s$ on p_{md}/p_{ms} obtained from Fig. 42. The points corresponding to equal temperatures provide a good fit to a straight line through the origin. This has the secondary

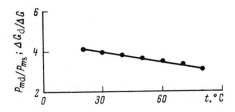

Fig. 42. Calculated pressure increase in a cavitation bubble during collapse in degassed water relative to the pressure in standing water.

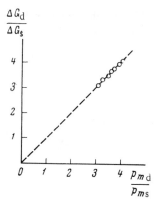

Fig. 43. Relation between the erosion and pressure at the time of collapse of cavitation bubbles.

implication that the cavitation damage to aluminum is linearly dependent on the pressure in the shock wave during the collapse of cavitation bubbles.

For a given acoustic pressure there exist well-defined bubble radii amenable to the formation of cavitation (see Fig. 5). In connection with the decrease in the gas content in the bubble in the case of degassing of the liquid, the number of nuclei and their radii also diminish, causing an increase in the cavitation threshold and a reduction in the size of the cavitation zone. With increasing temperature the pressure inside the bubbles increases, their radii increase, and they form new cavitation nuclei. Thus, degassing of the liquid induces a reduction in the cavitation zone with a concurrent increase in the shock wave intensity created by the cavitation bubbles, while an increase in the temperature leads to growth of the cavitation zone and a reduction of the shock wave intensity.

Consequently, if the size of a specimen subjected to cavitation damage exceeds that of the cavitation zone, an increase in temperature can result at first in an enhancement of the cavitation erosion due to the growth of the cavitation zone, but then, when the shock wave intensity drops considerably, the result will be a reduction in the erosion. This explains the temperature maximum observed by Bebchuk [56] in the cavitation damage when the dimensions of the specimens subjected to damage were greater than those of the cavitation zone formed by the tip of a nickel tube vibrator [57].

§ 3. The Hydrostatic Pressure

As shown in the preceding section, as well as in [58], a reduction in the gas content parameter δ by degassing of the liquid or lowering of the temperature affords a several-fold increase in the shock wave pressure p_m. However, a further increase in the pressure in the shock wave can be realized by the creation of an elevated static pressure [59]. Here $p_m = (P/81)\delta^3$ will grow not only linearly with P, but also due to the reduction of the parameter δ, also a function of P.

Under the influence of an elevated hydrostatic pressure the gas present in the cavitation nuclei, becoming partially dissolved, decreases their radii. But a reduction in the radius R_0 causes an increase in the cavitation threshold P_K and, consequently, a reduction in R_{max} [see Eq. (13)]. Knowing the variation of R_{max} for various hydrostatic pressures relative to the initial bubble radius R_{max0}, one can find the variation of δ relative to δ_0 at $P = P_0$

$$\delta_{re} = \frac{\delta}{\delta_0} = \frac{P_0 R_{maxs}^3}{P R_{max}^3}. \tag{43}$$

The maximum cavitation bubble radius R_{max} can be estimated from Eq. (13) or determined experimentally by filming under a pulsed light source (see Chap. 2, §3).

Our own experiments to determine the dependence of the shock wave intensity during the collapse of cavitation bubbles on the hydrostatic pressure were carried out with standing distilled water at a temperature of 23°C. We used a focusing concentrator in the form of a solid aluminum sphere 200 mm in diameter bonded to a piezoelectric mosaic, at the center of which was a water-filled spherical cavity ~8 mm in radius [60].

A schematic section of the experimental apparatus is shown in Fig. 44. The nomenclature is as follows: 1) aluminum sphere; 2) piezoelectric mosaic; 3) water-filled cavity; 4) aluminum cylinder placed in the focal spot for the assessment of cavitation erosion; 5) attachment for fixation of the cylinder in the focal spot (two thin needle springs pressed against recesses in the end surfaces). The required static pressure on the liquid was created by compressed air fed through the valve 6 in the upper cover plate 7. The latter also contained a glass window 8 for the ob-

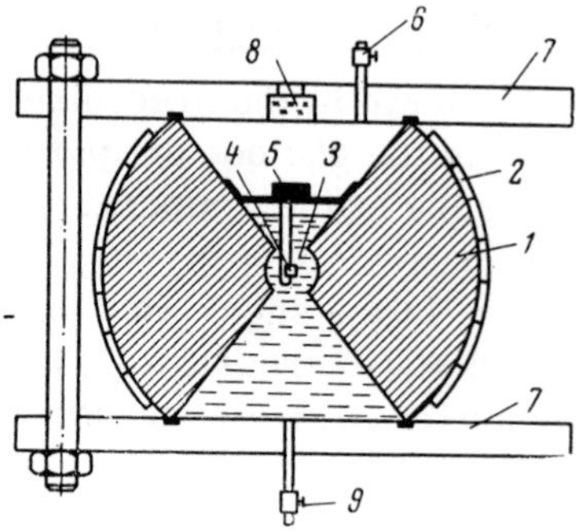

Fig. 44. Diagram of the experimental apparatus used to investigate the influence of hydrostatic pressure on cavitation processes.

servation of luminescence. The valve 9 in the lower cover plate was used to drain off the water, which was changed after each experiment.

The investigations were carried out at a frequency of 560 kc, which corresponded to one of the radial harmonics of the aluminum sphere. The diameter of the focal spot was 3.3 mm.

The dependence of the effective acoustic pressure P_c in the focal spot on the voltage U on the concentrator mosaic (1) and the extrapolation of that dependence on the assumption of zero cavitation (2) are shown in Fig. 45. The experimental dependence of the cavitation threshold P_K in the focal spot on the hydrostatic pressure P with an aluminum cylinder placed in the focal spot for the investigation of cavitation erosion is shown in Fig. 46. The onset of cavitation was registered with an oscilloscope connected to a piezoelectric element, which was bonded to the outer surface of the concentrator. The concentrator voltage, at which characteristic cavitation noise was detected, was determined in this case. The acoustic pressure corresponding to that voltage was found on the basis of curve 2 in Fig. 45.

Also shown in Fig. 45 are the variations of the acoustic pressure in the cavitation zone as a function of the voltage on the concentrator for various values of P (the circles correspond to the data of Fig. 46). The measurements were performed by means of a waveguide receiving unit [33] (frequency characteristic to 1 Mc) capable of operating at elevated hydrostatic pressures. The resulting value of P_c, as already mentioned in Chap. 2, §3, can be used as an approximate characteristic of the tensile stress in the cavitation zone, $P_t(R) \simeq P_c/2$.

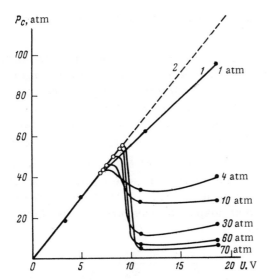

Fig. 45. Effective acoustic pressure in the focal spot versus the voltage on the mosaic transducer of the concentrator.

The experimental data make it possible from expression (43) and Eq. (13)

$$R_{max} \simeq \sqrt{\frac{2}{3} \frac{P_t(R)}{\rho_\kappa} \frac{T}{4} \left(\frac{2P_m - P_\kappa}{P_m} \right)}$$

to find the relative variation of the parameter δ_{re} as a function of P. Bearing in mind that in the measurements $P_m = \sqrt{2} P_c = \text{const}$, we obtain

$$\delta_{re} \simeq \frac{P_0}{P} \left[\frac{\sqrt{P_{t0}(R)}}{\sqrt{P_p(R)}} \frac{(2P_m - P_{\kappa 0})}{(2P_m - P_\kappa)} \right]^3 \tag{44}$$

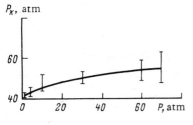

Fig. 46. Variation of the cavitation threshold as a function of the hydrostatic pressure.

(the zero subscript indicates the initial conditions, when $P = P_0 = 1$ atm).

The variation of the parameter δ_{re} as a function of the hydrostatic pressure as calculated from Eq. (44) is shown in Fig. 47. Curve 1 refers to the case when the voltage on the concentrator is 11 V, and curve 2 corresponds to a voltage of 18.7 V, which corresponds, as implied by Fig. 45, to $P_c = 66$ and 112 atm (for $P_c = 66$ atm and $P_0 = 1$ atm we have $R_{max} = 1$ and $\delta_{re} = 1$).

From the relative variation of δ_{re} shown in Fig. 47, one can find the ratio of the maximum pressure p_m in the cavitation bubble at an elevated hydrostatic pressure P to the pressure p_{m0} in the bubble at standard atmospheric pressure P_0. Invoking Eqs. (22) and (43), we obtain

$$\frac{p_m}{p_{m0}} \simeq \frac{P}{P_0} \frac{1}{\delta_{0m}^3}.$$

The dependences on the static pressure thus obtained are shown in Fig. 48. Curve 1 indicates the variation of the maximum pressure in the bubble with increasing static pressure for $P_c = 66$ atm (U = 11 V), and curve 2 shows the same relation for $P_c = 112$ atm (U = 18.7 V).

Inasmuch as the cavitation erosion is linearly dependent on the pressure in the bubble at the time of collapse, $\Delta G \sim p_m$ (see §2 of the present chapter), it was estimated in terms of the ero-

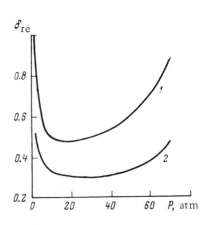

Fig. 47. Variation of the relative gas content parameter of the cavitation bubbles as a function of the hydrostatic pressure.

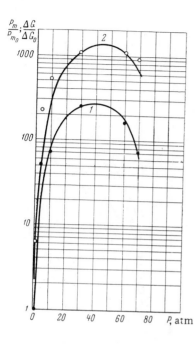

Fig. 48. Variation of the pressure at the time of collapse of a cavitation bubble and cavitation erosion as a function of the hydrostatic pressure.

sion of a small (2 × 2 mm) aluminum cylinder, which was smaller than the focal spot in diameter. The irradiation time of the cylinder, which was situated at the center of the focal spot, was 10 min. The weight loss ΔG, determined as the weight difference before and after the action of ultrasound, was a measure of the relative shock wave intensity created by the cavitation bubbles.

The dependence of the cavitation erosion ΔG of the aluminum cylinder on the voltage U on the concentrator for various values of the static pressure P is shown in Fig. 49. Each experimental point comprises the average of 3 to 10 measurements with a maximum scatter of 15%.*

We see from the figure, in particular, that at a static pressure equal to the atmospheric value the erosion curve has a maximum at a concentrator voltage of 18.7 V. There is every reason to suppose that this maximum, as indicated in Chap. 2, §5, is determined by the coincidence of the cavitation bubble collapse time

---

*For large U and P the cylinder was rapidly disintegrated, so that it became necessary to shorten the irradiation time to 2-5 min with an appropriate conversion.

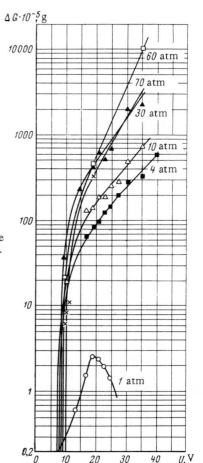

Fig. 49. Cavitation erosion versus the concentrator voltage U at various hydrostatic pressures.

and the acoustic half-period, i.e., $\tau = 0.5T$, where $\tau = R_{max} F(\rho/P)$. Hence, it follows that for a fixed bubble radius R_{max} an increase in the hydrostatic pressure will tend to diminish the collapse time. This will also be fostered by a reduction in the coefficient F, which drops somewhat with increasing P.

Consequently, by raising the hydrostatic pressure it is possible to shift the cavitation erosion maximum toward larger values of the transducer voltage, in fact roughly as $(1/P)^{\frac{1}{2}}$. Thus, the maximum decreases as the static pressure is increased.

Finally, it is apparent from Fig. 49 that with an increase in the hydrostatic pressure, as to be expected, the erosion and, hence, the shock wave intensity of the cavitation bubbles increase profoundly.

The dots and circles in Fig. 48 show the experimental values of the cavitation damage at elevated static pressures relative to atmospheric pressure, $\Delta G/\Delta G_0$. These values are obtained from the data of Fig. 49 for the same values of the concentrator voltage as those used to plot the analytical curves.

As evident in Fig. 48, the predicted theoretical increase in the shock wave intensity during the collapse of the cavitation bubbles in connection with an increase of the hydrostatic pressure is highly consistent with the experimental. Under the experimental conditions described this increase is more than 200-fold.

Other experimental studies [61, 62] also support the fact that an increase in the hydrostatic pressure induces growth of the cavitation erosion.

The studies described in this section corroborate the validity of the physical interpretation given to expression (22) for the determination of the pressure at the time of collapse of a cavitation bubble and demonstrate its excellent agreement with experiment.

Conclusion

Numerous investigations devoted to the study of individual aspects of the cavitation process and periodic surveys summarizing their results [1-5, 63] have made possible the incisive investigation of this complex process.

Nevertheless, many aspects of the cavitation process yet remain unclear. For example, the existing theory accounting for the behavior of a cavitation bubble in a liquid is clearly applicable only for the isolated bubble or for a small number of bubbles in a cavitation zone. In this case one is permitted to neglect the decrease in pressure near the bubble due to the discontinuity of the liquid and to assume that the tensile stress is equal to the acoustic pressure. In real circumstances, however, one is often compelled to deal with a cavitation zone consisting of a great many cavitation bubbles. In such a zone, as we have demonstrated, the active pressure is always lower than the acoustic pressure, the relation between them depending both on the cavitation bubble concentration and on the degree of coherence of their onset and collapse. Experiments have shown that in the collapse stage the pressure acting on the bubble is approximately equal to the hydrostatic pressure. Further experimental and theoretical research on this relatively unexplored effect should prove rewarding.

References

1.   A. D. Pernik, Problems of Cavitation, Sudpromgiz (1963).
2.   E. Webster, Cavitation, Ultrasonics, 1:39 (1963).
3.   H. G. Flynn, Physics of Acoustic Cavitation in Liquids, Physical Acoustics (W.P. Mason, ed.), Vol. 1B, Academic Press, New York (1964).
4.   M. G. Sirotyuk, Ultrasonic cavitation (review), Akust. Zh., 8(3):255 (1962).
5.   M. Kornfel'd, Elasticity and Strength of Liquids, Moscow—Leningrad (1951).
6.   Ya. B. Zel'dovich, Theory of the formation of a new phase, cavitation, Zh. Éksp. Teor. Fiz., 12(11-12):525 (1942).
7.   R. Esche, Untersuchung der Schwingungskavitation in Flüssigkeiten [Investigation of vibration-induced cavitation in liquids], Akust. Beih., 4:208 (1952).
8.   D. Messino, D. Sette, and F. Wanderligh, Statistical approach to ultrasonic cavitation, J. Acoust. Soc. Am., 35(10):1575 (1963).
9.   M. Strassberg, Onset of ultrasonic cavitation in tap water, J. Acoust. Soc. Am., 31(2):163 (1959).
10.   W. J. Galloway, An experimental study of acoustically induced cavitation in liquids, J. Acoust. Soc. Am., 26(5):849 (1954).
11.   V. A. Akulichev and V. I. Il'ichev, Spectral indication of the origin of ultrasonic cavitation in water, Akust. Zh., 9(2):158 (1963).
12.   W. Connolly and F. E. Fox, Ultrasonic cavitation thresholds in water, J. Acoust. Soc. Am., 26(5):843 (1954).
13.   I. T. Alad'ev (ed.), Handbook: Problems in the Physics of Boiling, Izd. "Mir" (1964).
14.   E. N. Harvey, R. D. McElroy, and A. H. Whiteley, On cavity formation in water, J. Appl. Phys., 18:162 (1947).
15.   R. T. Knapp, Cavitation and nuclei, Trans. ASME, 80:6 (1958).
16.   F. G. Blake, Tech. Mem. Acoustics Res. Lab., Harvard Univ., Cambridge, No. 9 (1949).
17.   Yu. A. Aleksandrov, G. S. Voronkov, V. M. Gorbunkov, N. B. Delone, and Yu. I. Nechaev, Bubble Chambers, Gosatomizdat (1963).
18.   W. E. Whyberew, G. D. Kinzer, and R. Gunn, Electrification of small air bubbles in water, J. Geophys. Res., 57(4):453 (1952).
19.   V. A. Akulichev, Hydration of ions and the cavitation resistance of water, Akust. Zh., 12(2):160 (1966).
20.   D. Lieberman, Radiation-induced cavitation, Phys. Fluids, 2(4):466 (1959).
21.   D. Sette, Sonic cavitation and ionizing radiation, Proc. Third Internat. Congr. Acoustics, Stuttgart (1959), Vol. I, p. 330 (1961).
22.   D. Sette and F. Wanderlingh, Nucleation by cosmic rays in ultrasonic cavitation, Phys. Rev., 125(2):409 (1962).

23. M. Bertoletti and D. Sette, On nucleation processes in ultrasonic cavitation and bubble chambers, Proc. Fourth Internat. Congr. Acoustics, Copenhagen (1962), Paper J26.

24. R. Macleay and L. Holroyd, Space-time analysis of the sonoluminescence emitted by cavitated water, J. Appl. Phys., 32(3):449 (1961).

25. B. E. Noltingk and E. A. Neppiras, Cavitation produced by ultrasonics, Proc. Phys. Soc., 63B(9):674 (1950); Proc. Phys. Soc., 64B:1032 (1951).

26. M. Minnaert, On musical air-bubbles and the sounds of running water, Phil. Mag., 16(7):235 (1933).

27. L. D. Rozenberg and M. G. Sirotyuk, Apparatus for the generation of focused ultrasound of high intensity, Akust. Zh., 5(2):206 (1959).

28. L. D. Rozenberg and M. G. Sirotyuk, A focusing radiator for the generation of superhigh-intensity ultrasound in 1 Mc, Akust. Zh., 9(1):61 (1963).

29. M. G. Sirotyuk, Cavitation strength of water and its distribution of cavitation nuclei, Akust. Zh., 11(3):380 (1965).

30. I. Schmid, Kinematographische Untersuchung der Einzelblasen-Kavitation [Motion picture investigation of the individual cavitation bubble], Acustica, 9(4):321 (1959).

31. A. T. Ellis, Techniques for pressure pulse measurements and high-speed photography in ultrasonic cavitation, Cavitation in Hydrodynamics, H.M.S.O., London (1956), 8-1-8-32; Discussion C1-C3.

32. E. V. Romanenko, Miniature piezoelectric ultrasonic receivers, Akust. Zh., 4(3):342 (1957).

33. L. D. Rozenberg and M. G. Sirotyuk, Factors limiting the acoustic power of a transducer operating in a liquid, Collected Papers of the All-Union Sci. Tech. Conf. Application of Ultrasonics in Industry, Ultrasonic Instruments for Measurement and Inspection, Moscow (1960), p. 157.

34. L. D. Rozenberg, Einige physikalische Erscheinungen, die in hochintensiven Ultraschallfeldern entstehen [Some physical phenomena occurring in high-intensity ultrasonic fields], Fourth Internat. Congr. Acoustics, Copenhagen, Vol. 2, p. 179 (1962).

35. M. G. Sirotyuk, Behavior of cavitation bubbles at high ultrasonic intensities, Akust. Zh., 7(4):499 (1961).

36. G. A. Khoroshev, Collapse of vapor-air cavitation bubbles, Akust. Zh., 9(3):340 (1963).

37. Rayleigh, On pressure developed in a liquid during the collapse of a spherical cavity, Phil. Mag., 34:94 (1917).

38. V. A. Akulichev, Pulsations of cavitation bubbles in the field of an ultrasonic wave, Akust. Zh., 13(2):170 (1967).

39. I. G. Mikhailov and V. A. Shutilov, A simple technique for the observation of cavitation in a liquid, Akust. Zh., 5(3):376 (1959).

40. M. G. Sirotyuk, Energetics and dynamics of the cavitation zone, Akust. Zh., 13(2):265 (1967).

41. M. G. Sirotyuk, Experimental investigation of the growth of ultrasonic cavitation at 500 kc, Akust. Zh., 8(2):216 (1962).

42.   V. A. Akulichev, Experimental investigation of an elementary cavitation zone, Akust. Zh., 14(3):337 (1968).

43.   G. W. Willard, Ultrasonically induced cavitation in water: a step-by-step process, J. Acoust. Soc. Am., 25(4):667 (1953).

44.   W. Güth, The formation of pressure waves by cavitation, Cavitation in Hydrodynamics, H. M. S. O., London (1956), 6, VII-X.

45.   V. F. Kazantsev, Motion of gas bubbles in a liquid under the action of the Bjerknes forces arising in an acoustic field, Dokl. Akad. Nauk SSSR, 129(1):74 (1959).

46.   P. D. Jarmen and K. J. Taylor, Some physical effects of acoustically induced cavitation in liquid helium and liquid nitrogen, J. Acoust. Soc. Am., 39(3):584 (1966).

47.   I. N. Kanevskii, Steady forces arising in a sound field, Akust. Zh., 7(1):3 (1961).

48.   M. G. Sirotyuk, Energy balance of an acoustic field in the presence of cavitation, Akust. Zh., 10(4):465 (1964).

49.   F. E. Borgnis, On the forces due to acoustic waves in the measurement of acoustic intensity, J. Acoust. Soc. Am., 25(3):546 (1953).

50.   W. G. Cady and C. E. Gittings, On the measurement of power radiated from an acoustic source, J. Acoust. Soc. Am., 25(5):892 (1953).

51.   I. P. Golyamina, Magnetostrictive ferrites as a material for electroacoustic transducers, Akust. Zh., 6(3):311 (1960).

52.   L. D. Rozenberg and M. G. Sirotyuk, Radiation of sound into a liquid in the presence of cavitation, Akust. Zh., 6(4):478 (1960).

53.   M. S. Plesset and S. A. Zwick, The growth of vapor bubbles in superheated liquids, J. Appl. Phys., 25(4):493 (1954).

54.   L. I. Ganeva and I. G. Golyamina, Properties of magnetostrictive ferrites at high temperatures, Akust. Zh., 9(4):413 (1963).

55.   Winkler, Chemischtechnische Untersuchungsmethoden [Chemical Engineering Research Methods], Vol. 1, Lunge, Berlin (1921), 5, p. 558.

56.   A. S. Bebchuk, On the cavitation destruction of solids, Akust. Zh., 3(1):90 (1957).

57.   A. S. Bebchuk, Investigation of the Cavitation Damage of Solids and Surface Films in a Sound Field, Candidate's Dissertation, Akust. Inst. AN SSSR, Moscow (1960).

58.   M. G. Sirotyuk, Effect of the temperature and gas content of the liquid on cavitation processes, Akust. Zh., 12(1):87 (1966).

59.   M. G. Sirotyuk, Ultrasonic cavitation processes at elevated hydrostatic pressures, Akust. Zh., 12(2):231 (1966).

60.   M. G. Sirotyuk, An ultrasonic focusing concentrator of solid material, Akust. Zh., 7(4):499 (1961).

61.   B. A. Agranat, V. I. Bashkirov, and Yu. I. Kitaigorodskii, Cavitation damage of metals and alloys in an ultrasonic field, Application of Ultrasonics in Machinery Construction, Minsk (1964).

62.   B. A. Agranat, V. I. Bashkirov, and Yu. I. Kitaigorodskii, Technique for increasing the efficiency of ultrasonic effects on processes in liquids, Ul'trazvuk. Tekh., 3:28 (1964).

63.   N. A. Roi, Onset and development of ultrasonic cavitation (review), Akust.
       Zh., 3(1):3 (1957).
64.   V. A. Akulichev and L. D. Rozenberg, Certain relations in a cavitation region,
       Akust. Zh., 11(3):287 (1965).

PART VI

THE CAVITATION ZONE

L. D. Rozenberg

Introduction

In the overwhelming majority of studies devoted to various physical aspects of ultrasonic cavitation (see, e.g., Part IV and portions of Part V of the present book), the authors are concerned with the behavior of the individual isolated cavitation bubble. In reality such a bubble is very rarely encountered, its existence being contingent upon a set of conditions that are difficult to realize. As a rule, even at acoustic pressures not too far above the cavitation threshold, an aggregate of cavitation bubbles appears at once, taking up a definite region of space, which we have chosen to call the cavitation zone. Inspite of the foregoing, even in Flynn's excellent survey [1] and Pernik's book [2] the bulk of the text refers to the behavior of an individual bubble, with only two or three pages given over to the cavitation zone.

It should not be forgotten, however, that all (or almost all) the experimental data have in fact been acquired for a cavitation zone. In all practical applications of ultrasonic cavitation, we are again always confronted by a cavitation zone. Hence, the investigation of the cavitation zone comprises a very timely problem.

Three problems come immediately to mind. First, the investigation of the properties of the cavitation zone as a specific entity involves the propagation in it of a primary sound wave, the transfer of its energy into the energy of cavitation bubbles, etc. Second, it is essential to clarify to what extent it is possible to extend the principles set down for an isolated cavitation bubble (see, e.g., Part IV) to bubbles "en masse," i.e., as part of the cavitation zone. Third, it must be ascertained how the average acoustical properties of the liquid occupying the cavitation zone vary relative to the acoustical properties of the aggregate liquid. With due regard to the physical substance and practical importance of the laws governing the cavitation zone, a systematic research program has been undertaken at the Acoustics Institute of the Academy of Sciences of the group of problems pertinent to this topic.

Of course, the most natural approach to the solution of the three problems set forth above (as well as many others) would be to start with rigorous equations applicable to the cavitation zone. However, even the compilation of such equations poses a formidable problem, not to mention their integration. Consequently, this kind of frontal attack on the solution of the problem is left for a later time. Nevertheless, we cite some preliminary endeavors in this direction at the end of the present part of the book. In the present state of the art, one is compelled in the investigation of the cavitation zone to be confined to special approximative and sometimes phenomenological approaches. Logically, of course, these special approaches, realized by any of a variety of methods, not only fails to give an overall picture of the effects in question, but sometimes can even result in rather inconsistent particular solutions. As a whole, however, the totality of special investigations affords a certain relatively orderly scheme from which one can infer a number of useful notions and sometimes even specific recommendations. It is to the description of this material that we now direct our attention.

Chapter 1

Phenomenological Treatment

§1. The Cavitation Index

The quantitative investigation of the diverse properties of the cavitation zone requires the introduction of suitable character- istic criteria. We shall invoke them as needed.

As the first quantity characterizing the degree of development of cavitation we introduce the cavitation index (or the cavitation development index) K [3]. Inside the cavitation zone we set off a certain volume V, which meets the following two requirements: 1) The linear dimensions of V must be small in comparison with the wavelength, so that the primary, cavitation-inducing sound pressure may be regarded inside that volume as constant in mag- nitude and coherent in phase; 2) the linear dimensions of V must be much larger than the dimensions of the cavitation bubble.

Now let the volume of all cavitation bubbles present inside V in the phase of maximum expansion be ΔV. We then define the cavitation index as the ratio

$$K = \frac{\Delta V}{V}.\tag{1}$$

It is readily seen that ΔV is proportional to the energy stored by all bubbles contained in the volume V at the time of maximum expansion. But the cavitation index K is a measure of the spatial density of that energy. As will be shown presently, the cavitation index characterizes a whole series of effects that take place in the cavitation zone.

In the steady state, under constant external conditions (static pressure, temperature, gas content, etc.), the cavitation index is a function of the field coordinates and may be thought of in the limit as a function of a point. In several cases one is interested in the average value over the entire cavitation volume $V_K \gg V$ of the index

$$\langle K \rangle = \frac{1}{V_{\kappa}} \int_{V_{\kappa}} K dV_{\kappa}. \tag{2}$$

It is clear from the definition that the value of K falls within the limits $0 \leq K \leq 1$. The lower limit is physically meaningful; it corresponds to the absence of cavitation. But also the upper limit, as will be shown below, is more than a mathematical artifice; it is a physical reality as well. In concluding this section, we note that the cavitation index can be averaged not only over the volume, but also over any cross section.

§ 2. Uniformity of the Cavitation Zone

Inasmuch as the nucleation of a cavitation bubble is dictated not only by the strictly deterministic sound pressure at a given point, but also by the chance presence at that point of a cavitation nucleus of particular dimensions, the distribution of bubbles in the cavitation zone is random. Due to convection, Brownian motion, and other factors, clearly, it may be assumed that the cavitation nuclei are uniformly distributed on the average throughout a volume of liquid at rest. However, this does not always imply that the volume distribution of the cavitation bubbles is also uniform on the average. As a matter of fact, as has been demonstrated in [4], the number of cavitation bubbles significantly exceeds the number of nuclei. The multiplication factor, i.e., the ratio of the number of cavitation bubbles in the steady development of cavitation to the number of nuclei contained in the same volume can vary from 10 [5] to 10^5 [4], depending on the frequency, intensity, and configuration of the primary sound field, as well as the gas content of the water. The multiplication process evolves around a collapsing bubble, as a result of which clusters of bubbles of one form or another appear. The presence of these clusters can upset the uniformity of the bubble distribution considerably; the number of bubbles in the clusters is great, while in the space between clusters it is small or even zero. The clusters can occur

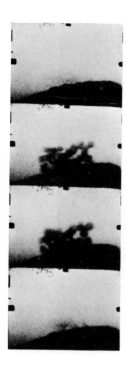

Fig. 1. Cluster of cavitation bubbles. High-speed filming at a rate of 50,000 frames/sec; $f = 15$ kc; $P_m = 2.5$ atm; ×20.

at an interface between a liquid and solid, where always, as a rule, cavitation nuclei are present. A typical cluster at the time of maximum expansion is clearly illustrated in Fig. 1, which was borrowed from [14]; it consists roughly of ten closely spaced bubbles, and in the neighborhood of the cluster, out to two or three times the diameter of the cluster, not a single bubble is observed. A similar pattern is observed in Fig. 29 of Part V (see also [6]); each light patch corresponds to a cluster of bubbles that are not visible separately, owing to their small size.

Also encountered on frequent occasion are clusters of cavitation bubbles having a peculiar filamentary form. Sometimes several filaments emanate and are called "spiders." An example of such a cluster is shown in Fig. 2 [7]. The causes of the formation of clusters of this configuration are not yet clear, but all experiments that have observed the filaments know that their configuration changes considerably with just a slight change of frequency. It may be postulated on this basis that the filaments are related to the interference structure of the near field of the radiator, as

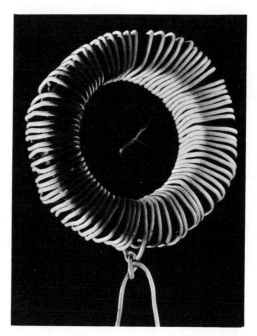

Fig. 2. Cavitation filaments in an annular ferrite
radiator.

this is the only factor which exhibits a strong response to a slight
frequency shift.

Consequently, the distribution of cavitation bubbles in the
zone is far from uniform in most instances. It is immaterial to
what extent this complicates affairs, other than the pure and simple fact that it often precludes any quantitative analysis of the
problems in which one is interested.

However, there are conditions under which the cavitation
bubble distribution in the zone may be regarded as essentially
uniform; this is true when the cavitation zone is formed near the
focus of a spherical focusing concentrator. Here, as a result, it
is entirely legitimate to treat the processes on the average. But
even in the aforementioned cases, when the bubble distribution in
the zone is plainly nonuniform, the investigation of the process
on the average is not altogether in vain, as it can at least shed
light on the qualitative aspect of the phenomena.

§3. Energy Losses in the Formation of the Cavitation Zone

Consider the model of a one-dimensional cavitation zone [8]. Let a plane ultrasonic radiator of sufficient radius R emit plane waves of length λ. The surface of the radiator has ideal wettability and does not carry any nuclei other than those already present in the liquid. It is clear that, with a uniform distribution of nuclei in the zone, the cavitation will be strongest at the radiator surface. A certain portion of the energy of the primary sound,* i.e., the sound produced by the radiator, will be lost in the formation of cavitation bubbles. Consequently, on moving away from the radiator surface, the primary sound and the cavitation it produces will be attenuated. This continues until the intensity of the primary sound falls below the cavitation threshold, at which time cavitation is halted. The distance at which that takes place is then the length of the cavitation zone. We will also assume that the following requirement holds

$$L < \frac{R^2}{\lambda},$$

i.e., over the extent of the entire cavitation zone the primary sound wave remains plane.

Now consider the energy losses in the formation of cavitation on the radiator axis (x axis) [9]. Referring to Fig. 3, let the sound intensity at the radiator surface in the plane x = 0 be I_0, and let it be independent of the variation of the average parameters of the

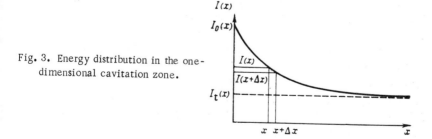

Fig. 3. Energy distribution in the one-dimensional cavitation zone.

---

*We introduce this term so as to distinguish the sound produced by the radiator from the sound generated by oscillating and collapsing cavitation bubbles.

medium due to the onset of cavitation (see Chap. 3). In the steady state, bubbles that have equilibrium radii (see Part IV) no smaller than a certain value R_t corresponding to the minimum primary wave threshold intensity I_t participate in the cavitation process. The condition for the onset of cavitation is $I \geq I_t$ and, of course, $I_0 \geq I_t$. Let us assume that the cavitating bubbles are uniformly distributed on the average in every cross section of the zone perpendicular to the radiator axis. We also neglect the linear and nonlinear absorption of sound during its propagation in the liquid. An energy flux $I(x)$ passes through a unit area perpendicular to the axis of the plane with coordinate x, and the flux perpendicular to the plane with coordinate $x + \Delta x$ is $I(x + \Delta x)$, where $I(x + \Delta x) < I(x)$ because some of the original energy goes for the formation of cavitation. The number of cavitating bubbles in our segregated volume element is

$$\Delta x \int_{I_t}^{I} \frac{dN\,(I)}{dI} dI,$$

where $N(I)$ is a function describing the concentration of stationary-cavitating bubbles at the sound intensity I. It is not to be confused with the function introduced in [10], which describes the concentration of nuclei.

As we stated above, every nucleus in the steady state forms a cluster comprising many bubbles. The function $N(I)$ can be determined experimentally either by direct counting with high-speed photographs or indirectly by measurement of the quantity ΔV and determination of the average cavitation bubble radius in the phase of maximum expansion, as proposed in [4] (see also Part IV, Chap. 2, §6). We shall assume that in the greater majority of situations (with the exception of ultrahigh intensities) the growth and collapse of the cavitation bubble occurs in just one period T of the primary sound (see Part V, Chap. 2, §5). Let the energy spent in the formation of one bubble from the nucleus be equal to $E_1 = E_1(I)$. Then the total energy lost by the wave in time T for the formation of all cavitation bubbles in the segregated elementary volume is

$$E_N = \Delta x \int_{I_t}^{I} E_1(I) \frac{dN\,(I)}{dI} dI.$$

However, it must also be equal to the variation of the intensity $I(x + \Delta x) - I(x)$ in the period T. Hence,

$$T\left[I\left(x+\Delta x\right)-I\left(x\right)\right] = -\Delta x \int_{I_t}^{I} E_1(I)\frac{dN\,(I)}{dI}\,dI. \tag{3}$$

Inasmuch as the functions $E_1(I)$ and $N(I)$ are continuous and differentiable (recall that they are obtained by averaging of the experimental data), the function $I(x)$ is also continuous and differentiable. Letting $\Delta x \to dx$, we obtain

$$-\frac{dI}{dx} = \frac{1}{T}\int_{I_t}^{I} E_1(I)\frac{dN\,(I)}{dI}\,dI, \tag{3a}$$

the integration of which, with regard for the initial condition $I(x) = I_0$ at $x = 0$, yields the desired relation for the intensity of the primary sound wave

$$\int_{I_0}^{I}\left[\frac{1}{T}\int_{I_t}^{I} E_1(I)\frac{dN\,(I)}{dI}\,dI\right]^{-1}dI = -x. \tag{4}$$

The specific form of the function $I(x)$ for a particular case can be obtained once we have analytic expressions for the functions $N(I)$ and $E_1(I)$. The situation with regard to the function $N(I)$ has been covered above. As for the function $E_1(I)$, so far we do not have any reliable experimental data. One can seek to determined it analytically by a rough schematic representation of the actual cavitation effect. However, as shown in §4, there is another, far more elegant way to determine $I(x)$.

An analogous investigation can be applied to a convergent spherical wave of the type that occurs in focusing concentrators, except that now it is required, in addition to the energy losses for the formation of the cavitation zone, to include the increase in intensity of the convergent wave as a result of focusing. It is difficult to say at the outset which of these two factors will be the prevalent one. The behavior of the cavitation zone in this case will be discussed in §5. For the moment we confine our perspective to a mathematical treatment of the process.

Assuming that the intensity in a convergent spherical wave grows inversely as the square of the radius of curvature of the wave front, we arrive quickly at an equation analogous to (3a), in the form

$$\frac{dI}{dr} + \frac{2}{r} I = \frac{1}{T} \int_{I_t}^{I} E_1(I) \frac{dN(I)}{d(I)} dI. \tag{5}$$

The initial conditions here are: $I = I_0$ ($I \geq I_t$) at $r = r_0$; $I = I_t$ (boundary of the cavitation zone) at $r = r_t$.

§ 4.   The Loss Function and Its

Application

The right-hand sides of Eqs. (3a) and (5) represent the energy lost per unit time in sustaining the cavitation zone in the steady state. Inasmuch as this is a function of I and I_t at a given frequency (for a given real liquid), we denote it by $C(I, I_t)$. As shown in [11] (see also Part V, Chap. 3), radiometric measurements can be used to determine experimentally the total power consumed in the formation of the whole cavitation zone. A series of photographs shown in Part V, Chap. 2 (Fig. 10 and accompanying text) shows the cavitation zone as having roughly the configuration of a sphere, whose radius grows with increasing voltage on the concentrator. Assuming that the distribution of the sound intensity and cavitation bubbles inside the cavitation zone is constant (this is considered somewhat more in detail in §5), we transform the results of [11] into a form better suited to the solution of our problem, recalling that

$$C(I, I_t) = \frac{W_\kappa}{V_\kappa}, \tag{6}$$

where W_K is the total power of the cavitation losses and V_K is the volume of the cavitation zone.

The dependence we seek is shown in Fig. 4 [9] for distilled water at a frequency of 500 kc. The points indicate the conversion of the experimental data according to Eq. (6), and the solid curve is an approximation:

$$C(I, I_t) = \begin{cases} A(I - I_t)^2, & I > I_t \\ 0 & , & I \leqslant I_t \end{cases} \tag{7}$$

For the given particular case $I_t = 900$ W/cm^2 and $A = 7 \cdot 10^{-3}$ cm/W.

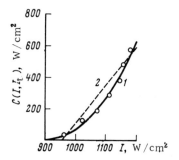

Fig. 4. Function $C(I, I_t)$ at 500 kc.
1) Square-law approximation; 2)
linear approximation; the circles
indicate the experimental points.

As an example of the application of the loss function, we refer back to the plane problem stated earlier. Recalling that

$$C(I, I_t) = \frac{1}{T} \int_{I_t}^{I} E_1(I) \frac{dN(I)}{dI} dI \qquad (6a)$$

and taking account of the approximation (7), we rewrite the equation in the form

$$\int_{I_0}^{I} \frac{dI}{A(I - I_t)^2} = -x, \qquad (8)$$

the solution of which is written as follows:

$$I(x, I_0) = \begin{cases} I_0 & , \quad I_0 \leqslant I_t \\ I_t + \dfrac{I_0 - I_t}{1 + xA(I_0 - I_t)} & , \quad I_0 > I_t. \end{cases} \qquad (9)$$

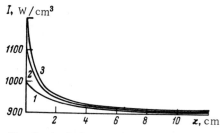

Fig. 5. Analytic intensity distribution on
the axis of a one-dimensional cavitation
zone.

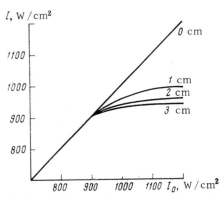

Fig. 6. Analytic behavior of the intensity in a one-dimensional cavitation zone as a function of the initial intensity.

Figure 5 shows a family of curves for I(x) for various values of I_0. Curve 1 corresponds to $I_0 = 1000$ W/cm^2, curve 2 to $I_0 = 1100$ W/cm^2, and curve 3 to $I_0 = 1200$ W/cm^2.

It is evident that the presence of cavitation causes a substantial intensity reduction. Already at a distance of a few centimeters from the radiator surface, the intensity of the primary field approaches the threshold value, and cavitation stops.

The same dependence is portrayed in Fig. 6 in different coordinates: $I(I_0)$ for various values of x, as indicated on the corresponding curves. The curves show that the intensity I at each point of space grows linearly as I_0 as long as $I < I_t$. With the onset of cavitation the growth of the curves slows down, departing further from the linear as the coordinate x increases. At a distance x = 3 cm the sound intensity after the onset of cavitation practically ceases its growth altogether. This means that the total increment of the original energy goes for the formation of the cavitation zone.

The situation is somewhat more complicated in the case of a convergent spherical wave. The substitution of (6a) into (5) yields the Riccati equation

$$\frac{dI}{dr} = A I^2 + \left(2 A I_t - \frac{2}{r}\right) I - A I_t^2, \tag{10}$$

which is not solvable in general form. For a rough estimate of the

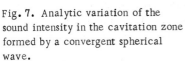

Fig. 7. Analytic variation of the sound intensity in the cavitation zone formed by a convergent spherical wave.

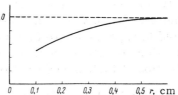

case one can replace the square-law approximation (7) with the linear approximation shown in Fig. 4 by the dashed line

$$C\,(I,\ I_t) = A_1\,(I - I_t).\qquad(7a)$$

The insertion of this expression into (5) leads to the following linear equation:

$$\frac{dI}{dr} + I\left(\frac{2}{r} - A_1\right) + A_1 I_t = 0.\qquad(10a)$$

The integral of this equation has a cumbersome form and need not be written out here. A curve computed on the basis of the integral, representing the ratio of the sound intensity in the convergent wave in the presence of cavitation to the intensity that would exist at that point in the absence of cavitation as a function of the distance r of the given point from the center of the focal spot, is shown in Fig. 7. It has been assumed here that the radius of the focal spot is 0.6 cm. It is apparent from the figure that, at a distance of 0.1 cm from the center of the focal spot, half of the total intensity of the primary sound is given up for the formation of cavitation. In fact, with regard for the fact that the losses in the formation of cavitation increase as the intensity squared, the intensity should fall off even more rapidly. This problem is attacked from another point of view in the next section.

§ 5. The Developed Cavitation Zone

(in a High-Intensity Focusing

Concentrator)

When we use a focusing concentrator, we are concerned with a relatively uniform zone far from the boundary with a solid surface and can therefore analyze the laws governing the cavitation zone in pure form, but this is not the only consideration that attaches special interest to the device. As will become apparent shortly, in focusing concentrators the ultrasonic energy is utilized far more efficiently than in plane radiators [12], so that it is reasonable to expect them to find more widespread use in the various applications of ultrasonics. Consequently, the more elaborate investigation of their sound field is also of practical urgency.

Consider the cavitation zone formed in an ultrahigh-intensity focusing concentrator operating at a frequency of 500 kc [13]. In concentrators of this type the cavitation zone occurs at distances much larger than a wavelength away from the radiating surface, and its onset does not elicit a reaction to the transmitted power (see Chap. 3). Therefore, the acoustic power is completely determined by the applied voltage U and is equal to $a U^2$, where for the given concentrator $a = 85$ W/kV^2.

The cavitation zone itself, as evident from Fig. 22 of Part V, has a roughly spherical form. As the voltage U, applied to the concentrator, is increased, the diameter of the sphere increases, and its center shifts in the direction of the radiating surface. The dependence of the diameter D_K of the cavitation zone on the applied voltage is shown in Fig. 8. The first thing we notice is that the dimensions of the zone are larger than the incident wavelength. As far as the dependence $D_K(U)$ is concerned, it approaches a square law (see, e.g., Fig. 23 of Part V).

Let us examine this problem more closely. As we are aware, in a convergent spherical wave the sound pressure increases as $1/r$ as the wave approaches the center of curvature (toward the principal focus of the system) at distances greater than the wave-

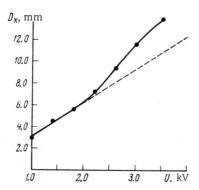

Fig. 8. Diameter of the cavitation zone versus the concentrator voltage.

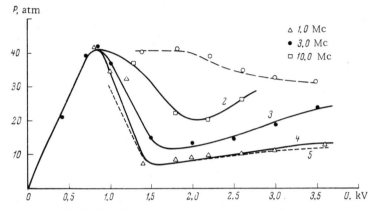

Fig. 9. Sound pressure versus concentrator voltage.

length, i.e., where diffraction effects are not yet pronounced. Consequently, with increasing voltage on the radiator or, equivalently, the sound pressure P_s on the radiator surface, the radius r_K of the wave front, at which the pressure is equal to the cavitation threshold P_K, will vary according to the law

$$P_{\text{к}} = P_s \frac{R_s}{r_{\text{к}}}, \tag{11}$$

where R_s is the radius of the spherical casing of the radiator.

In other words, proceeding from these considerations, the radius (or diameter) of the cavitation zone must for $r_K \ll \lambda$ increase as the voltage on the radiator increases, as illustrated by the dashed line in Fig. 9. Of course, the pressure at the boundary of the cavitation zone in this case must remain constant and equal to P_K. Actually, however, as the figure reveals, the true growth is somewhat faster (solid curve in Fig. 8). If now, relying on the validity of relation (11), we calculate the actual value of P_K at the boundary [9], it will have the form shown in Fig. 9 (curve 1). It is obvious that P_K stays constant up to U = 1.8 kV, but for larger values of U it gradually diminishes.

The reason for the diminution, clearly, is to be found in the fact, as shown in Part V, Chap. 2, §6 (p.   ) that inside the cavitation zone there is a continual process of multiplication and coagulation of the cavitation bubbles, the cavitation threshold de-

creasing somewhat, because in the steady state the role of the
cavitation nuclei begins to be taken over by so-called equilibrium
bubbles (see Part IV, Chap. 2, p. 221), whose volume and gas content are larger than those of the nuclei proper. Due to the presence of acoustic streaming, which elicits vigorous mixing, these
equilibrium bubbles are ejected toward the boundary of the cavitation zone, slightly lowering the cavitation strength of the liquid.
The pressure distribution inside the cavitation zone [8] is shown
in Fig. 10 for various voltages on the concentrator. Whereas the
diffraction structure of the spot is dinstinctly seen at a subthresh-
old voltage (U = 0.7 kV), above the cavitation threshold the entire
picture washes out, and the pressure distribution inside the zone
becomes essentially uniform. The pressure distribution curves
inside the cavitation zone as depicted in Fig. 13 of Part V have an
analogous form.

As mentioned, a convergent spherical wave propagating inside the cavitation zone, on the one hand, loses some of its energy
and, therefore, some of its intensity in the formation of cavitation
bubbles. On the other hand, this intensity is magnified by focusing.
Not to be overlooked is the possibility of the two processes com-
peting with one another. Any increase in the intensity above the
threshold value causes a disruption of the cavitation nuclei, and
this in turn induces a decrease in the intensity to the threshold

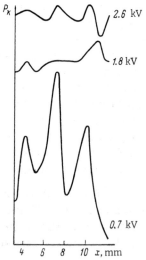

Fig. 10. Sound pressure distribution in the focal region
of a concentrator at various
voltages.

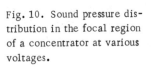

value and often, considering the foregoing remark about the role of equilibrium bubbles, to values considerably below threshold.

These considerations qualitatively explain the constancy of the sound pressure inside the cavitation zone.

Curves 2-4 in Fig. 9, which characterize the variation of the sound pressure at the center of the focal spot with variation of the radiator voltage, appear rather contradictory in this connection. However, here it is essential first of all to understand clearly that the interpretation of the results of measurements of the variable (sound) pressure in the cavitation zone is by no means a simple problem. As a matter of fact, in addition to the remaining (i.e., not given over to the formation of cavitation bubbles) portion of the primary sound energy, the sound receiver also reacts to cavitation noise, which has a broad spectrum. This is plainly evinced in the same Fig. 9, which shows the sound pressures recorded by receivers having different upper limits to the frequency band: 1.0, 3.0, and 10.0 Mc. Curve 2, which corresponds to the last receiver, is twice (and for voltages of about 1.5 kV, three times) as high as curve 4, which was recorded with a low-frequency receiver. It may be assumed with a certain approximation that the low-frequency receiver measures only the primary sound pressure.

The notion of a certain effective cavitation layer existing on the outer entrant boundary of the cavitation zone, wherein all the losses are concentrated, has been introduced in [14]. Assuming that after the transversal of this layer the remaining sound energy is focused, we can compute the pressure variation in the center of the zone. At voltages above 1.5 kV the result (curve 5 of Fig. 9) coincides very well with the experimental curve 4 recorded with a receiver having an upper frequency limit of about 1.0 Mc.

§ 6. Experimental Determination of the Cavitation Index

As implied by the definition (1), the cavitation index can be written, incorporating the above notation, in the form

$$K = \frac{1}{V_\kappa} \int_{I_t}^{I} V_1(I) \frac{dN\,(I)}{dI}\,dI, \tag{12}$$

where $V(I)$ is the maximum volume of one cavitation bubble in the phase of maximum expansion, where the bubble in question is formed from a nucleus at which cavitation occurs at an intensity equal to I.

The functions $V_1(I)$ and $dN(I)/dI$ are not known at this time, although in principle they can be determined experimentally for each specific case. For this we can use, say, high-speed motion pictures of the cavitation zone (or a small part thereof); in the subsequent processing of the frames we can obtain the functions $V_1(I)$ and $dN(I)/dI$ and calculate K from Eq. (12). This technique is suitable for measurement of the local values of the cavitation index in small volumes, although even in this case, besides the relatively complex apparatus required, the processing of the results is difficult.

A second technique for measuring K emerges directly from Eq. (1) and reduces to a measurement of ΔV by dilatometric methods. It is far simpler and yields an answer right away, but only permits one to measure the averaged value $\langle K \rangle$ over a relatively large volume.

High-speed motion pictures have been used in [14]. The experiments were carried out at a frequency of 15 kc in a cylindrical ceramic barium titanate focusing concentrator. The films were made with an SSKS-3 camera at a speed of 200,000 frames/sec with a magnification of five times on the film. Since the SSKS-3 camera has a low acuity, for stabilization in the space of the filmed cavitation zone a brass wire 1.5 mm in diameter was inserted into the focal region of the radiator, and the cavitation zone was produced at the end of the wire. The sound pressure was measured with a calibrated miniature receiver about 4 mm in radius placed on the radiator axis at a distance of 4 or 5 mm from the investigated zone.

Some frames obtained by high-speed photography to be used for calculation of the sizes and number of cavitation bubbles are shown in Fig. 11. The radius of the investigated focal zone was 1 or 2 mm.

Passing over the details of the procedure by which the experimental data were processed, we give the final results. The dependence of the sound pressure in the liquid on the voltage delivered

Fig. 11. High-speed film of a cavitation zone. The film speed is 200,000 frames/sec; f = 15 kc; P_m = 2.5 atm; ×20.

to the radiator is shown in Fig. 12a; cavitation sets in at a voltage of about 135 V. It can be verified that at the time of onset of cavitation the linear growth of the pressure slows down and ostensibly tends to a constant value. After the voltage on the radiator exceeds 200 V the pressure begins once again to increase, roughly linearly, but with a lesser slope.

It might be supposed that the slowing-down of the pressure rise is associated with the occurrence of cavitation bubbles on the surface of the receiver; however, the fact that the receiving surface is coated with a film of epoxy resin about 1 mm thick precludes that possibility, because, as shown in the next chapter, such a distance eliminates the possibility of interaction. The pressure reduction is more likely attributable to a reaction of the radiator in

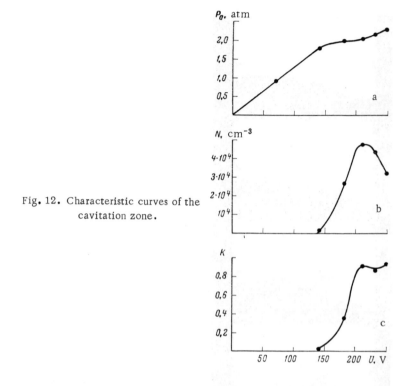

Fig. 12. Characteristic curves of the cavitation zone.

the given instance to the variation of the acoustical properties of the medium, because (in the particular case considered) the cavitation voids occur near the radiator axis, i.e., at distances smaller than half the wavelength from the radiating surface. This problem will be analyzed more in detail in Chap. 3.

The variation of the bubble density N in the given volume, counted directly in the film frames, is illustrated in Fig. 12b. The most striking fact is that N goes through a maximum in the vicnity of 200 V, and its decrease coincides with the secondary increase of the pressure. In seeking to explain this occurrence, we refer to Fig. 12c, which shows the dependence of the cavitation index calculated from the same frames. Rising sharply from the cavitation threshold, K attains values close to one. This implies that, in the phase of maximum expansion, almost the entire cavitation zone is filled with bubbles, which displace the aggregate liquid (see Fig. 11). With a further increase in the acoustic power of the primary field the bubble volumes increase even more, and since the aggregate liquid is already practically non-

existent, they begin to displace one another outside the volume V
in question. Here the number of bubbles in the given volume, i.e.,
their concentration, of course, falls off, and the cavitation index
is almost invariant. It is readily imagined that the complete displacement of the liquid from the cavitation zone by bubbles under
these conditions must lead to powerful microstreaming, both
around the cavitation bubbles and near the boundary of their cluster. Nor is the possibility excluded that the cavitation filaments
mentioned in §2 of this chapter could represent filamentary clusters
with a cavitation index close to one.

Next we consider the second technique for measuring the
cavitation index. As already stated, it relies on a measurement
of the total volume ΔV of liquid displaced by the bubbles.

The method is based on the dilatometric method proposed in
[15] for determining the cavitation threshold and is described in
detail in Part V, Chap. 2, §6. Although, as we said, the method
yields the averaged value $\langle K \rangle$, in spherical focusing concentrators the cavitation zone is small and has a fairly regular configuration and reasonably sharp boundaries; outside this zone,
cavitation is absent. Consequently, the dilatometric method can
be used to ascertain the average value $\langle K \rangle$ in the cavitation zone
in this case. Using the data of [4] (see also Fig. 8), we can deduce [14] the dependence of $\langle K \rangle$ on the voltage applied to the concentrator; the result is depicted in Fig. 13.

Now we perceive a different picture. The average cavitation
index for the cavitation zone at first increases rapidly, reaching
a value of 0.15 at a voltage of 1.8 kV, then begins to diminish.
This fits nicely within the framework of the hypothesis advanced
in [14] (see also Part V, Chap. 2) to account for the mechanism of
the multiplication of cavitation bubbles. It is postulated that
during collapse some of the bubbles disintegrate into fragments,

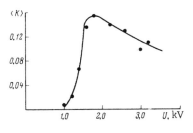

Fig. 13. Cavitation index in the focal
region of a high-intensity focusing concentrator.

each of which acts as a new nucleus. However, in addition to the increase in the number of bubbles, their number is diminished by coagulation. For every value of the primary sound field, the equilibrium state is established after a few periods. But, as shown in Part V (p.313), at concentrator voltages exceeding 1.8 kV the shape of the bubble oscillations begins to approach a sine wave, the number of "implosion" events decreases, and the number of fragments is reduced accordingly. Consequently, the equilibrium number of bubbles per unit volume falls off very abruptly, as is well evinced in Fig. 28 of Part V. The volume of each bubble grows with increasing primary sound intensity, whereupon the result is a gradual decrease of $\langle K \rangle$, as see in Fig. 13.

Not to be forgotten, however, is the fact that the local values of K inside the cavitation zone can still differ from the average value $\langle K \rangle$, nor is the possibility to be excluded that K can tend to unity at isolated points of the zone.

Chapter 2

Oscillations of a Bubble in a Cavitation Zone

§ 1. Statement of the Problem

Part IV of the present book is devoted to the oscillations of an individual cavitation bubble. The material presented there reveals that we are now in a position to describe the behavior of such bubbles in an alternating (acoustic) pressure field with due regard for a whole series of subtle effects and attendant processes [16] through the vehicle of nonlinear differential equations.

These equations, of course, are not amenable to integration in general form, but can be solved by numerical techniques. In [14] a great many particular solutions of the equations have been obtained by numerical integration on a digital computer, leading to a number of important conclusions and the formulation of some general principles.

The logical question arises, to what extent can these concepts, deduced for the individual bubble, be extended to a bubble as part of the cavitation zone. Given an affirmative answer, we could then avail ourselves of an entire battery of qualitative and, sometimes, quantitative principles. Otherwise, we would be placed in a very troublesome situation to the extent that, as suggested in the introduction to the present part, even the formulation of any kind of general equations to characterize the behavior of the cavitation zone in general with regard to its microstructure entails tremendous, often insurmountable difficulties.

Indeed, in this case we are compelled to consider a large number of qualitatively new phenomena: the multiplication and coagulation (coalescence) of cavitation bubbles, terminating in the buildup of a kind of dynamic equilibrium; the interaction of neighboring (or more distant) bubbles, i.e., a variation in the nature of their pulsations due to shock waves and the acoustic radiation of nearby bubbles; variations of the average acoustical properties of the medium as the result of trapped gas inclusions and, ensuing from this, variations of the sound pressure inside the cavitation zone at a fixed amplitude on the part of the radiator oscillations; microstreaming developed in the vicinity of each bubble, as well as in the cavitation zone as a whole, etc.

It is highly probable that the foregoing list will grow larger with time. We have only begun to delve into the complex aggregate of interrelated processes that attend the cavitation zone. However, even the effects enumerated above are sufficient indication that the strong temptation to borrow the results obtained in Part IV of the present book, as well as the survey [1], is justifiable, although certainly seldom without some preliminary arguments and at least crude estimates. We need only remind ourselves that, in accepting these arguments and estimates independently of one another, we are presupposing the admissibility of their ultimate superposition, i.e., linearity of the process in the cavitation zone.

It has been firmly established, however, that such a presupposition is inconsistent with reality. Hence, the only significant criterion of the validity of a particular estimate or any combination of such estimates at a given stage is their experimental confirmation.

What then can be stated right now about such crude estimates? Information regarding the multiplication of cavitation bubbles is presented in Part V, Chap. 2 (p.309). Considerations of the interaction between neighboring bubbles will be set forth in the next section of the present chapter; finally, Chap. 3 of this part will be given over to the variations of the average properties of the medium.

The problem of microstreaming in the cavitation zone remains wide open at the present time, apart from the cursory remark in §3.

§ 2.   The Interaction of Adjacent

Bubbles

In the process of its expansion a cavitation bubble, agitating the liquid, creates a radially divergent flow. The velocity of this flow in contact with a neighboring bubble affords a potential criterion of interaction. Thus, we imagine two identical bubbles expanding in phase and lying in close proximity to one another (Fig. 14). If we denote the time-varying bubble radius by $R(t)$, the radial velocity of its surface will be $\dot{R}(t)$. The flow generated by one bubble will act on the wall of the second expanding bubble. If this action is small, it may be assumed that each bubble oscillates without any "awareness" of the presence of the other one, and the principles deduced for the oscillation of the individual bubble may be applied to the bubble in question. Otherwise, the bubbles will interact.

Let us plant the coordinate origin at the center of one of the bubbles. Let the velocity of its wall be $\dot{R}_1$, and let the velocity of the resulting radial flow be $v_1(r)$, where r is the instantaneous coordinate. We shall assume that the interaction is negligible when $v_1(a - R_2) \ll \dot{R}_2$, where a is the distance separating the centers of the bubbles and $\dot{R}_2$ is the velocity of the surface of the second bubble. Inasmuch as $R_1 = R_2 = R$ and $v_1 = v_2 = v$, the noninteraction condition may be written

$$\frac{v}{R} \ll 1, \tag{13}$$

i.e., the flow velocity created by the first bubble at the wall of the second must be considerably smaller than the velocity of the wall of the second bubble. With regard for the symmetry of the effect, this condition may be expressed in simpler fashion; at the time of arrival of the flow at the wall of the neighboring bubble the flow velocity must diminish by one order of magnitude relative to $\dot{R}$.

Fig. 14. Diagram pertaining to the interaction of two bubbles.

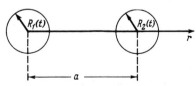

For the calculation of the flow velocity created by the expanding bubble, we can use an expression of the type, for example, given in [17]:

$$v(r) = \psi_1 \frac{1}{r^2},\tag{14}$$

where the function ψ is determined by the initial conditions. In our case these conditions are: $v = \dot{R}$ at $r = R$, whereupon

$$\psi_1 = \dot{R}R^2.$$

Substituting this condition into (13), we obtain

$$v(r) = \dot{R}\left(\frac{R}{r}\right)^2.\tag{14a}$$

Then condition (13) assumes the form

$$\left(\frac{R}{r}\right)^2 \gg 1,\tag{13a}$$

or, assuming that the sign $\gg$ in the given instance denotes reduction by one order of magnitude,

$$r = 3R.\tag{13b}$$

As we know, during oscillations of the bubble, R varies between the limits $R_0 < R \leq R_m$, where R_0 is the initial equilibrium radius and R_m is the radius of the bubble at the instant of maximum dilation.

Clearly, condition (13b) will be the most stringent for $R = R_m$. Then for the distance between bubbles a, we obtain

$$a \geqslant 4R_m.\tag{15}$$

The latter condition is easily translated into terms of the cavitation index K. Thus, converting to a single bubble, $\Delta V = \frac{4}{3}\pi R_m^3$, and $V_K = \frac{4}{3}\pi(a/2)^3 = \frac{4}{3}\pi(2R_m)^3$, so that, recalling the definition (1), for the most critical value of the cavitation index such that the bubbles may be assumed noninteracting, we obtain

$$K_{cr} = \frac{\Delta V}{V} = 0.125.\tag{15a}$$

It is important to note that this quantity depends very strongly (as the cube) on the bubble radius specified in condition (13b). If,

for example, we assume that not only the magnitude of the inter-
acting flows (13), but also the interaction time is important, and
choose $R = 0.75R_m$, thereby admitting the contingency of a certain
interaction over a small final time interval, then the critical value
of the cavitation index increases by a factor of 2.4, attaining values ~0.3.

It is also necesary to recognize the fact that, in general, the
bubbles in the cavitation zone do not oscillate in phase; this means
that different bubbles can attain their maximum radius at different
times, and with a certain scatter. Logically, this also tends to
ameliorate condition (15a).

§ 3. Experimental Estimation of the
Interaction Criterion

In order to assess the validity of condition (15a), one can use
the results described in Parts IV and V of the present book. Thus,
in Fig. 8 of Part V we find the experimentally recorded time dependence of the radius of a cavitation bubble located at about two
diameters from the surface of the radiator producing the sound
field. Not too far from the bubble is seen (slightly out of focus)
another bubble. The distance between the two bubbles is difficult
to determine precisely, but it clearly conforms to condition (13b).
Also shown in the same figure is the corresponding curve calculated
on the assumption that the bubble oscillates in isolation. The occurrence of the calculated and experimental results is very good.

An example of another character is shown in Fig. 12 of Part
IV. There the calculation and experimental observation of the
pulsations of a cavitation bubble at the edge of the cluster shown
in Fig. 11 are compared; the bubble in question is indicated by an
arrow for one of the observations. The solid curves in Fig. 12
correspond to the solutions of the Herring – Flynn equation, and
the dashed curves correspond to the solutions of the Noltingk –
Neppiras equation for a single bubble at sound pressures of 1.75,
2.00, and 2.75 atm.

The experimental results, indicated by dots, agree very well
with the calculated results, except for the last point ($\omega t \approx 5$) on
the curve for a sound pressure of 2.75 atm. The absolute value
of the bubble radius in this case is large enough, $R \approx 10^{-2}$ cm, to
warrant the suspicion of an error in its measurement; in order to

fit the analytic curve the bubble radius should be an order of magnitude smaller. Of course, it is extremely precarious to draw any sort of inferences on the basis of one experimental point, but in the light of the excellent agreement of all the other experimental points with the analytic data it may be assumed that the rate of collapse in this case is considerably smaller than the analytic value. The curve drawn through the last three points intersects the horizontal axis at a much smaller angle than the analytic curve.

What, then, are the values of the cavitation index in the given case? All the data necessary to answer this question are to be found in Fig. 12 of the present part, wherein the sound pressure (a) and the corresponding cavitation indices (c) are presented as a function of the same argument, the voltage on the radiator. A sound pressure of 1.75 atm corresponds to a cavitation index of about 0.04, and a pressure of 2.0 atm corresponds to an index of 0.35, but at pressures above 2.3 atm the cavitation index fluctuates between values of 0.85 and 0.95.

The first quantity fits condition (15a) perfectly, but the second somewhat exceeds the milder requirement $K = 0.3$. For the third curve the value of K is equal to an average of 0.9, corresponding to condition (13) for $R \approx \frac{1}{2} R_m$. It is appropriate at this point to recall that for this curve (see above) it is in fact admissible to assume the existence of a discrepancy between the analytic and experimental data in the region of collapse of the bubble. Clearly, the relatively slow process of expansion of the given bubble is the same as for the isolated bubble. The agreement of the maximum radius R_m with the analytic value indicates that kinetic energy stored by the bubble is independent of the presence of neighboring bubbles. But with the initiation of the contraction process, which proceeds at an ever-growing rate, the time comes when the liquid flow through the comparatively narrow spaces between the bubbles gains such large velocities that the resulting Bernoulli pressure near the bubbles decreases, and the process of collapse is retarded. Should the mechanism just described be subsequently confirmed, the investigation of microstreaming in the cavitation zone will pose a more pressing problem than it does now.

Not wishing to gamble on conclusive inferences based on the results of a single experiment, we must nevertheless assert that condition (15a) is extremely severe and that clearly in the majority of practical situations the principles deduced in Part IV for an individual pulsating bubble are little affected by the presence of neighboring bubbles in the cavitation zone.

Chapter 3

Variation of the
Average Acoustical Characteristics
of the Medium

§ 1. The One-Dimensional Cavitation
Zone

In the overwhelming majority of cases, except at high mega-cycle frequencies, where cavitation is not normally observed, the radii of the individual bubbles and their coalesced clusters turn out to be considerably smaller than the wavelength of the primary sound field. This entitles us to treat the liquid and the gas inclusions trapped in it as an ostensibly new medium with equivalent acoustical characteristics that differ from the acoustical character-istics of the aggregate liquid. This fact was first brought to general attention in [18].

For the precise analysis of the effect, one must necessarily take account of its nonlinear character. The pulsations of the vapor-gas bubbles, which act the part of weakening inclusions, cannot, strictly speaking, be represented as linear oscillations of an amplitude small in comparison with the average stable radius. As a first approximation, however, we shall adopt just such a model. In this case the average values of the fundamental acous-tical characteristics of the equivalent medium, its density and com-pressibility, over one period may be written in the form

$$\rho_{\kappa} = \rho_0 (1 - \overline{K}) + \rho_v \overline{K};$$

$$\beta_{\kappa} = \beta_0 (1 - \overline{K}) + \beta_v \overline{K}, \tag{16}$$

where ρ_0 and β_0 are the density and compressibility of the aggregate liquid, ρ_v and β_v are the density and compressibility of the vapor-gas mixture in the bubbles, and $\overline{K}$ is the average cavitation index corresponding to the average bubble radius, around which the oscillations take place. The logical choice for this average radius is half the maximum radius of the bubble, $\overline{R} = R_m/2$. Then the average cavitation index for one period, $\overline{K}$, may be expressed in terms of the cavitation index:

$$\overline{K} \simeq 0.1 \, K. \tag{17}$$

The time-average wave resistance of the equivalent medium may be written in the form

$$\overline{\rho_{\kappa} c_{\kappa}} = \rho_0 c_0 \left[\frac{(1 - \overline{K}) + \rho_v/\rho_0 \overline{K}}{(1 - K) + \beta_v/\beta_0 \overline{K}} \right]^{1/2}. \tag{18}$$

Allowing for the fact that for water $\rho_v/\rho_0 \ll 1$, $\beta_v/\beta_0 \approx 10^4$, and $K \ll 1$, we find that expression (18) takes the form

$$\overline{\rho_{\kappa} c_{\kappa}} = \rho_0 c_0 \left[\frac{\beta_0}{\overline{K} \beta_v} \right]. \tag{18a}$$

We need merely recall that this expression is invalid for $K \to 0$ and, in particular, for $K \ll 0.05$. In this case it is impossible to neglect the one in the denominator, and it is required to use the

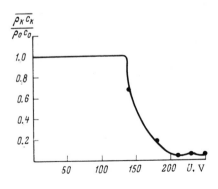

Fig. 15. Variation of the effective wave resistance of the medium in cavitation.

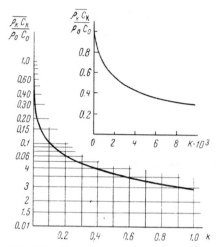

Fig. 16. Effective wave resistance of
the medium versus the cavitation index.

formula

$$\overline{\rho_{K}c_{K}} = \left[\frac{1}{1+\overline{K}\beta_{v}/\beta_{0}}\right]^{1/2}.$$ (18b)

For an approximate calculation we use the data given in Fig. 12 for the parameters of the local cavitation zone (cavitation cluster). Taking $K = K(U)$ from there and taking account of relation (17), we obtain the dependence of the ratio $\rho_{K}c_{K}/\rho_{0}c_{0}$ on the voltage U applied to the radiator. This dependence is illustrated in Fig. 15. It is evident that for highly developed cavitation, i.e., for a cavitation index close to unity, the wave resistance of the equivalent liquid is a fraction of one tenth of the wave resistance of the aggregate liquid, in this case water.

Results similar in order of magnitude are obtained for the calculation of the wave resistance in the focal region of a focusing concentrator operating at 500 kc [9, 14]. These analytic value exhibit poor agreement with the experimental results, bearing in mind, however, that the latter were obtained indirectly under a number of assumptions.

Although the foregoing linearized estimate is rather crude, it gives some idea of the order of magnitude involved. We therefore present the relation (18b) for water, assuming that $\beta_{v}/\beta_{0} = 10^{4}$. This dependence is illustrated in Fig. 16. For convenience, its initial portion, corresponding to very small values of K, is also shown on a blown-up scale. It is apparent that the wave resistance varies considerably, even for small values; for example, for K = 0.001 it decreases 30%, while for K = 0.003 it is cut in half. Clearly, this effect cannot be ignored.

§ 2. Experimental Investigations of the

Equivalent Wave Resistance in Cavitation

As already mentioned, the first experimental study of the effect discussed in the present chapter was carried out in [18]. The experiment was conducted at a frequency f = 28 kc in a container with sound-absorbing walls, filled with degassed water having a residual air content of 0.8%. The oscillations were generated by a ferrite resonance-mode radiator rigidly connected to another ferrite transducer designed to monitor the amplitude of the emitting surface. By recording the frequency characteristics of this dual system at a constant particle velocity when loaded in water and in air, it was possible to determine both the losses in the system and the acoustic load presented by the cavitating water. The curve shown in Fig. 17 [18] reveals that the effective resistance decreases to 30% of its original value. Also investigated in the same paper was the influence of the decrease of $\overline{\rho_K c_K}$ for pulsed sound at a repetition rate of 12.5 cps. It turned out that, for a radiator particle velocity of 15.7 cm/sec at a frequency of 28.0 kc and a pulse duration of 3 msec, $\overline{\rho_K c_K}$ decreases to a value of $8 \cdot 10^5$, for a 5-msec pulse to $7 \cdot 10^5$, for an 8-msec pulse to $6.5 \cdot 10^5$, and for ∞ (i.e., continuous sound) to $6 \cdot 10^5$.

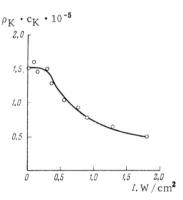

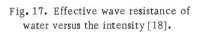

Fig. 17. Effective wave resistance of water versus the intensity [18].

The same problem was studied in [19]. The principle of the method was the same as that used in [18], except that the particle velocity was measured by five miniature piezoelectric pickups, whose readings were averaged, and the acoustic load on the radiator was determined with a specially designed measurement bridge; the magnetostrictive radiator, with a radiating surface of 8.5×8.5 cm^2, operated at a frequency of 21.4 kc. In addition, the alternating pressure acting in the cavitating liquid was measured with miniature receivers shielded against cavitation damage [20].

It was mainly established that, in the range of particle velocity amplitudes from 0 to 40 cm/sec, the shape of the surface oscillations remained sinusoidal or very close to it. The shape of the sound pressure curve, on the other hand, changed from sinusoidal in the subcavitation regime to very complex in the case of well-developed cavitation. Consequently, it was reliably verified that the observed nonlinear effects were indeed due to variation of the properties of the liquid.

As stated, a cavitating liquid is a nonlinear medium; hence, the conventional concept of wave impedance as the ratio of the instantaneous values of the pressure to the particle velocity becomes inapplicable. Recognizing, however, that the particle velocity of the emitting surface remains sinusoidal, we can stick to the notion of the average radiation resistance regarded as the ratio of the total radiated power to the square of the radiator particle velocity:

$$\overline{R}_{\mathrm{rad}} = 2\,\frac{W_a}{v_m^2} = 2\,\frac{I \cdot S}{v_m^2}. \tag{19}$$

We need only remember that, as opposed to the linear case, the quantity $\overline{R}_{\mathrm{rad}}$ is not a constant of the medium, but a function of v_m or K.

If we refer the quantity thus obtained to the radiating surface S, we obtain the same quantity as that which we called the average wave resistance of the equivalent medium:

$$\overline{\rho_{\kappa}c_{\kappa}} = \frac{\overline{R}_{\mathrm{rad}}}{S} = 2\,\frac{I}{v_m^2}. \tag{20}$$

The results obtained in [19] are shown in Fig. 18 (curve 1). The quantity v_m^2 is plotted on the horizontal axis, and the vertical axis represents $\overline{\rho_K c_K}$. Plotted on a second scale on the hori-

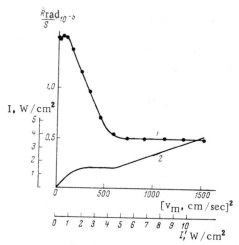

Fig. 18. Effective wave resistance and actual radiated power versus the particle velocity of the radiator surface.

zontal axis is $I' = (v_m^2/2)\rho_0 c_0$. This is the value that the radiated
intensity would have if cavitation did not occur. The limiting val-
ue to which the ratio $\overline{\rho_K c_K}/\rho_0 c_0$ decreases is, as in [18], approximately 30%; this corresponds, as apparent from Fig. 16, to an
average value of K = 0.01 for the zone as a whole. However, in
departure from the data of [18], it is obvious that after reaching
this value at $v_m = 25$ cm/sec it then remains constant.

Once we have the value of $\overline{\rho_K c_K}$ as a function of v_m, we can
use Eqs. (19) and (20) to find the values of W_a and I, respectively,
which differ only by a constant factor S. Curve 2 of Fig. 18 shows
the variation of I; the appropriate scale is given on the vertical
axis. If the wave resistance of the medium was to remain constant, we would have to obtain a straight line. Actually, however,
a more peculiar type of dependence results; prior to the onset of
the cavitation threshold (in the given case at $I_t \approx 0.9$ W/cm^2) the
curve is a straight line; then with the initiation of cavitation it
rises more slowly, and after attaining a maximum value of $I \approx$
1.5 W/cm^2 it levels off.

Over the entire descending interval of the wave resistance
the radiated intensity does not change; only after the stabilization
of curve 1 at $v_m > 25$ cm/sec does the intensity again begin to

rise along a straight line, but with a considerably smaller (about one third) slope.

Of course, the data presented here and the subtle peculiarities of the curve, such as the strict constancy of I in the interval 10 cm/sec < v_m < 25 cm/sec, are typical only of the conditions of the given experiment; yet the general behavior of the curves affords a characteristic of the process as a whole. For I > I_t the inequality I < I' always holds; for example, in the given specific case for I' = 12 W/cm^2 the value of I is only about 4 W/cm^2. This indicates the error of estimating the power output of an ultrasonic radiator by measuring it in the subcavitation regime and extrapolating the result with respect to the voltage applied to the radiator. Unfortunately, this is the kind of fallible practice that we encounter repeatedly in the literature.

In order to compute the instantaneous value of the sound pressure or its effective value, we must be able to describe all the processes associated with the oscillations of the bubbles in the cavitation zone; but at the present time we are not able to do so. Nevertheless, we can introduce the concept of the effective sound pressure as the reaction of the nonlinear medium to the linear oscillations of the radiator

$$P_{\text{eff}} = \frac{v_m}{2} \rho_K c_K.$$ (21)

All nonlinear properties of the medium are now contained in the quantity $\overline{\rho_K c_K}$, which we know on the basis of experimental measurements, and we obtain a function of the particle velocity of the radiator surface. The pressure thus calculated is given in Fig. 19 (curve 1). This function has two extrema, between which is a descending interval corresponding to the development stage of the cavitation process. In the same figure curves 2 and 3 represent two experimental curves recorded with a miniature probe at two different points of the cavitation zone. In lieu of a reliable calibration the scale of the vertical axis is made arbitrary, although, qualitatively the behavior of these curves is consistent with the one calculated on the basis of Eq. (21); there are at least two extrema with descending intervals between them in the interval of cavitation development.

We note in conclusion that the effect described here is in fact the reason for the deviation of the sound pressure buildup from the linear in the case treated in Chap. 1, §6 (see Fig. 12a).

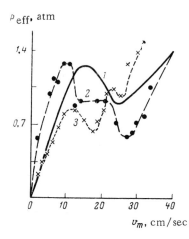

Fig. 19. Effective sound pressure in
the cavitation zone versus the particle velocity of the radiator [19].

§ 3.   Compressibility of Cavitation Bubbles

In §1 we adopted a linear model for the oscillations of cavitation bubbles in the form of small pulsations about a certain average radius. However, even within the scope of this interpretation, a bubble filled with a vapor-gas mixture cannot always be
regarded merely as an additional compressibility. In particular, the fundamental resonance of the radial modes of the bubble
is determined, as we are well aware, by the compressibility of
the vapor-gas mixture and the additional mass of the liquid.

Therefore, the bubble behaves as a compressibility element
only at frequencies below the first resonance. At resonance its
reactance components cancel one another, and it becomes a pure
active load. At frequencies above resonance (but below the next
resonance) it represents a mass reactance, because the additional mass surrounding the bubble has the effect of shielding its
compressibility, and it becomes exempt from the propagation of
the sound wave. Consequently, even the linear compressibility
model produces an increase in the compressibility only at lower
frequencies than the resonance frequency of the bubbles, which is
determined by their radii and gas content.

In reality the cavitation bubble oscillations analyzed in detail
in Part IV bear a far more complicated nature and are described
by nonlinear differential equations. One might seek to investigate
their true compressibility by using a linear analogy and taking

account of the variation of the maximum radius (and, hence, the "resonance" frequency) with the variation of the external variable pressure. In the given instance, however, there is another more workable approach, which, although it does not lead to analytic expressions, gives greater confidence in the validity of the argumentation.

The fact is that the numerical solutions of the nonlinear differential equations describing the oscillations of cavitation bubbles permit one to obtain a function expressing the time variation of the bubble radius $R = R(t)$ for any specified conditions. Once the bubble radius is known, it is a simple matter to determine its volume at every given instant. Of course, the compressibility of the bubble is simply

$$\beta_v(t) = -\frac{1}{V(t)}\frac{dV(t)}{dp}.\tag{22}$$

The average compressibility over the period T, which determines the variation of $\overline{\rho_K c_K}$ (recalling that ρ_K changes very little in practice in all cases except when the cavitation indices are

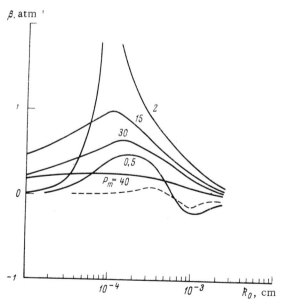

Fig. 20. Compressibility of a cavitation bubble versus the equilibrium radius.

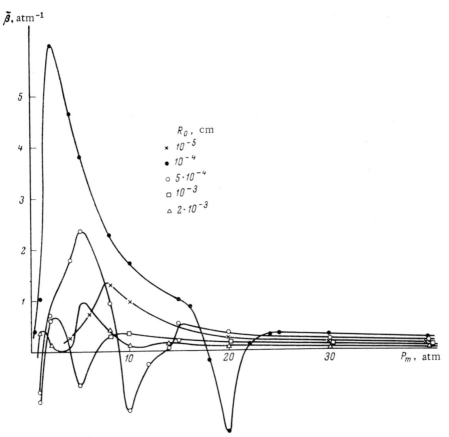

Fig. 21. Compressibility of a cavitation bubble versus the sound pressure.

large), may be written in the form

$$\overline{\beta}_v = \frac{1}{T} \int_0^T \beta(t)\, dt. \qquad (23)$$

With the function R(t) at our disposal we can readily determine the values of β_v numerically. Below we give, as an example,
the computed values of β_v at f = 500 kc for various sound pressure and various values of the equilibrium bubble radius.*

---

*The computations were carried out by V. A. Akulichev on the Minsk-2 electronic
digital computer.

The dependence of the compressibility on the equilibrium radius is shown in Fig. 20, and its dependence on the variable sound pressure for various values of the equilibrium radius between the limits of 10^{-5} and $2 \cdot 10^{-3}$ cm is shown in Fig. 21. It is apparent in these figures, especially in Fig. 21, that the compressibility of the individual bubble can vary over very wide limits, decreasing from large values to zero and assuming negative values.

Unfortunately, we do not have access to reliable data on the distribution of equilibrium bubbles, nor are there any precise methods for determining it. Indirect information on the predominant bubble radii can be obtained from cavitation noise spectrograms, as described in Part IV, Chap. 3, §3, but these data are inadequate for conversion from the quantity $\overline{\beta}_V$ to Eqs. (16) or (18a).

Some considerations relevant to the more rigorous investigation of the influence of bubble compressibility on the properties of the cavitation zone will be presented in Chap. 5.

Chapter 4

The Cavitation Efficiency

§ 1. Fundamental Definitions

A certain portion of the energy of the primary sound field is utilized for the creation of the cavitation zone. We denote the ratio of the spent energy to the total energy of the primary field by $\varkappa$ and call it the coefficient of cavitation utilization of acoustic energy [12]:

$$\varkappa = \frac{E_\kappa}{E},\qquad(24)$$

where E_K is the energy spent in the formation of cavitation per unit volume and E is the input primary energy density.

It is clear that not all of the energy E can be tapped for the formation of cavitation, because for $E < E_t$ cavitation stops, so that always $\varkappa < 1$.

Some of this energy is recovered in the collapse of bubbles in various forms, chiefly in the form of shock waves capable of producing mechanical erosion and as cavitation noise; furthermore, some of the energy is spent in the formation of active chemical radicals, luminescence, etc.

For the industrial application of ultrasound mainly cavitation erosion is used, namely for ultrasonic cleaning, emulsification, etc. If we denote the energy transmitted per unit volume to shock waves by E_M, then the ratio

$$\varepsilon = \frac{E_M}{E_\kappa}\qquad(25)$$

is a measure of the erosive cavitation activity. It is clear from the definition that $\varepsilon < 1$. It should be remembered, however, that the erosive activity is determined not only by the amount of energy transferred, but also by its rate of release, i.e., by the peak power values.

It is well known that mechanical damage requires definite values of the mechanical stresses or strains above threshold. Thus, for electrical breakdown, for example, it is necessary to exceed the threshold values of the electrical voltage. The mechanical stresses are determined by the shock wave pressure, and the latter in turn depends specifically on the peak power carried by the shock wave. If the pressure of the shock wave is less than the critical value capable of producing mechanical damage, then, of course, damage will not ensue, regardless of the values of ε. But if the power of the shock wave is such that the pressure exceeds the threshold strain values, damage will take place, and it is in this case that the volume of the damage will be proportional to ε, all other conditions being equal.

The sound field energy going for the formation of the cavitation zone is lost during the entire expansion phase of the cavitation bubble. As the many curves presented in Part IV indicate, this time is somewhat less than the period T of the primary sound. The error is not too appreciable if we set that time equal to T.

As for the time in which the energy stored by the bubble is given up, a few additional remarks are in order. Strictly speaking, energy is given up during the entire collapse time τ_m of the bubble (see part IV) and in the very beginning of the expansion phase of the ultimately compressed bubble. But the rate of collapse of the bubble in this period varies so sharply, from zero at the beginning of compression to very large values at the end of the compression period and the beginning of expansion, that at the instant of maximum contraction some of the energy stored in the bubble will already have been extracted for the creation of microstreaming, heating of the bubble interior, the generation of luminescence, and possibly the formation of chemically active radicals. The remainder, and clearly the greater part, of the energy is transmitted into the surrounding liquid, the instantaneous power (i.e., the rate of energy transfer) becoming greater when the rate of expansion of the bubble is maximal, i.e., at the very beginning of the expansion phase. In the course of expansion this rate falls off very rapidly. If we assume for the sake of simplification that the rate is constant at the beginning of expansion, after which it rapidly drops to zero, then the period τ_0 of existence of the large velocity will, together with the energy spent in the formation of a shock wave (as characterized by the parameter ε), determine the peak power and, hence, the quantity of principal concern in the given case, namely the pressure in the shock wave. If we denote the period-average power spent in the formation of the total cavitation zone by W_K, then according to (24) it is equal to

$$W_{\kappa} = \frac{E_{\kappa}}{T} = \varkappa \frac{E}{T} = \varkappa W, \qquad (26)$$

where W is the primary acoustic power delivered by the radiator into the volume occupied by the cavitation zone. The power extracted by all shock waves generated in the cavitation zone is equal to

$$W_{sh} = \frac{E_{\text{M}}}{\tau_0} = \frac{E_{\kappa}}{\tau_0} \varepsilon = \frac{\varkappa \varepsilon}{\tau_0} E. \qquad (27)$$

Comparing Eqs. (26) and (27), we readily verify that the quantity characterizing the erosive efficiency of the sound field may be written in the form*

$$\eta = \frac{W_{sh}}{W} = \varkappa \varepsilon \frac{T}{\tau_0}. \qquad (28)$$

The quantities $\varkappa$ and ε, as indicated, are always smaller than one, although in some cases they can be close to that value. The ratio T/τ_0 can fluctuate over very wide limits. For nearly sinusoidal oscillations of the cavitation bubble, which can take place either in the case of bubbles having a very large gas content or in the case when the Rayleigh (i.e., the total) collapse time τ_m of the bubble is close to $T/2$ [Eq. (21); see also Part IV], the ratio T/τ_0 is on the order of unity. If, however, the amount of gas in the bubble is very small and $T/\tau_m > 1$, the ratio T/τ_0 can attain very large values.

---

*It is important to note that here the coefficient η does not correspond to the value of η in [12], but differs from it by the amount of the factor T/τ_0.

In other words, the individual cavitation bubble or cavitation zone representing an accumulation of such bubbles may be regarded as a special kind of power transformer, in which the cumulative energy is liberated rather slowly in a very short period of time, so that the instantaneous power is many times the average power delivered by the radiator into the cavitation zone.

It is self-evident that analogous arguments are applicable to any other form of cavitation efficiency; hence, expressions of the type (28) can be deduced. These expressions will always contain the factor $\varkappa$ characterizing the transfer of energy from the primary sound field into stored energy in the cavitation zone. Where the other coefficients are concerned, they will be determined by the particular efficiency involved, for example, chemical, luminescent, etc.

It is important to note in conclusion that although the factors $\varkappa$, ε, and T/τ_0 involved in η do not depend directly on one another, they can be functions of the same variables, such as the intensity and configuration of the primary sound field, distribution of the cavitation nuclei, temperature, static pressure, gas content, etc. A variation of one of the latter can induce a variation of any of the factors and, if so, in different directions. Therefore, in the investigation of the possible effects of one of the above factors on η, one should not overlook the possibility of concomitant variations of the other two.

§ 2. Experimental Determination of $\varkappa$

The most universal coefficient involved in the expression for any cavitation efficiency is $\varkappa$. The definition (24) implies the possibility of determining it by the radiometric measurement techniques already familiar to us (see Part V, Chap. 3). The readings P_0 of a radiometer without a protective film are proportional to E, and with a protective film to intercept the steady flow $P' \approx E - E_K$. Hence,

$$\varkappa = \frac{P_0 - P'}{P_0}.$$

The behavior of the dependence of $\varkappa$ on the voltage on the winding of a magnetostrictive ferrite radiator operating at 20 kc is shown in Fig. 22. It is clear that $\varkappa$ begins to depart from zero

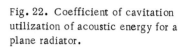

Fig. 22. Coefficient of cavitation
utilization of acoustic energy for a
plane radiator.

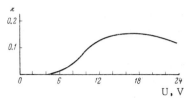

after passing the cavitation threshold (U ≈ 5 V) and increases to
0.2 in the working mode, acquiring a gentle maximum at about
16 V.

Considerably more favorable conditions are realized in focusing devices. Thus, for an ultrahigh-intensity radiator operating
at 500 kc [13] the general path of the dependence of $\varkappa$ on the radiator voltage, as evident from Fig. 23 (curve 1), has the same
character as for a ferrite radiator; the absolute value of $\varkappa$ in
the working mode (about 1.8 kV), however, attains values of 0.6-0.7,
i.e., it is three times the value for a conventional plane radiator.

The reasons for this gain have already been discussed in
Chap. 1, §5. They bear on the fact that, in the case of a plane
concentrator, the intensity of the propagating wave diminishes
continuously and is used for the formation of cavitation bubbles.

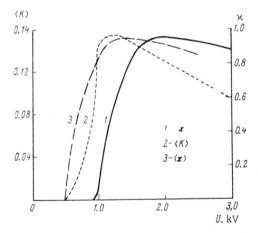

Fig. 23. Coefficient of cavitation utilization of
acoustic energy for a spherical focusing radiator.

The diminished intensity can induce cavitation only if the wave encounters a nucleus of large radius in its path, but this occurrence is quite rare. In focusing systems, on the other hand, the intensity in the wave propagation process constantly grows due to the focusing effect; hence, the total number of cavitating bubbles becomes considerable.

It is appropriate at this point to reflect on the consideration that the growth of the coefficient $\varkappa$ still does not imply an increase in η. In fact, as we know [21], the magnitude of the erosion in the given concentrator has a maximum at a voltage of about 2 kV, after which it begins to fall off steeply. This means that, even though the amount of sound energy going for the formation of cavitation decreases only slightly with an increase in voltage above 2 kV, as evident from Fig. 23, the other coefficients involved in the expressions for η begin to decrease rapidly. Therefore, in order to obtain the maximum erosion in a given concentrator it is necessary to operate at voltages no greater than 1.8 to 2.0 kV.

Curiously, the behavior of the curve for $\varkappa = \varkappa(U)$ in this case is strikingly reminescent of the behavior of $\langle K \rangle$ as a function of the same applied voltage (see Fig. 13); the latter curve is transferred to Fig. 23 with a change of vertical scale (curve 2). The fact that the two curves have a horizontal shift is probably attributable to the different properties of the water used in the two experiments. If the curve for $\varkappa$ is moved to the left until the cavitation thresholds coincide (curve 3), fully satisfactory agreement results, fostering the assumption that the relation between $\langle K \rangle$ and $\varkappa$ has an almost-linear character.

Also related to the cavitation index is the free enthalpy G of the cavitation bubbles [14]. The increment of the free enthalpy is equal to [22]

$$\Delta G = -S \Delta T + V p + \mu \Delta n,$$

where S and V are the entropy and volume of the subsystem, ΔT and Δp are the temperature and pressure increments, μ is the chemical potential of the vapor per unit volume, and Δn is the relative volume change.

It may be assumed in the expansion process of the cavitation bubble that $\Delta T = \Delta p = 0$. Then the free enthalpy of one bubble is

$$G_1 = \mu \Delta n,$$

and the free enthalpy of all cavitation bubbles per unit volume is expressed as

$$G = \mu \int_{I_t}^{I} V_1 \frac{dN\,(I)}{dI}\, dI = \mu K,\tag{29}$$

i.e., is a linear function of the cavitation index.

§ 3. The Measurement of the Acoustic Energy Cavitation Efficiency

In order to meet practical requirements, several methods have been proposed for the assessment of cavitation efficiency. The majority of the recommendations entail the application of various kinds of test objects, both mechanical and chemical. In studies carried out at the Acoustics Institute of the Academy of Science of the USSR, the weight loss of a small aluminum cylinder placed at a test point of the cavitation domain was measured in order to estimate the cavitation erosion [23]. Somewhat later [24] it was proposed that the shape and intensity of the cavitation zone be assessed in terms of the damage to a polished glass surface and in terms of the damage to a light-sensitive photographic layer. Very recently [25] a special miniature instrument was developed in which a photocell is used to measure the total area of the punctures formed in a piece of aluminum foil exposed to cavitation. Of course, this is quicker than the tedious methods of irradiating and weighing a small sample, but it is far less accurate, because if a puncture is formed at a particular site, the repeated action of cavitation on that site can no longer be ascertained. Moreover, for a large number of puctures the minute pieces of foil left between the punctures will simply fall out, creating the illusion of an increase in the observed erosion. This could very conceivably explain the square-law time dependence obtained for the erosion in [25], as opposed to the many investigations on different materials in [26], which showed that this dependence should have a linear behavior.

One of the most sensitive methods based on the chemical effects of cavitation involves the decomposition of carbon tetrachloride [27] with an estimation of the quantity of free chlorine released, in terms of the degree of coloration of the reaction product of free chlorine with ortho-tolidine.

In [28] the chlorine release was determined from the amount
of iodine that it displaced from a solution of potassium iodide.
This investigation is also interesting in that it corroborates the
linear time dependence of the amount of chlorine released and
demonstrates a correlation between that amount of chlorine and
the efficiency of ultrasonic cleaning.

A basic shortcoming of all chemical methods is the unaccept-
ability of the existence of complete correlation between the cavitation erosion and the chemical action of cavitation as a proven fact.
The example cited in [28] is merely a special case. Moreover,
in the series of investigations [29] the presence of two spontane-
ously alternating distinct cavitation phases is verified, one of
which is responsible for chemical effects and luminescence. Unfortunately, in these investigations it is not proved that the second
phase is responsible for mechanical damage; yet the authors make
claims to that effect. In any case, however, this study prohibits
the use of chemical techniques without special preliminary tests
for the given specific conditions.*

There are also purely acoustical methods. They too must
be resorted to with extreme caution, because the sound field in the
cavitation zone is a very intricate combination of the primary
sound and the cavitation noise, which, in addition to discrete components, also contains white noise [8]. Evidently, this white noise
constitutes the sum of the spectra of the shock waves formed by
cavitation bubbles. If this is true, then the white noise intensity
must be determined both by the number of bubbles and by the
slope of the shock wave front. However, these are the very factors responsible for cavitation damage.

Consequently, the white noise intensity and magnitude of the
cavitation erosion can be expected to have a correlation. This
has been confirmed experimentally in [30]. At a frequency of 8.1
kc the cavitation damage to an aluminum cylinder was measured
in three different liquids at three distances from the end of the
vibrator. Also measured at the same points was the sound pressure, in which, according to test measurements, white noise sub-

---

*In a recently published paper, Ibisi and Brown [J. Acoust. Soc. Am., 41(3):568 (1967)]
have demonstrated experimentally that the temperature dependence of the erosion
and chemical reactions in cavitation are totally different.

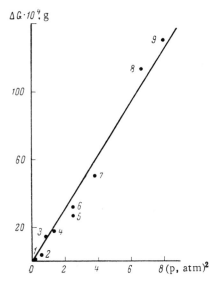

Fig. 24. Relation between cavitation erosion and white noise.

stantially dominated. The results are exhibited in Fig. 24, which shows the sound pressure as a function of the weight loss of the sample. It is apparent that all the points are well grouped about a straight line, whose slope is a measure of the cavitation strength of the sample. The experimental conditions are summarized in Table 1. The numerical entries correspond to the label numbers of the points in Fig. 24.

Comparatively recently a foreign company developed and began the manufacture of a special device based on the same principle [31], and apparently it has been successful in solving the stated problem.

Medium	Distance, mm		
	2.25	1.5	0.5
Acetone	1	2	3
Water + OP-10	4	5	6
Water	7	8	9

§ 4.   Estimation of T/τ_0

The intervals τ_0 are so small and the velocities of the bubble walls at the time just preceding collapse or the beginning of expansion so great that they cannot even be read on a digital computer. It is necessary, therefore, to resort to what are analytically rather crude estimates based on a series of approximations. We postulate first of all, as several authors have done, that the collapse of a cavitation bubble is an adiabatic process and that its expansion is isothermal. Then we can invoke the first integral of a Rayleigh-type equation with the presence inside the bubble of a certain quantity of gas, whose pressure at the initial time, i.e., at $R = R_{max}$, is P_g. This integral has the form [32]

$$\tfrac{3}{2}\rho\dot{R}^2 = P_a\left[\left(\frac{R_{max}}{R}\right)^3 - 1\right] - P_g\frac{\left[\left(\frac{R_{max}}{R}\right)^3 - \left(\frac{R_{max}}{R}\right)^\gamma\right]}{1 - \gamma} ,$$

where R is the instantaneous bubble radius, P_a is the pressure acting in the medium at the initial instant of collapse, and γ is 4/3 for the adiabatic process.

Here the maximum and minimum radii are joined by the relation

$$\left(\frac{R_{max}}{R_{min}}\right)^{3(\gamma-1)} = (\gamma - 1)\frac{p_a}{p_g} = \frac{p_a}{3p_g}. \tag{30}$$

The maximum pressure in the bubble at the time of its contraction, i.e., for $R = R_{min}$, is

$$p_{max}' = p_g\left[\frac{p_a(\gamma - 1)}{p_g}\right]^{\frac{\gamma}{\gamma-1}} = \frac{p_a^4}{81p_g^3}. \tag{31}$$

For the expansion phase, when it may be assumed that $\gamma = 1$, the main equation, disregarding surface tension and assuming that the pressure inside the compressed bubble is much larger than the pressure in the medium (at infinity), i.e., $p_g' \gg p_a$, will have the form [33]

$$\tfrac{3}{2}\rho\dot{R}^2 = p'\left(\frac{R_{min}}{R}\right)^3 \ln\left(\frac{R_{min}}{R}\right)^3 - p'\left[1 - \left(1 - \frac{R_{min}}{R}\right)^3\right], \tag{32}$$

where p' is the pressure inside the bubble. At time t = 0 we find
$R = R_{min}$, $\dot{R} = 0$, and $p' = p'_{max}$.

Analyzing Eq. (32) at the extremum, we readily verify that $\dot{R}$
has a maximum at $R = R_1 = R_{min} e^{1/3}$, and $\dot{R}$ passes through a
minimum.

If we assume that $\ln (R/R_{min})$ varies only slightly, we can
obtain an approximate solution of Eq. (32) for the initial expansion phase in the form*

$$R = 8.75 \left(\frac{p'_{max} \cdot R^3_{min}}{\rho} \right)^{1/5} \cdot t^{2/5}. \tag{33}$$

Let us assume that the maximum power is radiated in the time
interval from t = 0 to t = τ_0, where τ_0 is the time at which the condition $\dot{R} = \dot{R}_{max}$ is met, i.e., $R = R_1$. Then from (33) we have

$$\tau_0 = \frac{1.7 R_{min}}{\sqrt{p'_{max}/\rho}}. \tag{34}$$

For a comparison of this time with the oscillation period we
refer to the Rayleigh collapse time for an empty bubble, equal to

$$\tau_m = 0.9 R_{max} \sqrt{\frac{2\rho}{3p_a}}, \tag{35}$$

recalling from the many curves given in Part IV that $\tau_m \approx T/4$.
Comparing (34) and (35), we obtain

$$\frac{T}{\tau_0} \approx 2.6 \frac{R_{max}}{R_{min}} \sqrt{\frac{p'_m}{p_a}}. \tag{36}$$

Now, using (30) and (31), we finally obtain

$$\frac{T}{\tau_0} \approx 0.1 \left(\frac{p_g}{p_a} \right)^{5/2}. \tag{37}$$

The active pressure p_a, i.e., the pressure acting on the bubble
at the very first moment of collapse, is composed of the constant

---

*This solution was obtained by Yu. Ya. Boguslavskii.

static pressure p_0 and the instantaneous value of the variable pressure $p_m \cos \omega t$. We write this in the form

$$p_a = p_0 + \alpha p_m,$$

where $-1 \leq \alpha \leq 1$. The quantity α depends on several factors, such as the equilibrium radius of the cavitation bubble, static pressure, variable pressure amplitude, etc. Our problem, however, does not involve the analysis of these dependences at the given time. We rewrite relation (37) in the form

$$\frac{T}{\tau_0} \approx 0.1 \left(\frac{p_0 + \alpha p_m}{p_g} \right)^{5/2}. \tag{37a}$$

This exhibits all the possibilities for controlling the ratio T/τ_0 for the individual bubble, viz., increasing p_0, increasing p_m, reducing p_g, and selecting those relations which will yield the maximum values of α. The third approach is trivial, but difficult to realize, because in the presence of cavitation it is impossible to keep air from coming in contact with the water. The fourth approach has not been studied and it is difficult at this time to say anything definite about it.*

We are left with the first two approaches. At first glance they seem quite adequate, because p_0 and p_m enter into a linear combination. It must be realized here, however, that in the presence of a cavitation zone the variations of the two variables will affect the cavitation index differently. An increase in p_0 will diminish both the number of cavitation bubbles and the volume of each one individually. In the limit, for very large values of p_0, cavitation will simply be totally suppressed. For values of p_0 not too large, it is hard to say beforehand how this will be felt in the cavitation erosion, whether it will decrease monotone to zero or, as p_0 begins to increases, whether it will elicit a rapid growth of the peak power of the shock waves, this effect overlapping the reduction in erosion due to the smaller number of active bubbles. In either case, the growth of p_0 will increase the cavitation efficiency as the 5/2 power. It must be added that, as first pointed out in [33], the growth of p_0 will promote an increase in the solubility of the gases and, hence, an increase in p_g, which is contained in the denominator in the same power 5/2.

---

*The possibility of controlling the value of α has been considered very recently in [45].

But then an increase in p_m will increase the cavitation index, because in this case the number of active bubbles, as well as their maximum radii, increase. Consequently, an increase in p_m must be expressed in a power greater than 5/2. The limit of the growth of p_m is, as we know, set by the aforementioned condition $\tau_m < T/2$ (see Chap. 5), which does not allow for expression (37), inasmuch as the initial equations correspond to a single collapse of a bubble under the assumption that p_a is constant.

For a more rigorous analysis of relation (37a), it would be necessary also to take account of the fact that p_0 and p_m are not independent.

Thus, the reduction in the number of "working" nuclei and, hence, of cavitation bubbles, per se, due to an increase in the static

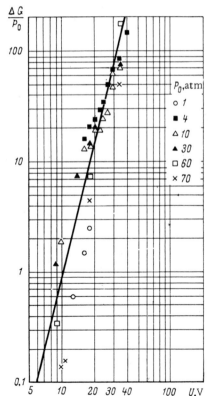

Fig. 25. Cavitation erosion versus the static and variable pressures.

pressure leads to an increase in the equivalent wave impedance
of the medium, as asserted in Chap. 3. An increase in the wave
impedance, on the other hand, with a constant particle velocity of
the radiating surface (or, equivalently, for a fixed radiator voltage), causes an increase in the acoustic power delivered into the
medium and an increase in the sound pressure p_m. However, we
do not have access to any data by which it would be possible to
include this effect quantitatively. Therefore, in the present discussion, we assume that the variable pressure p_m is proportional
to the voltage u on the radiator and is independent of the static
pressure p_0.

For the experimental verification of the considerations expressed above with regard to the dependence of the cavitation
erosion on p_0 and p_m, we make use of the data acquired in [34]
(see Fig. 49 or Part V), which represent the results of an investigation of mechanical erosion over a broad range of values of
p_0 and u. On the horizontal axis we plot the voltage on the radiator and assume that the sound pressure p_m is proportional to it.
On the vertical axis we plot $\Delta G/p_0$, where ΔG is the absolute value
of the cavitation erosion. If we choose a log scale for both axes,
then, as seen in Fig. 25, the points provide a fairly good fit to a
straight line,* from which it follows that

$$\Delta G \sim p_0 \cdot p_m^4. \tag{38}$$

It is easily verified that the power exponent of p_0 is indeed
smaller whereas the power of p_m is larger than 5/2 over a wide
range of variation of p_0, p_m, and ΔG.

We can assume on this basis that the given mechanism for
the transformation of power into the slow buildup of energy by
the cavitation bubble and its rapid withdrawal not only does take
place, but also makes a large contribution to the erosive efficiency
of the cavitation bubbles.

---

*The figure does not contain the last three points corresponding to p_0 = 1 atm, as
they do not meet the condition $\tau_m < T/2$.

Chapter 5

Analysis of Bubble Oscillations

§ 1. Introductory Considerations

For the solution of the problem of the behavior of a cavitation zone in the next-higher approximation, it is required to give consideration to the actual oscillations of the cavitation bubbles, variations in the initial properties of the medium, etc. As already mentioned, the problem in general is exceedingly complex, so that the attempts at such an analysis to date have been imbued with many simplifying assumptions. The first attempts were aimed at analyzing cases of hydrodynamic cavitation, and not until some time later were they extended to cavitation induced by ultrasound.

For the solution of the problem it is first of all necessary to build a mathematical model, i.e., a system of hydrodynamic equations adequately describing the flow of a liquid containing cavitation bubbles. One of the first such models was formulated in' [35], in which it was postulated that the bubbles present in the liquid are compressed adiabatically and that the pressure in the water is equal to the pressure inside the bubbles. In [36] a diffusion model of a cavitating liquid was formulated. However, in the foregoing theories, no attention was given to the dynamics of the individual bubbles; hence, they are only of historical interest.

In [37] for the first time a model of a medium was proposed which could describe a cavitating liquid; in the model the expansion and contraction of the bubbles obeyed the ordinary hydrodynamic equations. The following assumptions were incorporated:

1) The cavitation nuclei represent spheres of equal radii, distributed uniformly throughout the entire volume of an incompressible liquid ρ_0, so that N such nuclei are ascribed to a unit volume of the liquid.

2) Cavitation bubbles are generated in the liquid the moment the pressure falls below a certain value p_b.

3) The pressure inside the cavitation bubbles during expansion and contraction remains constant and equal to p_b, i.e., it is supposed that there is no gas inside the bubble and that the vapor can be condensed.

4) The radial motion of all the bubbles proceeds according to the same law as the motion of the unit cavitation void, with a constant pressure inside the infinite volume of the incompressible liquid.

This assumption permits the use of the equation

$$R\ddot{R} + {}^3/_2 \dot{R}^2 = \frac{p_b - p}{\rho_0}. \tag{39}$$

5) The liquid-vapor mixture is regarded as a homogeneous medium with a density equal to the average density of the mixture, computed according to the formula

$$\rho = \frac{\rho_0}{1 + b(R^3 - R_0^3)}, \tag{40}$$

where

$$b = \frac{4}{3}\pi N.$$

Equations (39) and (40) establish a relation between the pressure and density

$$p = p\left(\rho, \frac{d\rho}{dt}, \frac{d^2\rho}{dt^2}\right). \tag{41}$$

Now the motion of the cavitating liquid in Euler coordinates is described by the system

$$\frac{du}{dt} = -\frac{1}{\rho}\operatorname{grad}p; \quad \rho = \frac{\rho_0}{1 + b(R^3 - R_0^3)}; \quad b = \frac{4}{3}\pi N,$$

$$\frac{d\rho}{dt} + \rho \operatorname{div}\mathbf{u} = 0; \quad p = p_b - \rho_0\left[R\frac{d^2R}{dt^2} + \frac{3}{2}\left(\frac{dR}{dt}\right)^2\right]. \tag{42}$$

The expression for the free energy and entropy is written in the form

$$F(R, \dot{R}, T) = \frac{3}{2} bR^3 \left(\frac{dR}{dt}\right)^2 - \frac{b}{\rho_0} p_{cr} R^2 + \varphi(T),$$

$$S = -\frac{\partial F}{\partial T}.$$

In [38] the development of the cavitation zone in a rarefaction wave near a piston moving with an alternating velocity was investigated. The piston, which is bounded on the left by a semi-infinite tube filled with a compressible liquid, begins at time t = 0 to move to the left according to a certain law x = P(t) < 0 relative to the initial coordinate x = 0. A rarefaction wave moves to the right of the piston into the liquid. As soon as the pressure in the wave drops to a value p_b, the cavitation nuclei begin to expand. Neglecting the density variation of the aggregate liquid relative to the variation of the average density in the medium, we solve the system of equations (42) in Lagrange coordinates for one-dimensional motion

$$\frac{\partial u}{\partial t} = -\frac{1}{\rho_0}\frac{\partial p}{\partial \xi}; \quad \rho = \frac{\rho_0}{1 + b(R^3 - R_0^3)}; \quad \rho = \frac{4}{3}\rho N;$$

$$\frac{\partial\left(\frac{1}{\rho}\right)}{\partial t} = \frac{\partial u}{\partial \xi}; \quad p = p_b - \rho_0 [R\ddot{R} + {}^3/_2 \dot{R}^2], \tag{43}$$

where ξ is the Lagrange coordinate under the following boundary conditions:

$$\text{for } \xi = 0 \quad u = u_b,$$
$$\text{for } \xi = ct \quad p = p_b, \ \rho = \rho_0, \ R = R_0, \ u = 0. \tag{44}$$

Under the boundary conditions (44) a particular solution was deduced in [38] corresponding to a certain one-parameter family of laws governing the motion of the piston:

$$p = p_b - b\rho_0 c^2 (R^3 - R_0^3);$$

$$u = -bc (R - R_0^3);$$

$$\frac{dR}{dt} = \sqrt{\frac{1}{3} bc^2 R^3 \left[1 - \left(\frac{R_0}{R}\right)^3\right]^2 + \chi R^{-3}};$$

$$t - \frac{\xi}{c} = \int_{R_0}^{R} \frac{dR}{\sqrt{\frac{1}{3} bc^2 R^3 \left[1 - \left(\frac{R_0}{R}\right)^3\right]^2 + \chi R^{-3}}}. \tag{45}$$

Here the parameter is a certain abstract quantity $\chi = \dfrac{\dot{u}_0^2}{9b^2 R_0 c^2}$ in which $\dot{u}_0$ is the initial acceleration of the piston.

The equation of motion of the piston can be determined by means of the relation

$$\chi = \int_0^t u\,(\xi = 0,\, t)\,dt = -bc \int_0^t (R_n^3 - R_0^3)\,dt.$$

The function $R_n(t)$ is found from the last equation of the system (45) for $\xi = 0$. From the first equation of (45) and the third equation of (43) we have

$$p_b - p = \frac{c^2 \rho_0^2}{\sigma} - \rho_0 c^2. \tag{46}$$

Equation (46) corresponds to a so-called Chaplygin gas. We know, of course, that in this case a progressive wave propagates without any change of waveform. Consequently, the profile of the cavitation rarefaction wave in this case will remain unchanged.

The next step was undertaken in [39], in which the following system of equations in Lagrange coordinates was adopted as the starting point:

$$\frac{\partial u}{\partial t} = -\frac{1}{\rho_0}\frac{\partial p}{\partial \xi}\,; \quad \rho = \frac{\rho_0}{1 + b\,(R^3 - R_0^3)}\,; \quad b = \frac{4}{3}\pi N;$$

$$\frac{\partial\left(\dfrac{1}{\rho}\right)}{\partial t} = \frac{1}{\rho_0}\frac{\partial u}{\partial \xi}\,; \quad p = p_b - f(R,\,\dot{R},\,\ddot{R},\,a,\,b) \tag{47}$$

under the same boundary conditions as (44). The new idea was the fact that the fourth equation of the system (47) now presupposed an arbitrary functional dependence between the pressure in the liquid and the variation of the bubble radius; the parameters a and b are constants.

On the basis of the general relation deduced by Riemann between the velocity, density, and pressure in a traveling plane wave of finite amplitude, namely the familiar expression [40]

$$u = \int \frac{c\,d\rho}{\rho} = \int \frac{dp}{\rho c}\,, \tag{48}$$

one can obtain the fundamental expressions for the velocity and pressure, as well as the equation of state:

$$u = -bc\ (R^3 - R_0^3);$$
$$p = p_b - \rho_0\ c^2 b\ (R^3 - R_0^3) = p_b + \rho_0 C u;$$
$$p_b - p = \frac{c^2 \rho_0^2}{\rho} - \rho_0 c^2. \tag{49}$$

It is essential to point out here that relations (49), in the case of a cavitation rarefaction wave, provided one neglects the compressibility of the aggregate liquid, are obtained directly from the Riemann solution without any presumption as to the specific form of the relation between the pressure and time variation of the bubble radius.

The fourth equation of (47) must be particularized in order to find the dependence of the velocity and pressure in the cavitation rarefaction wave on the coordinates and time.

Later in [39] the following expression was adopted for the fourth equation of the system (47)

$$p = p_b - \tfrac{3}{2}\rho_0 \dot{R}^2 \left[1 - \frac{4}{3}\frac{R}{a}\right] - \rho_0 R\ddot{R}\left(1 - \frac{R}{a}\right) - \frac{1}{2}\rho_0\frac{R^4}{a^4}\dot{R}^2, \tag{50}$$

which was first obtained by Rayleigh [41]. Here the parameter a is the average distance $a \approx 1/\sqrt[3]{N}$ between the cavitation nucleus and the point at which the values of the pressure and velocity characterizing the Lagrangian particle ξ at time t are determined.

Omitting the highest-order small term in (50), $\frac{1}{2}\rho_0\frac{R^4}{a^4}\dot{R}^2$, we have

$$p = p_b - \tfrac{3}{2}\rho_0 \dot{R}^2\left(1 - \frac{4}{3}\frac{R}{a}\right) - \rho_0 R\ddot{R}\left(1 - \frac{R}{a}\right). \tag{51}$$

In this case we obtain, along with relations (49),

$$t - \frac{\xi}{c} = \int_{R_0}^{R} \frac{\sqrt{1 - \frac{R}{a}}\ dR}{\sqrt{\frac{1}{3}bc^2R^3\left[1 - \left(\frac{R_0}{R}\right)^3\right]^2 + \chi R^{-3}}}. \tag{52}$$

The velocity and pressure in the cavitation rarefaction wave as a function of the coordinates and time are determined by substitution into (49) of the value obtained for $R = R[t - (\xi/c)]$ from Eq. (52).

The equation of motion of the piston is obtainable from the expression

$$\chi = \int_0^t u\,(t,\ \xi = 0)\,dt = -\int_0^t (R_n^3 - R_0^3)\,dt. \tag{53}$$

The function $R_n(t)$ is determined from (52) for $\xi = 0$. As $a \to \infty$ this expression goes over to the solution found in [38].

§ 2. The Cavitation Zone in a Plane Wave

For the solution of this problem it is required, along with the system of equations given in §1, to include the density variation of the medium due to the onset of cavitation bubbles. The appropriate system of equations has the form

$$\frac{du}{dt} = -\frac{1}{\rho}\,\mathrm{grad}\,p; \quad \rho = \frac{\rho_l}{1 + b\,(R^3 - R_0^3)}; \quad b = \frac{4}{3}\pi N;$$

$$p = p_b - \rho_0\left\{R\ddot{R} + \tfrac{3}{2}\dot{R}^2 - \frac{1}{Rc}\,\frac{d}{dt}\,[R^2\dot{R}^2]\right\}; \tag{54}$$

$$\frac{d\rho}{dt} + \rho\,\mathrm{div}\,\mathbf{u} = 0; \quad p = p\,(\rho_l),$$

where N is the number of cavitation bubbles per unit volume of liquid, R_0 is the equilibrium radius of the nuclei, R is the instantaneous radius of the cavitation bubbles, ρ_l is the density of the ideal liquid in the presence of the sound-induced perturbation, ρ_0 is the equilibrium density of the ideal liquid, p, ρ, u are the pressure, density, and velocity of the cavitating liquid, p_b is the critical pressure at which the growth of cavitation bubbles is initiated, and c_0 is the velocity of sound in the liquid at the bubble boundary at the pressure p_b. When $\rho_l = \rho_0$ the first four equations of (54) correspond to the system postulated in [37].*

The expression for the sound wave energy in a medium with cavitation bubbles, i.e., in a medium described by the system of equations (54), may be written in the form

$$\int_v \frac{1}{2}\,\frac{c^2\rho_l^2}{\rho_0} - bp_b(R^3 - R_0^3) + \frac{3}{2}\,\rho_0 b R^3 \dot{R}^2 - \frac{2\rho_0}{c_0}\,bR^3\dot{R}^3 -$$

$$- \rho_0 b\,(R^3 - R_0^3)\,\rho_l\left(\frac{p_0}{\rho_0^2} + \frac{\rho_0 u^2}{2}\right)dv, \tag{55}$$

---

*Instead of the Rayleigh equation used in [37], the Kirkwood–Bethe equations [43], which are analogous and take account of the compressibility of the aggregate liquid at small Mach numbers $\dot{R}/c_0 < 1$, are adopted as the fourth equation in the system (54).

where the integrand is the sound energy density in the cavitating liquid. The first term of the integrand of (55) is the potential energy of the sound wave per unit volume of the ideal liquid. The second, third, fourth, and fifth terms in (55) describe the variation of the energy per unit volume of the cavitating liquid due to transmission of the sound wave in the presence of bubbles. In particular, the second term is the work done reversibly in the expansion of the bubbles under the pressure p_b; the third term is the kinetic energy of the liquid motion; the fourth term is the sound wave energy radiated by the pulsating bubbles; the fifth term is the potential energy of the cavitating liquid due to compressibility of the bubbles in the ideal liquid; and the sixth term is the kinetic energy of the liquid in the sound wave.

The average total energy flux through the given closed surface in the cavitating liquid is equal to

$$\oint [p_m u + \overline{b(R^3 - R_0^3) \, p_0 u} - \overline{ub(R^3 - R_0^3) \, p_b} +$$
$$+ \overline{{}^3/_2 u \rho_0 b R^3 \dot{R}^2} - \overline{\rho_0 \frac{2u}{c_0} b R^3 \dot{R}^3}] ds, \tag{56}$$

where p_m is the amplitude value of the pressure in the sound wave.

In general the law of energy conservation, as we know, has the form [40]

$$\overline{\frac{\partial}{\partial t} \left(\rho \, \frac{u^2}{2} + \varepsilon \right)} = - \operatorname{div} \left\{ \rho u \left(\frac{u^2}{2} + \omega \right) \right\}, \tag{57}$$

where the bar denotes time averaging. In order to obtain the law of conservation of sound energy in the cavitating liquid, it is necessary to substitute the integrands of (55) and (56) into Eq. (57).

Let us analyze further the system (54). Substituting the third equation of that system into the second, by means of the fourth and fifth we obtain

$$3bR^2 \frac{dR}{dt} + \frac{[1 + b(R^3 - R_0^3)]}{c^2} \left\{ \frac{dR}{dt} \frac{d^2R}{dt^2} + R \frac{d^3R}{dt^3} + 3 \frac{dR}{dt} \frac{d^2R}{dt^2} - \right.$$
$$\left. - \frac{dt}{d} \left[\frac{1}{Rc_0} \frac{d}{dt} (R^2 \dot{R}^2) \right] \right\} = \rho_l \, [1 + b(R^3 - R_0^3)] \operatorname{div} \mathbf{u}. \tag{58}$$

The first term on the left-hand side of expression (58) characterizes the compressibility of the cavitating liquid, a quantity that is related only to the volume variation of the bubbles, and the second term determines the compressibility of the ideal liquid.

Let T be a certain characteristic time, for example, the half-period of the sound wave, and let $\bar{R}$ be the characteristic bubble radius (the radius to which the bubble grows in the period T during expansion). Then we have two limiting cases.

1) The first term on the left-hand side of (58) is much larger than the second

$$\frac{3b\bar{R}^3}{T} \gg \frac{\bar{R}^2}{c^2 T^3}, \quad \text{i.e.,} \quad 3bc^2 \bar{R} T^2 \gg 1.$$

This implies that the compressibility of the ideal liquid is negligible in comparison with the compressibility of the cavitating liquid due to the volume variation of the bubbles. Then $\rho_l \equiv \rho_0$, so that the motion of the cavitating liquid is described by the first four equations of the system (54); this is precisely the case treated in [37] and [38].

2) The first term on the left-hand side of (58) is much smaller than the second

$$\frac{3b\bar{R}^3}{T} \ll \frac{\bar{R}^2}{c^2 T^3}; \quad 3bc^2 T^2 \bar{R} \ll 1.$$

This case is typified by the fact that the compressibility of the cavitating liquid due to the volume variation of the bubbles may be neglected relative to the compressibility of the ideal liquid. The implication here is that, as mentioned in Chap. 3, §3, the additional mass of the liquid isolates the bubble compressibility, which becomes inoperative and ceases to exert any real influence on the sound propagation process.

Then $\rho = \rho_l$, and, therefore, the third and fourth equations may be dropped from the system of equations (54); the motion of the liquid in the sound wave is described by the system of equations for the ideal liquid, i.e., the first, second, and fifth equations of the system (54).

We next consider the propagation of a traveling plane sound wave under the condition $3bc^2 T^2 \bar{R} < 1$, taking account of the cavitation-induced losses. Using the integrand of (56) and relation (57) and neglecting the highest-order small terms, we obtain

$$\frac{\overline{\partial}}{\partial t} \frac{3}{2} \rho_0 b R^3 \dot{R}^2 = -\frac{\partial}{\partial x} \overline{p_m u}. \tag{59}$$

Next, using the familiar relations, we write

$$I = \overline{p_m u} = \frac{\overline{p_m^2}}{\rho c}, \quad R \approx \sqrt{\frac{2}{3} \frac{\sqrt{\overline{p_m^2}}}{\rho_0}} \ t;$$

we know from [10] that

$$b = \alpha (I - I_t) \quad \text{for} \quad I > I_t,$$

where I_t is the threshold intensity for the onset of cavitation and α is a constant to be determined experimentally.

Now

$$\frac{\partial}{\partial t} \frac{3}{2} \rho_0 b R^3 \dot{R}^2 = 2 \frac{c_4^{5/4} \alpha T^2}{\rho_0^{1/4}} (I - I_t) (I^{5/4} - I_t^{5/4}) \approx K (I - I_t)^2,$$

where

$$K = 2 \frac{c^{5/4} \alpha T^2 I_t^{1/4}}{\rho_0^{1/4}}.$$

and after some straightforward manipulation we obtain

$$\frac{dI}{dx} = -k (I - I_t)^2. \tag{60}$$

Integrating Eq. (60) under the condition that the average sound energy flux density in the plane x = 0 is equal to I_0, we obtain the dependence of the average sound energy flux density of the primary wave on the coordinate x in a one-dimensional cavitation zone

$$I(x) = I_t + \frac{I_0 - I_t}{1 + kx(I_0 - I_t)}. \tag{61}$$

Note that expression (61) coincides perfectly with Eq. (9), which was derived on the basis of an analysis of the phenomenological theory of the one-dimensional cavitation zone with the use of experimentally determined quantities.

§3. Cavitation in a Convergent Spherical Wave

Consider the propagation of a convergent spherical sound wave [44] under the condition $3bc^2T^2\bar{R} < 1$, taking account of the losses induced by cavitation. As the preceding section reveals, in the case of a convergent spherical wave we can write the conservation of energy with regard for cavitation-induced losses:

$$\overline{\frac{\partial}{\partial t} \frac{3}{2} \rho_0 b R^3 \dot{R}^2} = \frac{1}{r^2} \frac{\partial}{\partial r} (r^2 \overline{p_m u}) \tag{62}$$

(the bar indicates time averaging), where p_m and u are the pressure and velocity in the sound wave.

Next, borrowing the data of [10], we have

$$b = \alpha (I - I_t), \tag{63}$$

where $b = \frac{4}{3} \pi N$, I_t is the cavitation threshold intensity, and α is a coefficient to be determined experimentally.

We know, of course, that [2]

$$R \approx \sqrt{\frac{2}{3} \rho_0 \sqrt{\overline{p_m^2}}} \; t. \tag{64}$$

Using (63) and (64), we obtain

$$\overline{\frac{\partial}{\partial t} \frac{3}{2} \rho_0 b R^3 \dot{R}^2} \approx 2 \frac{c^{5/4} \alpha T^2}{\rho_0^{1/4} 4} (I - I_t) I^{5/4},$$

where T is the period of the sound wave.

Next we put $I^{1/4} \approx I I_t^{1/4}$. In place of (62) we write

$$\overline{\frac{\partial}{\partial t} \frac{3}{2} \rho_0 b R^3 \dot{R}^2} \approx K_1 I_t^{1/4} (I^2 - I I_t), \tag{65}$$

where $K_1 = \dfrac{\alpha T^2 c^{5/4}}{2 \rho_0^{1/4}}$.

From (64) and (65) we finally obtain the equation

$$\frac{dI}{dr} + \frac{2}{r} I = K_1 I_t^{1/4} (I^2 - I I_t), \tag{66}$$

which is more conveniently written in the form

$$\frac{dI}{dr} + \left(\frac{2}{r} + K I_t^{5/4} \right) I - K_1 I_t^{1/4} I^2 = 0. \tag{66a}$$

Equation (66a) is a first-order nonlinear ordinary differential equation, called the Bernoulli equation. Integrating it under the condition that the average sound energy density at $r = r_K$

is equal to I_t, we obtain the dependence of the average sound energy flux density of the primary convergent spherical wave on the coordinate r in a one-dimensional cavitation zone:

$$I(r) = \frac{\dfrac{r_K^2}{r^2} e^{-K_1 I_t^{5/4}(r-r_K)}}{\left[\dfrac{I}{I_t} - \displaystyle\int_{r_K}^{r} \dfrac{K I_t^{1/4} r_K^2}{r^2} e^{-|K_1 I_t^{5/4}(r-r_K)} dr\right]} \cdot \qquad (67)$$

We first examine the case when $K_1 I_t^{5/4} < 1$, which corresponds to weakly ·developed cavitation. Replacing the exponential in (67) by one, we obtain

$$I(r) \approx \frac{I_t \dfrac{r_K^2}{r^2}}{1 + I_t^{5/4} K_1 r_K^2 \left(\dfrac{1}{r} - \dfrac{1}{r_K}\right)} \cdot \qquad (67a)$$

For $r \ll r_K$ we infer from (67a)

$$I(r) \approx \frac{I_t \dfrac{r_K^2}{r^2}}{I_t^{5/4} K_1 r_K^2 \dfrac{1}{r}} \approx \frac{1}{I_t^{1/4} K_1 r} \cdot \qquad (67b)$$

Hence, it is clear that the sound wave intensity increases as $1/r$. But if there are no cavitation losses, i.e., if it is allowed to set $K_1 = 0$ in Eq. (67a), then the intensity grows as $1/r^2$, as it should in the spherical wave case.

In the second case, corresponding to strongly developed cavitation, when $K_1 I_t^{5/4} \gg 1$, the exponential in (67) is a rapidly decaying function, so that, taking the slowly varying factor with $K_1 I_t r_K^2/r^2$, outside the integral, we obtain

$$I(r) \approx \frac{I_t \dfrac{r_K^2}{r^2} e^{-K_1 I_t^{5/4}(r-r_K)}}{1 + \dfrac{K_1 I_t^{5/4} r_K^2}{r^2 K_1 I_t^{5/4}} \left(e^{-K_1 I_t^{5/4}(r-r_K)} - 1\right)}, \qquad (68)$$

whereupon for $r \ll r_K$

$$I(r) \approx \frac{I_t \dfrac{r_K^2}{r^2}}{\dfrac{K_1 I_t^{5/4} r_K^2}{r^2 K_1 I_t^{5/4}}} \approx I_t, \qquad (69)$$

i.e., the intensity of the convergent spherical sound wave in the cavitation zone does not grow, but stays equal to I_t for all time.

This result is entirely consistent with the assumptions set forth in Chap. 1, §5, on the basis of the phenomenological analysis, and with the experimental results in Fig. 10.

Let us examine in closer detail the physical mechanism of the cavitation-induced losses, using the example of a rarefaction wave.

During the propagation of the rarefaction wave, beginning with some isobaric surface on which the pressure is equal to p_b, the nuclei begin to cavitate. Near the cavitation bubbles a pressure drop takes place, which can be determined from the Rayleigh formula

$$ p = p_b = -\,^3/_2 \rho_0 \dot{R}^2 \left(1 - \frac{4}{3}\frac{R}{a}\right) - \rho_0 R\ddot{R}\left(1 - \frac{R}{a}\right) - \frac{1}{2}\rho_0 \frac{R^4}{a^4}\dot{R}^2, \quad (50) $$

where a is the distance from the center of the bubble to the point at which the pressure p is determined. Inserting into (50) the expressions

$$ R \approx \sqrt{\frac{2}{3}\frac{p_\infty - p_b}{\rho_0}}\,t, \qquad \dot{R} \approx \sqrt{\frac{2}{3}\frac{p_\infty - p_b}{\rho_0}}, \qquad \ddot{R} \approx 0, $$

where p_∞ is the pressure at infinity, and omitting the last term in (50) as the one of highest-order smallness, we obtain the expression

$$ p \approx p_b - \rho_0 \frac{(p_\infty - p_b)}{\rho_0}\left(1 - \frac{4}{3}\frac{R}{a}\right), \quad (50a) $$

which determines the pressure in the liquid as a function of the distance from the bubble. It is apparent that $p = -p_\infty$ as $a \to \infty$, $p \approx -\,^5/_9 p_\infty$ for $a = 3R$, etc.

Clearly, this pressure drop in the liquid, computed according to Eq. (50a) for all bubbles whose distribution density is n, produces a loss of energy from the primary sound wave. The amount of energy going for the formation of cavitation is exactly equal to the kinetic energy of motion of the liquid due to the expansion and collapse of the cavitation bubbles, i.e., to $\,^3/_2 bR^3\dot{R}^2$.

On the other hand, as we are aware [38], the quantity $\frac{3}{2} bR^3\dot{R}^2$ is equal to

$$\int_{R_0}^{R} (p_b - p)3bR^2 dR,$$

i.e., to the work done by the sound field in the expansion and collapse of the bubbles.

This means that the overwhelming portion of the energy absorbed and transferred by the cavitation bubbles exists as kinetic energy of the additional mass of the liquid.

§ 4. Acoustic Streaming Induced by the Formation of Cavitation

As stated several times already (see, e.g., Parts III and V of the present book), the energy losses associated with the formation of cavitation, as any other losses, result in a reduction of momentum and are thus a stimulus for the onset of Eckart acoustic streaming. Relying on the materials of the present chapter, §2 in particular, we can calculate the streaming velocity and compare it with the experimental measurements [46].

The simplest place to start is the equation for the conservation of momentum in the form

$$\rho_0 \frac{\partial u}{\partial t} + \rho_0 (u, \nabla) u = -\frac{\partial E}{\partial x}, \tag{70}$$

where u is the streaming velocity, E is the sound energy density, and ρ_0 is the density of the aggregate liquid.

For steady-state motion ($\partial u / \partial t = 0$) and the one-dimensional problem, Eq. (70) is simplified:

$$\rho_0 u \frac{du}{dx} = -\frac{dE}{dx}. \tag{70a}$$

We use Eq. (59) to determine the right-hand side of the above expression. We at once obtain

$$-\frac{dE}{dx} = \frac{3\rho_0}{2c_0} \overline{\frac{d}{dt} bR^3 \dot{R}^2}. \tag{71}$$

Recall that $R = R(t)$ is the instantaneous value of the cavitation bubble radius; $b = \frac{4}{3}\pi N$, where N is the number of bubbles per unit volume. The bar, as before, indicates time averaging.

Recognizing the admissibility of the assumption [2]

$$R \approx \sqrt{\frac{2}{3} \frac{p_m}{\rho_0}} \, t$$

in the expansion stage (where p_m is the sound pressure amplitude and t is the time measured from the beginning of expansion) and assuming that in the steady state the number of cavitation bubbles is invariant, from (71) we obtain

$$-\frac{dE}{dx} = kE^{5/4}, \tag{71a}$$

where $k = 0.3 \dfrac{c_0^{5/4} b}{f^2 \rho_0^{1/4}}$; and f is the frequency.

Integrating (71a) in order to find the function $E(x)$ in explicit form, substituting it into (70a), and integrating with regard to the boundary condition at $x = 0$, viz., $u = 0$ and $I = I_t$, we obtain

$$u = \left\{ \frac{2I_t}{\rho_0 c_0} \left[1 - \frac{1}{(\frac{1}{4}kxI_t^{1/4} + 1)^4} \right] \right\}^{1/2}, \tag{72}$$

where I_t is the cavitation threshold.

The resulting equation is valid for the case of a cavitation zone formed by a traveling plane wave. However, as indicated above, in this case it is very difficult to realize a cavitation zone with a sharply delineated leading edge in a free liquid. We therefore rely on the fact that a convergent spherical wave transforms to a plane wave near the focal region (see, e.g., [47]). In this case a cavitation zone is formed with an "entrant" surface, and inside this zone propagates a wave that scarcely differs from a plane wave.

It is most convenient to use the data of [11] and [9] (see also, respectively, Part V, Chap. 3 and the present part, Chap. 1, § 5), which were obtained for a focusing concentrator operating at 500 kc. In [11] the hydrodynamic head F of acoustic streaming was measured, while in [9] the characteristic diameter D_K of the

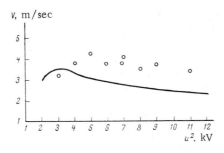

Fig. 26. Acoustic streaming velocity due to
the losses of sound energy in the formation
of cavitation.

cavitation zone was measured as a function of the electrical voltage
u applied to the concentrator. The flow velocity can be found from
the simple relation

$$u = \sqrt{\frac{2F}{\rho_0 S_K}},$$ (73)

where $S_K = \pi D_K^2/4$.

The circles in Fig. 26 indicate the values of the acoustic
streaming velocity, converted from the experimental data according to Eq. (73), and the solid curve represents the result of a calculation according to Eq. (72) under the condition $x = D_K$. The
latter implies that we are allowing for the streaming velocity at
exit from the cavitation zone.

The reasonable agreement of the absolute values attests to the
fact that the strong Eckart-type acoustic streaming observed in a
cavitation zone does in fact stem from the energy losses in the
formation of cavitation.

References

1. H. G. Flynn, Physics of Acoustic Cavitation in Liquids, Physical Acoustics
(W. P. Mason, ed.), Vol. 1B, Academic Press, New York (1964).
2. A. D. Pernik, Cavitation Problems, Sudpromgiz (1963).
3. V. A. Akulichev and L. D. Rozenberg, Certain relations in a cavitation region,
Akust. Zh., 11(3):287 (1965).
4. M. G. Sirotyuk, Energetics and dynamics of the cavitation zone, Akust. Zh.,
13(2):265 (1967).

5.   V. A. Akulichev, Experimental investigation of an elementary cavitation zone, Akust. Zh., 14(3):337 (1968).

6.   M. G. Sirotyuk, Experimental investigation of the growth of ultrasonic cavitation at 500 kc, Akust. Zh., 8(2):216 (1962).

7.   I. P. Golyamina, Ferrite magnetostrictive radiators, Sources of High-Intensity Ultrasound (L. D. Rozenberg, ed), Vol. 1, Plenum Press, New York (1969), pp. 165-222.

8.   L. D. Rozenberg, Einige physikalische Erscheinungen die in hochintensiven Ultraschallfeldern entstehen [Some physical effects in high-intensity ultrasonic fields], Fourth Internat. Congr. Acoustics, Vol. 2, Copenhagen (1962), p. 179.

9.   V. A. Akulichev, L. D. Rozenberg, and M. G. Sirotyuk, Certains rélations dans le champ de la cavitation ultra-sonore [Certain relations in the field of ultrasonic cavitation], Proc. Fifth Internat. Congr. Acoustics, E64, Liège (1965).

10. D. Messino, D. Sette, and F. Wanderlingh, Statistical approach to ultrasonic cavitation, J. Acoust. Soc. Am., 35(10):1575 (1963).

11. M. G. Sirotyuk, Energy balance of an acoustic field in the presence of cavitation, Akust. Zh., 10(4):465 (1964).

12. L. D. Rozenberg, Estimation of the cavitation efficiency of acoustic energy, Akust. Zh., 11(1):121 (1965).

13. L. D. Rozenberg and M. G. Sirotyuk, Apparatus for the generation of focused high-intensity ultrasound, Akust. Zh., 5(2):206 (1959).

14. V. A. Akulichev, Investigation of the Onset and Development of Acoustic Cavitation, Candidate's Dissertation, Akust. Inst. AN SSSR (1966).

15. I. G. Mikhailov and V. A. Shutilov, A simple technique for the observation of cavitation in a liquid, Akust. Zh., 5(3):376 (1959).

16. V. A. Akulichev, Yu. Ya. Boguslavskii, A. I. Ioffe, and K. A. Naugol'nykh, Radiation of finite-amplitude spherical waves, Akust. Zh., 13(3):321 (1967).

17. V. P. Korobeinikov, R. S. Mel'nikova, and E. V. Ryazanov, Theory of Point Detonation, Moscow (1961).

18. Y. Kikuchi and H. Shimizu, On the variation of acoustic radiation resistance in water under ultrasonic cavitation, J. Acoust. Soc. Am., 31(10):1385 (1959).

19. L. D. Rozenberg and M. G. Sirotyuk, On the radiation of sound in the presence of cavitation, Akust. Zh., 6(4):478 (1960).

20. E. V. Romanenko, Ultrasonic receivers and methods for their calibration, Sources of High-Intensity Ultrasound (L. D. Rozenberg, ed), Vol. 2, Plenum Press, New York (1969), p. 187.

21. M. G. Sirotyuk, Behavior of cavitation bubbles at high ultrasonic intensities, Akust. Zh., 7(4):499 (1961).

22. A. Sommerfeld, Thermodynamik und Statistik (Thermodynamics and Statistics), Wiesbaden (1952).

23. A. S. Bebchuk, On the cavitation damage of solids, Akust. Zh., 3(1):90 (1957).

24. Y. Olaf, Oberflachenreinigung mit Ultraschall [Ultrasonic surface cleaning], Acustica, 7(5):253 (1957).

25. A. E. Crawford, The measurement of cavitation, Ultrasonics, 2(3):120-123 (1964).

26. L. D. Rozenberg, On the physics of ultrasonic cleaning, Ultrasonic News, 4(4):16 (1960).

27. A. Weissler, Ultrasonic cavitation measurements by chemicals, Proc. Fourth Internat. Congr. Acoustics, J32 (1962).

28. Shin Pin Liu, Chlorine release test for cavitation measurements, J. Acoust. Soc. Am., 36(5):1019A (1964).

29. M. Degrois and I. Badilian, Influence du phénomène de relaxation, produit au sein des solutions soumises d'un rayonnement ultrasonore, sur le rendement des éffets chimiques et la luminescence [Influence of relaxation phenomena in ultrasonically irradiated solutions on the efficiency of chemical and luminescence effects], Compt. Rend., 254:231 (1962).

30. A. S. Bebchuk, Yu. Ya. Borisov, and L. D. Rozenberg, On cavitation erosion, Akust. Zh., 4(4):361 (1958).

31. R. M. Boucher and B. Polansky, The measurement of ultrasonic cavitation, IEEE Symp. Ultrasonics, Los Angeles (1964).

32. I. A. Roi, Initiation and development of ultrasonic cavitation, Akust. Zh., 3(1):3 (1957).

33. Yu. A. Aleksandrov, G. S. Voronov, V. M. Gorbunkov, N. V. Delone, and Yu. N. Nechaev, Bubble Chambers, Gosatomizdat (1963).

34. M. G. Sirotyuk, Ultrasonic cavitation processes at elevated hydrostatic pressures, Akust. Zh., 12(2):231 (1966).

35. J. Ackeret, Experimentalle und theoretische Untersuchungen über Hohlraumbildung im Wasser [Experimental and Theoretical Studies of Cavitation in Water], Eidgenöss. Materialprüfungsanstalt, E. T. H., Zurich (1930).

36. L. A. Éinshtein, Trudy TsAGI, No. 584 (1946).

37. B. S. Kogarko, A model of a cavitating liquid, Dokl. Akad. Nauk SSSR, 137(6):1331 (1961).

38. B. S. Kogarko, One-dimensional nonsteady motion of a liquid with the onset and development of cavitation, Dokl. Akad. Nauk SSSR, 155(4):779 (1964).

39. Yu. Ya. Boguslavskii, Onset and development of a cavitation rarefaction wave, Akust. Zh., 13(4):538 (1967).

40. L. D. Landau and E. M. Lifshits, Mechanics of Continuous Media, GTTI (1954).

41. Rayleigh, On pressure developed in a liquid during the collapse of a spherical cavity, Phil. Mag., 34:94 (1917).

42. Yu. Ya. Boguslavskii, Propagation of sound waves in a liquid during cavitation, Akust. Zh., 14(2):185 (1968).

43. R. H. Cole, Underwater Explosions, Princeton Univ. Press (1948).

44. Yu. Ya. Boguslavskii, Cavitation zone in a convergent spherical sound wave, Akust. Zh., 14(3):463 (1968).

45. F. A. Bronin, Investigation of the Cavitation Damage and Dispersion of Solids in a High-Intensity Ultrasonic Field, Dissertation, Moskov. Inst. Stali i Spetsial'nykh Splavov, Moscow (1967).

46. Yu. Ya. Boguslavskii and Yu. G. Statnikov, Mechanism of the generation of acoustic streaming in a sound field and calculation of its velocity in a cavitation zone, Fourth All-Union Conference on Acoustics, BV1, Moscow (1968).

47. L. D. Rozenberg, Ultrasonic focusing radiators, Sources of High-Intensity Ultrasound (L. D. Rozenberg, ed), Vol. 1, Plenum Press, New York (1969), pp. 223-309.

Index